Evaluation of Environmental Contaminants and Natural Products:

A Human Health Perspective

Edited by
Ashita Sharma
Department of Civil Engineering, Chandigarh University,
Gharuan, Mohali, India

Manish Kumar
Department of Biology, S.D. College Barnala, Punjab, India

Satwinderjeet Kaur
Department of Botanical and Environmental Sciences,
Guru Nanak Dev University, Amritsar, Punjab, India

&

Avinash Kaur Nagpal
Department of Botanical & Environmental Sciences,
Guru Nanak Dev University, Amritsar, Punjab, India

Evaluation of Environmental Contaminants and Natural Products:

A Human Health Perspective

Editors: Ashita Sharma, Manish Kumar, Satwinderjeet Kaur and Avinash Kaur Nagpal

ISBN (Online): 978-981-14-10963

ISBN (Print): 978-981-14-1095-6

need for a court order if at any point you breach any terms of this License Agreement. In no event will any delay or failure by Bentham Science Publishers in enforcing your compliance with this License Agreement constitute a waiver of any of its rights.

3. You acknowledge that you have read this License Agreement, and agree to be bound by its terms and conditions. To the extent that any other terms and conditions presented on any website of Bentham Science Publishers conflict with, or are inconsistent with, the terms and conditions set out in this License Agreement, you acknowledge that the terms and conditions set out in this License Agreement shall prevail.

Bentham Science Publishers Pte. Ltd.
80 Robinson Road #02-00
Singapore 068898
Singapore
Email: subscriptions@benthamscience.net

BENTHAM SCIENCE

CONTENTS

PREFACE

In the last century, soaring economic developments had led to mushrooming of industries and ill planned urbanization. The plethora of byproducts and/or waste products of industrialization acted as environmental contaminants in the biosphere. Currently, addition of some xenobiotics to our environment and increase in concentration of existing natural elements in biosphere has resulted in the deterioration of ecosystem, alterations in biogeochemical cycles, degradation of gene pool thus, affecting the health of flora and fauna existing on the earth. The health effects associated with exposure to these contaminants include various degenerative diseases including cancer. Increase in cancer incidences and deterioration of human health have led to the awakening of scientific community and efforts are being made to formulate policies and framework associated with environmental contaminants.

The present book has been conceptualized keeping all of the above issues into consideration. The book is broadly divided into two sections; Section A covers the toxic effects of environmental contaminants on human health, tools to analyze the pollutants and ways to mitigate them. Section B focuses on the protective effects of various plant products against the toxicity induced by environmental pollutants. Section A was mainly edited by Dr. Ashita Sharma while the main editor for Section B was Dr. Manish Kumar, concerning their respective areas of expertise.

In the first chapter of the book, Savita *et al.* have summarized various pollutants that have been added to the environment by anthropogenic activities. The chapter also summarizes various effects of the pollutants on flora, fauna and environment. The Second chapter of our book by Chattopadhyay summarizes various carcinogenic pollutants in our ambient environment and their mechanism of action. The chapter holds a descriptive approach to the carcinogens and their effects. The next two chapters of the book by Kaur T. *et al.* and Kaur M. *et al.* focus on various tools which help in evaluating the toxicity of pollutants so as to identify various carcinogenic and/or toxic chemicals in our surroundings. These chapters have carefully summarized the common techniques adopted for the screening of pollutants. Pollutants like pesticides and biphenyls, pertaining to their wide application and severe toxic effects, have gained attention of the scientists and researchers throughout the world. The advent of green revolution and the demand of increased food production have resulted in injudicious use of agrochemicals, which may have increased the crop production but at the same time have caused many toxic effects to the consumers. Pesticides are one of most common contaminants which effect large section of population, hence these holds importance over the other pollutants. So, in the next chapter, Yadav and Sharma throw light on the toxic effects of pesticides on human population. Biphenyls are the emerging pollutants which have increased in biosphere due to various anthropogenic reasons. The increase in urbanization and ill planned disposal has made this pollutant spread drastically in the ambient environment. The Chapter sixth of the book by Sharma *et al.* reflects on various sources and effects of biphenyls. Also, the chapters have tried to provide the remediation strategies for both the pollutants. Before the pollutant reaches to the human beings and showcase the effects, it can be curbed in the environment. The next chapter by Kalia and co-workers, thus focuses on the emerging technology which identifies an emerging strategy to mitigate the pollutant from environment employing unique group of microorganisms called Actinomycetes.

Since ancient times, the natural phytochemicals in plants are known to exhibit medicinal properties. Thus, the power of these phytoconstituents was also evaluated against the deadly contaminants and in many cases these phytochemicals proved to be the curative nectars. The next section of the book throws light on protective benefits of plants products against degenerative diseases including cancer. The subsequent chapters by Kaur P. *et al.* and Kaur S. *et al.* cover modulatory and protective effects of plants products and nutraceuticals. The second last chapter of book by Kumar *et al.*, establishes an important link between caspases and plant products in cancer chemoprevention. In the last chapter, Patel and Bhattacharya have focused on important diagnostic tools related to cancer.

This version is a concise handbook for researchers and academicians to understand the issues related to various environmental toxins and protective agents which can safeguard gene pool against these toxins.

Ashita Sharma
Department of Civil Engineering,
Chandigarh University, Gharuan,
Mohali, India, 140413

&

Manish Kumar
Department of Biology,
S.D. College Barnala,
Punjab,
India 148101

FOREWORD

Environmental contaminants are constantly increasing in the environment due to the ill planned economic development. Increase in degenerative diseases as a result of continuous exposure to environmental contaminants is posing a serious threat to mankind. While living in such an environment, it has become an absolute necessity to find ways to enhance the protective nature of human metabolic response against the exposure to carcinogens. Various phytochemicals were considered as nectar against many degenerative diseases in ancient civilizations. Modern science is also making efforts to understand the protective effects of natural products. There is a need to compile the information about pollutants which can damage mankind and the protectants to safeguard our existence. The book is a unique and well-arranged compilation of these important aspects.

The first section of the book covers various sources and effects of emerging contaminants in our ambient environment. Environmental contaminants are associated with an increase in incidences of various types of cancers. The chapter entitled "Environmental risk and cancer" covers all such pollutants in detail. Due to the emergence of various contaminants, it has also become important for scientific community to assess the toxicity of various chemical agents present in the environment. Next chapters of the book suggest various plant and bacterial bioassays which could be used to analyze the toxic potential. Injudicious urbanization and economic development have increased the concentrations of some toxins into the environment which could have been prevented if judiciously used; pesticides and biphenyls are perfect examples. In the second section, editors have tried to compile the information about various natural products and their protective effects against carcinogenic potential of environmental contaminants. The metabolic link between plant products and cancer chemoprevention is also summarized well in one of the chapters in the book. The volume of *Environmental Contaminants* and *Natural Products* is a concise and well-organized handbook for researchers working in the field.

Renu Wadhwa
National Institute of Advanced Industrial Science & Technology (AIST)
Tsukuba Science City - 305 8565
Japan

List of Contributors

Mr. Ajay Kumar is currently working as Research Scholar in the Department of Botanical and Environmental Sciences, Guru Nanak Dev University, Amritsar, India. He is pursuing his Ph.D. on cancer chemopreventive potential of natural products.

Dr. Anu Kalia is presently working as an Assistant Professor in the Department of Soil Science, College of Agriculture, Punjab Agricultural University, Ludhiana, Punjab, India.

Dr. Anuradha Sharma is an Associate Professor, Botany Department, Hindu College, University of Delhi, New Delhi. She works in the area of cell biology.

Dr. Ashita Sharma has completed Ph.D. from the Department of Botanical and Environmental Sciences, Guru Nanak Dev University, Amritsar, Punjab, India. Her areas of specialization are genotoxicity studies, heavy metal analysis, environmental monitoring and bioremediation. At present she is working as Assistant Professor at Chandigarh University.

Dr. Avinash Kaur Nagpal, Professor in the Department of Botanical and Environmental Sciences, Guru Nanak Dev University, Amritsar, Punjab, India. Her areas of specialization are antigenotoxic studies, plant tissue culture, plant databases and environmental monitoring.

Dr. Gurpreet Kaur is an Assistant Professor, Crop Protection Division, Department of Agriculture, Khalsa College, Amritsar, India. She is working in the field of Mycology and Plant Pathology.

Dr. Harsimran Kaur is presently working as Assistant Professor in PG department of Agriculture in Khalsa College, Amritsar, India. She has expertise in the field of plant tissue culture and natural plant products.

Dr. Inderpreet Kaur is presently working as an Assistant Professor in the Department of Chemistry, Centre for Advanced Studies, Guru Nanak Dev University, Amritsar, Punjab, India. Her areas of specialization are electrochemistry, sensors, surface modified adsorbents, water analysis, heavy metal analysis and environmental monitoring.

Dr. Indranil Chattopadhyay is an Assistant Professor in the Department of Life Sciences, Central University of Tamil Nadu, India. His areas of research expertise are cancer genetics, genomics and environmental mutagenesis.

Dr. Jatinder Kaur Katnoria is currently working as an Associate Professor in the Department of Botanical and Environmental Sciences, Guru Nanak Dev University, Amritsar, Punjab, India. Her areas of specialization are environmental toxicology, heavy metal analysis and environmental monitoring.

Ms. Kritika Pandit is currently working as Research Scholar in the Department of Botanical and Environmental Sciences, Guru Nanak Dev University, Amritsar, India. She is pursuing her Ph.D. on cancer chemopreventive potential of natural products.

Dr. Madhurama Gangwar is Senior Microbiologist in the Department of Microbiology, College of Basic Sciences and Humanities, Punjab Agricultural University, Ludhiana, Punjab, India.

Ms. Mandeep Kaur is a Research Scholar in the Department of Botanical and Environmental Sciences, Guru Nanak Dev University, Amritsar, Punjab, India. Her areas of specialization are genotoxicity evaluation and environmental monitoring.

Dr. Manish Kumar is currently working as an Assistant professor in the Department of Biology, S.D. College Barnala, Punjab, India His research activities are focused on evaluation of cancer chemopreventive properties *viz.* antioxidant, antigenotoxic, antimutagenic, COX-2 inhibitory and anticancer activities of medicinal plants.

Dr. Paramjeet Kaur is currently working as an Assistant Professor in Botany at PG Department of Botany, Khalsa College-Amritsar, India. Her areas of research expertise are genetic toxicology and cancer chemoprevention.

Ms. Priyanjali Bhattacharya is pursuing Ph.D. from Vellore Institute of Technology, VIT, Vellore, India. She is currently working on research topic "Mutational Profiling of DNA Repair Genes in Hematologic Malignancies".

Dr. Rajneet Kour Soodan is Ph.D. from the Department of Botanical and Environmental Sciences, Guru Nanak Dev University, Amritsar, Punjab, India. Her areas of specialization are environmental toxicology, plant and animal bioassays, and antioxidative enzyme analysis.

Dr. Renu Bhardwaj is a Professor in the Department of Botanical and Environmental Sciences, Guru Nanak Dev University, Amritsar, India. Her field of research includes plant stress physiology, plant metabolomics, chemical ecology and ecophysiology, phytoremediation and plant growth regulation.

Ms. Sakshi Sharma is a Research Scholar in the Department of Botanical and Environmental Sciences, Guru Nanak Dev University, Amritsar, Punjab, India. Her areas of specialization are botany and environmental monitoring.

Ms. Sandeep Kaur is currently working as Research Scholar in the Department of Botanical and Environmental Sciences, Guru Nanak Dev University, Amritsar, India. She is pursuing her research on cancer chemopreventive potential of natural products.

Dr. Saroj Arora is a Professor in the Department of Botanical and Environmental Sciences, Guru Nanak Dev University, Amritsar, India. Her areas of expertise are natural plant products and environmental toxicology.

Dr. Satwinderjeet Kaur is currently working as a Professor in the Department of Botanical and Environmental Sciences, Guru Nanak Dev University, Amritsar, India. Her areas of expertise are cancer chemoprevention and natural plant products.

Dr. Satyawati Sharma is a Professor in the Centre for Rural Development and Technology, Indian Institute of Technology Delhi, Hauz Khas, New Delhi, India. She has a wide area of research including waste management, soil science and agronomy.

Dr. Savita is an Assistant Professor in Botany Department, Hindu College, University of Delhi, New Delhi. Her areas of expertise are plant tissue culture and plant pathology.

Ms. Sneh Rajput is currently working as Research Scholar in the Department of Botanical and Environmental Sciences, Guru Nanak Dev University, Amritsar, India.

Ms. Sonal Yadav is a Ph.D. Research Scholar in the Centre for Rural Development and Technology, Indian Institute of Technology Delhi, Hauz Khas, New Delhi, India.

Ms. Sukhjinder Kaur is working in the Department of Microbiology, College of Basic Sciences and Humanities, Punjab Agricultural University, Ludhiana, Punjab, India.

Dr. Tajinder Kaur is currently working as Assistant Professor, Crop Protection Division, Department of Agriculture, Khalsa College, Amritsar, India. Her areas of expertise are water quality management and abiotic stress management.

Dr. Trupti Patel is an Associate Professor at Vellore Institute of Technology (VIT) in the department of Integrative Biotechnology, School of Bio Sciences and Technology (SBST), India. She works in the area of cancer genetics, with an emphasis on cancer cytogenetics and molecular diagnostics. She also works on natural products to explore their potential as cancer therapeutic enhancers.

Dr. Vaneet Kumar is a Research Scholar in the Department of Botanical and Environmental Sciences, Guru Nanak Dev University, Amritsar, Punjab, India. His areas of specialization are genotoxicity evaluation and environmental toxicology.

Dr. Vanita Chahal is currently working as an Assistant Professor at Department of Sciences, Kamla Nehru College for Women, Phagwara, Punjab, India. Her areas of specialization are environmental genotoxicity, soil pollution, heavy metal contamination and multipesticide analysis.

Ms. Varinder Kaur is currently working as Research Scholar in the Department of Botanical and Environmental Sciences, Guru Nanak Dev University, Amritsar, India. She is pursuing her research on cancer chemopreventive potential of natural products.

Mr. Vivek Chopra is an Assistant Professor in Botany Department, Hindu College, University of Delhi, New Delhi. His areas of expertise are algal and environmental biotechnology.

CHAPTER 1

Environmental Contaminants: Sources and Effects

Savita*, **Vivek Chopra** and **Anuradha Sharma**

Botany Department, Hindu College, University of Delhi, New Delhi, India

Abstract: Environmental contaminants are those substances which are present in the environment above the permissible limits of concentration, which adversely alters the environment and is toxic to the human, animal and plant health. Due to industrialization and overuse of chemical fertilizers, our environment has become contaminated with various types of contaminants. The contaminants include solid, liquid and gaseous substances which are produced by human activity for short-term economic benefits at the cost of long-term ecological benefits for humanity. The sources of contaminants may be point source or non-point source. The common sources of contaminants are fossil fuels, industries and industrial accidents, oil spills, mining, ammunitions and agents of war *etc*. The contaminants may be natural or xenobiotic (man-made) in nature. The common environmental contaminants are polyaromatic hydrocarbons, heavy metals, pesticides organic solvents and inorganic solvents *etc*. Accidental release of these contaminants in the environment leads to evolution of new diseases to human health and as well as mass death of population. The present chapter deals with the various types of contaminants, sources and effects of environmental contaminants.

Keywords: Chemical Contaminants, Contaminants, Environment, Electronic Waste, Industrial Accidents, Natech Accidents, Oil Spills, PAM, Pesticides, Pollution, RDS and Ammunitions and Agents of War, Smog, USTs, Waste Disposal.

INTRODUCTION

Environment is the complex of biotic and abiotic factors (Physical and chemical) which interact with each other. The environment helps in sustaining human life by providing the natural resources and particular ways of life. Our modern life style and our dependency on technology have leaded us towards a highly industrialized state at the cost of the natural environment, which is one of the main reasons for an ecological crisis we are facing today [1]. Because of industrialization our environment has become contaminated with various types of chemicals.

---
* **Corresponding author Savita:** Botany Department, Hindu College, University of Delhi, New Delhi, India; Tel:+91-9716482372; E-mails: savitagndu@gmail.com; savita14.du@gmail.com

Ashita Sharma, Manish Kumar, Satwinderjeet Kaur & Avinash Kaur Nagpal (Eds.)

Environmental contaminants are those substances which are present in the environment above the permissible limits of concentration, which adversely alters the environment and is toxic to the human, animal and plant health. These compounds may either originate from man-made sources (industrial waste, wastewater discharge, thermal and nuclear power plants and agricultural run-off) or from natural sources (odor-causing chemicals produced by algal and bacterial blooms and volcanic eruptions). The man-made chemicals are called as xenobiotic compounds which are foreign to the biosphere. They are produced mainly through the waste discharged from industries, pesticides, insecticides, landfilling and mining. Common examples of xenobiotic compounds are petroleum hydrocarbons, heavy metals, pesticides lead, organic solvents and inorganic solvents. Contaminated soil, air and water affect human health adversely through direct contact. Accidental release of these contaminants in the environment leads to evolution of new diseases to human health and as well as mass death of population. Present chapter will cover the following contents:

1. Types of environmental contaminants.
2. Sources of environmental contaminants.
3. Effects of environmental contaminants.

TYPES OF ENVIRONMENTAL CONTAMINANTS

Depending upon the Existence of Environmental Contaminants in Nature, there are two Types of Contaminants

Quantitative Contaminants: Those substances which occurs normally in the environment but their concentration gets increased due to various anthropogenic factors *e.g.* carbon dioxide.

Qualitative Contaminants: Those substances which are not found naturally in the environment but are produced by man for example fungicides, pesticides, insecticides and chemical fertilizers. Qualitative contaminants are further classified depending upon various parameters like their form, persistency, disposability and their chemical composition.

Depending upon their Form in which they Persist in the Environment

Primary Contaminants: The contaminants which are present naturally into the environment or emitted directly from a source and persist in the form in which they were added to the environment are called as primary contaminants. Smoke, fumes, dust, ash, soil particles and volatile organic compounds *etc.* are the common examples of primary contaminants.

Secondary contaminants: Primary contaminants react in the atmosphere to form secondary contaminants. Formation of acid rain by reaction of sulfur dioxide/nitrogen oxides with water and nitrogen dioxides, PAN formation in the presence of sunlight by reaction of hydrocarbons (HC) with nitrogen oxides (NOx).

Depending upon their Natural Disposal

Biodegradable contaminants: Those contaminants which can be quickly degraded or broken down into simpler or harmless substances by natural means are called biodegradable contaminants. Common biodegradable contaminants are urine, faecal matter, sewage, cattle dung, animal bones, agriculture residues, paper, wood, cloth, leather, wool, vegetable stuff *etc.*

Non-biodegradable contaminants: Those contaminants which either cannot be broken down into harmless substances or degraded very slowly in nature are called non-biodegradable contaminants. Examples of non-biodegradable contaminants are heavy metals (mercury, arsenic and lead), glass objects, DDT, silver foils, plastics, polythene, insecticides, pesticides *etc.*

Depending upon whether Manmade or Natural

***Natural and xenobiotic contaminants*:** Chemical compounds which occur naturally in the environment are called natural contaminants while those which are not found naturally in the environment and are foreign to the biosphere are called xenobiotic contaminants. These compounds are generally non-degradable and called as recalcitrant. Aromatic halogenated compounds, polyaromatic hydrocarbons and azo dyes are the common examples of xenobiotic contaminants [2, 3].

Depending upon their Chemical Composition

Organic contaminants: Organic contaminants are the carbon-based chemicals, which enter into the water table and water supply through agricultural runoff and waste disposal from industries. The examples of organic contaminants of environment are organic solvents, DDT, Dieldrin, PCP, PAHs, RDX and TNT.

Inorganic contaminants: Inorganic contaminants are mineral-based compounds such as lead, mercury, antimony, selenium, asbestos *etc.* Inorganic contaminants can occur naturally in water, or they may get into water through agricultural runoff, direct discharge from chemical industries, leaching from mining sites and decay of natural deposits.

SOURCES OF ENVIRONMENTAL CONTAMINANTS

The sources of contaminants may be point source or non-point source. If pollutants are being released from a single, identifiable source of pollution, such as discharge from industries, sewage treatment plant and drain, then the contaminants are called as point source. Non-point sources of contamination is also called as 'diffuse' contamination which is attributed to multiple sources like land runoff, drainage, precipitation, seepage *etc*. The sources and effects of environmental contaminants are summarized in Table **1**. The common sources of contaminants are as follows:

Table 1. Sources and effects of contaminants (soil, water and air) on human health

Contaminant of soil and water	MCL (maximum contaminant level) mg/L	Effects	Sources of contaminants	References
Microorganisms				
Cryptosporidium	-	Watery diarrhea, weight loss, stomach pain, vomiting and nausea.	Sewage sludge and solid waste discharge.	[14]
Giardia lamblia	-	Giardiasis disease having symptoms: Diarrhea, vomiting, cramps, malnutrition and growth failure in infancy.		[15]
Legionella	-	Sever kind of pneumonia (Legionnaire's Disease).	Water systems like, hot-tub, mist sprayers, air conditioners *etc*.	[16]
Disinfection Byproducts				
Bormate	0.010	Can induce cancer.	Produced during disinfestation of drinking and swimming pool water.	[17]
Chlorite	1.0	Nervous system impairments and anemia in infants and young children.		
Haloacetic acid (HAA)	0.060	Can induce cancer.		
Total Trihalomethanes (TTHMs)	0.10	Can induce cancer and also adversely affects the functioning of liver, kidney and central nervous system.		
Inorganic contaminants				

(Table 1) cont.....

Contaminant of soil and water	MCL (maximum contaminant level) mg/L	Effects	Sources of contaminants	References
Antimony	0.006	Irritation of the skin, eyes, and lungs. High levels of antimony in water can cause stomach pain, stomach ulcers diarrhea, and vomiting; altered electrocardiograms.	Fumes from burning of oil fuels, petroleum refineries; volcanic eruptions, forest fires *etc.*	[18]
Arsenic	0.010 as of 01/23/06	Cancer, cardiovascular and peripheral vascular diseases; anemia, leukopenia, eosinophilia, developmental anomalies, diabetes, hearing loss and portal fibrosis.	Runoff of Mining sites, agricultural runoff from glass & electronics wastes	[19]
Asbestos	7 MFL	Highly carcinogenic and can develop benign intestinal polyps, lung cancer and mesothelioma (mesothelium cell line cancer).	Building demolition, decay of asbestos cement in water bodies, fire proofing and insulation.	[20]
Barium	2	Cardiovascular and kidney diseases, metabolic, neurological, and mental disorders.	Discharge from metal refineries mining sites, drilling wastes, and erosion of natural deposits	[21]
Beryllium	0.004	Person exposed to Berylium becomes sensitized to beryllium and later on develops Berylliosis or chronic beryllium disease (CBD).	Discharge from metal refineries mining sites, drilling wastes, and erosion of natural deposits	[22]
Cadmium	0.005	Bone demineralization or bone damage, Kidney damage, impaired lung functioning and increased risk of lung cancer and the Itai-Itai-disease.	Runoff from cadmium plating, waste batteries, paints, alkaline batteries, copper alloys and plastics, erosion of natural deposits, discharge from metal refineries.	[23]
Chromium	0.1	Cancer in humans.	Discharge from Industries like steel and pulp mills and erosion of natural deposits.	[24]

(Table 1) cont.....

Contaminant of soil and water	MCL (maximum contaminant level) mg/L	Effects	Sources of contaminants	References
Cyanide	0.2	Seizers, cardiac arrest and permanent nerve damage if a person survives after exposure to cyanide.	Discharge from metal, steel, plastic and fertilizer factories.	[25]
Fluoride	4.0	Bone disease; Children may get mottled teeth.	Discharge from fertilizer and aluminum factories; natural deposit erosion.	[26]
Lead	0.015	Kidney problems and high blood pressure in adults, delays in physical or mental development in infants and deficit in attention span and learning abilities in children;	Erosion of natural deposits, sewage sludge, waste batteries, paints, munitions *etc*.	[27]
Mercury	0.002	Health effects depend upon the age of person, duration of exposure and the form of mercury. Mercury is a neurotoxin which affects the central nervous system and causes neuromuscular changes, irritability, fatigue, tremors, headaches, dysarthria, hearing and cognitive loss, emotional changes, hallucinations, cardiovascular problems and death.	Discharge from metal refineries, and factories; runoff from landfills, mining sites and croplands and erosion of natural deposits.	[28]
Nitrates and nitrites	10 and 1.0 respectively	Risk of cancer in older women.	Runoff from agriculture; natural deposit erosion; leaching from septic tanks and sewage treatment plants.	[29]
Selenium	0.05	Diarrhea, fatigue, Dermatologic effects (nail and hair loss, nails discoloration or brittleness and dermatitis), joint pain, and nausea.	Discharge from mines and petroleum refineries; erosion of natural deposits.	[30]

(Table 1) cont.....

Contaminant of soil and water	MCL (maximum contaminant level) mg/L	Effects	Sources of contaminants	References
Thalium	0.002	Fatigue, headache, nausea and vomiting, muscle and joint pain, numbness of fingers and toes, cranial nerve abnormalities, altered mental status.	Discharge from electronics, glass, and drug factories; industrial combustion of coal; bottom sediments and flood plain soil, leaching from mining and ore-processing sites.	[31]
Radionuclides				
Alpha particles	15 picocuries per Liter	May cause lung and liver cancer.	Alpha decay of heavy radioactive atoms, ternary fission, accelerator (cyclotrons and synchrotrons), erosion of natural deposits of radioactive minerals.	[32, 33]
Beta particles and Radium	4 milligrams per year and 5 pCi/L respectively	May cause bone cancer.	Decay of radioactive minerals that may emit photons and beta radiation.	[33]
Uranium	30 ug/L	May cause lung cancer and affects the normal functioning of kidney.	Natural deposits erosion and leaching from radium and uranium mining sites.	[33]
Gaseous Contaminants				
Carbondioxide (CO_2)	-	Global warming.	Burning of fossil fuels, and transportation.	[34 - 37]
Sulphur dioxides and nitrogen oxides	-	Formation of acid rain.	Combustion of coal and petroleum, burning of fossil fuels, volcanic eruptions and industrial processes.	

(Table 1) cont.....

Contaminant of soil and water	MCL (maximum contaminant level) mg/L	Effects	Sources of contaminants	References
Carbon monoxide	-	Sudden illness or death, lung diseases.	Produced by the incomplete combustion of natural gas, coal or wood; heating systems, water heaters, gas cooking ranges and portable generators.	[34 - 37]
Non-methane volatile organic compounds	-	Non-methane volatile organic compounds are considered as carcinogens that may cause leukemia with prolonged exposure.	Fossil fuels, vehicle exhaust, and industries.	
Particulate matter	-	Lung diseases.	Produced from burning of fossil fuels, volcanoes, dust storms, forest fires, sea spray, power plants *etc.*	
Chlorophlorocarbons (CFCs)	-	Causes harm to ozone layer.	released from air conditioners, refrigerators, aerosol sprays, *etc.*	
Ammonia	-	Ammonia can cause bronchiolar and alveolar edema, and airway destruction. Exposure to higher concentrations can cause immediate burning of nose, throat and respiratory tract. Exposure to lower concentrations can cause nose, throat irritation and coughing.	Emitted from agricultural processes.	
Radioactive pollutants	-	Risk of cancer.	Produced by nuclear power plants, nuclear bomb testings, diagnostic centers and hospitals, war explosives.	[38]

Fossil Fuels

Dead remains of living organisms form the fossil fuels under high pressure and heat. Fossil fuel formation is a long-term process which takes millions of years to be completed. Coal, crude oil and natural gas are the three major types of fossil fuels. Combustion of fossil fuels releases various gases like carbon dioxide, nitrogen oxides and sulfur dioxide. Burning of gasoline and diesel releases greenhouse gases like carbon dioxide (CO_2), methane (CH_4), nitrous oxide (N_2O), and hydro fluorocarbons (HFCs) into the environment, which are also responsible for global warming and climate change.

Oil Spills

Oil spill pollution refers to the spill of crude oil or oil distilled products such as diesel fuels, jet fuels, kerosene, gasoline, Stoddard solvent, hydraulic oils, lubricating oils *etc*. which results in negative effect on environmental health and health of living organisms. Oil spills mainly pollute the surface of the water, air and soil. Oil spill pollution may comprise of a variety of organic compounds like paraffins (alkanes), isoparaffins (isoalkanes), aromatics, cycloalkanes, alkenes and alkynes. Other compounds which are present in crude oil and oil discharges include sulfur, nitrogen and oxygen atoms. Torray canyon (super tanker) oil spill was one of the first major oil spills. The super tanker was carrying 1,20,000 ton oil in it when it hit the reef off the Coast of Cornwall. More than 15000 sea birds and huge number of aquatic animals were killed by this oil spill.

Mining

Mining is the extraction of any material that cannot be produced either artificially in laboratory and factory or through agricultural processes. Mining includes the extraction of geological materials which are generally non-renewable resources from the earth like petroleum, natural gas, gemstones, limestone, metals, coal, rock salt, potash, gravel *etc*. Besides providing essential raw materials for various industries, mining is a major source of contamination of surface and groundwater. During the process of mining many oxidation reactions occur such as acidic solutions are formed by the oxidation of iron salts and sulfur-containing materials in the presence of water. The major contaminants produced at the mining sites are heavy metals, selenium, thalium, barium, beryllium, arsenic, asbestos, antimony, mercury and cyanide produced by the processing of ores.

Industries and Industrial Accidents

Industries don't use the raw products cost effectively instead they produce a large number of byproducts in the form of gas, liquid or solid. The gases such as NO,

NO_2, SO_2, SO_3, Cl_2, CO, CO_2, fumes of H_2SO_4 and other acids, *etc.* are directly liberated into the atmosphere; the water used as cooling agent who also contains harmful chemicals is directly discharged into rivers without prior treatment, solid wastes are illegally dumped on the land. The common most polluting industries include the cement, distillery, oil refineries, paper and pulp industry, agro-chemicals such as pesticides, fertilizers, herbicides, insecticides *etc.*, iron and steel industry, pharmaceuticals, sugar, textiles, thermal power plants, and tanneries. The wide variety of contaminants enters into soil air and water and disturbs the natural eco-system affecting the health of living organisms. Chemical accidents which happen by natural events or act of god, such as earthquakes, floods, lightning *etc.*, are called as "Natech" accidents. Due to Natech accidents hazardous substances are released from refineries, overland oil, gas pipelines, oil and gas storage, which are responsible for sever environmental pollution, fatalities, injuries, and economic losses [4 - 7]. Some of the toxic substances which are released in the environment by industrial accidents are acrylonitrile, ammonia, chlorine, carbon disulphide, hydrogen cyanide, hydrogen sulphide, hydrogen fluoride, hydrogen chloride, methyl isocyanate, Phosgene, sulphur dioxide and sulphur trioxide.

Transport

Ozone, particulate matter, smog-forming emissions and greenhouse gases such as nitrogen oxides, carbon monoxide, and carbon dioxide are the major contaminants of air which are produced by passenger vehicles and heavy duty trucks. Cars, buses and trucks emit the primary pollutants in the environment in the form of gases, which react in the environment to form the secondary contaminants like smog, fog acid rain. These secondary contaminants are responsible for asthma and bronchitis, increase the risk of lung and other cancer. Major percentage of greenhouse gas (GHG) emissions of the world is from transportation.

Corrosion of Underground Storage Tanks

Underground storage tanks (USTs) are generally made-up of network of steel pipes which has 10% volume underground. They are generally used for the storage of petroleum products (gasoline, diesel, kerosene). Over time steel corrodes and causes leakage of USTs containing hazardous liquids which may contaminate millions of gallons of ground water and large areas of soil. After leakage from the tank, petroleum products leached out into soil and enter into the ground water.

Intensive Farming and Agrochemicals

In the recent past the use of agrochemicals has increased many folds to fulfill the requirement of food for the growing population. Agrochemicals include the fertilizers, pesticides, insecticides, fungicides and herbicides. The most serious problems is the contamination of groundwater, surface water and drinking water supply, rivers, lakes and ponds from agrochemicals due to leaching of chemicals, agricultural runoff and floods [8 - 11].

Road Debris

Road deposited sediments (RDS) are derived from emission from the various transportation means, vehicle tires, brakes, body frames, asphalt road surfaces, road railings, fences, deicing salt *etc*. RDS consist of metals (zinc, cadmium, mercury, lead *etc.*) metalloids (arsenic, silicone, boron, asbestos, antimony, selenium *etc.*) and polycyclic aromatic hydrocarbons (PAH) [12]. Due to heavy rainfall and floods, RDS enter into the rivers, lakes and drinking water supplies and it severely affects the water quality and life of aquatic flora and fauna.

Waste Disposal

Waste refers to those materials which are not usable further with any kind of processing. Waste is created by all life forms but the waste created by modern life style of humans and the industries cannot be cope up by nature. The waste may be hazardous and non-hazardous. The hazardous waste can be classified as the biological, chemical and radioactive waste, which generally originates from various sources like house hold materials, industries, hospitals and clinics, construction sites, nuclear and thermal power plants, and agriculture [13]. Untreated waste, leads to contamination of air, water and soil.

Nuclear Wastes

Nuclear waste is a material which has no further use and that contains radionuclides. Radionuclides are unstable elements which disintegrate spon-taneously and emit radiations. Nuclear waste is created by nuclear power plants, laboratories, hospitals and diagnostic centers, military weapons production and nuclear bomb testing, mining, and other industrial uses. The natural radiations or the background reactions include the radioactive elements like uranium, thorium, radon, potassium, carbon, and radium. These elements can be traced in the rocks, soil, and water. Man-made radiations include refining of plutonium, mining of thorium, strontium-90 and Iodine-131. The Chernobyl nuclear power plant disaster which happened in 1986 in Ukraine exposed the environment with Strontium-90 and Iodine-131 (Both are radioactive materials), which is responsible for the increased risk of thyroid and other cancers.

Discharge of Sewage Sludge

Sewage sludge is produced as a byproduct by sewage treatment plant, which generally contains micro-organisms, toxic materials like pesticides, and heavy metals, nanomaterial and hormones. Treated sewage sludge is also known as bio-solids and is applied to agricultural cropland as fertilizers. Sewage sludge is a rich source of Nitrogen, Phosphorus, calcium, organic matter and other micronutrients such as boron, copper, iron, magnesium, molybdenum, zinc *etc*. The qualities of soil improve by the application of sewage sludge to agricultural. However, sewage sludge has become controversial as a fertilizer; because it contains heavy metals and pathogenic microorganisms which may enter into the water sources and the heavy metals may become accumulated in the top soil. The heavy metals that are present in the sludge exist in the form of oxides, hydroxides, sulfides, silicates, *etc*., and they also bound with the organic matter of the sludge.

Landfill and Illegal Dumping

A landfill is a site where waste materials are filled into pits by burial. It is also called as, dump, garbage dump or dumping ground. The illegal dumping or the fly dumping is the dumping of waste illegally without using the authorized rubbish dump. Because of landfilling and illegal dumping, soil and ground water both have become contaminated. Soil contaminants enter into groundwater aquifers through leaching. Various types of hazardous contaminants originate from the dumping site whose common sources are batteries, waterproofing agents, epoxies, lead based paint, asbestos and commercial cleaning compounds, gypsum, shingles, ceiling tile, insulation and vinyl floor covering. Decomposing wastes also generate methane and other gasses.

Coal Ash

Coal combustion residuals (CCR) or coal ash is produced by burning of coal. Coal naturally concentrates lead and zinc and other heavy metals during its formation. Heavy metals are lesser in degree in comparison with lead and zinc. Due to burning of coal most of the heavy metals become concentrated in the ash except mercury. Coal ash and slag usually contains sufficient lead, zinc, cadmium, nickel, barium, boron, chlorine, cobalt, beryllium, manganese, molybdenum, chromium, copper, sulphate, bromide, antimony, aluminium, vanadium and variable concentrations of polynuclear aromatic hydrocarbons (PAHs). These toxic elements and PAHs are responsible for increased risk of cancer and nervous system impairments such as developmental delays, impaired bone growth in children, behavioral problems and cognitive deficits, lung and kidney diseases, heart damage, gastrointestinal illness, reproductive problems, respiratory distress and birth defects.

Electronic Waste

Electronic devices have become integrated part of our daily life. When the electric appliances reach to the end of their useful life, they are considered as electronic waste. If electronic waste is dumped with municipal waste then it releases 1000s of different toxic substances such as lead, zinc, lithium, germanium, nickel, silicone, tin, mercury, arsenic, cadmium, copper, aluminum, selenium, and hexavalent chromium at the dumping site. These toxic substances contaminate the soil and later contaminate the groundwater too through leaching from soil.

Ammunitions and Agents of War

Ammunitions are the materials fired, scattered, dropped, from any weapon, as bombs or rockets, shrapnel, bullets, or shells fired by guns. The improper disposal of munitions can contaminate soil and water. Some hazardous substances used in the formation of munitions are 2-ADNT, DNB, DNA, HMX, RDX, TNT, TNB, and tetryl. They are also known as nitro-aromatic munition compounds. These compounds are persistent and remain present in soil, sediment, and surface water or groundwater at the sites where they were produced or processed for example when an organism is shot with the lead bullet, it remains in the carcass of the organism that was never retrieved. The carcass or the gut of these organisms remains open in the environment and they can be consumed by the scavengers like carnivores. In this way lead enters into the food chain and causes lead poisoning.

EFFECTS OF ENVIRONMENTAL CONTAMINANTS

With the increase in pollutant concentrations environmental damage increases linearly. A definite threshold in pollutant concentration must be achieved to induce any adverse effect in the environment. Biodegradable pollutants show such type of effect. An important type of the effect where pollutant produces severe effects while interacting with other pollutant instead of acting individually is known as synergistic effect. A good example of this effect is smoking in the presence of asbestos, which increases the chances of causing lung cancer at a higher rate.

Effects of various contaminants on environment and organisms are as follows:

Gaseous Pollutants

The major pollutants like SO_2, HNO_3 and O_3 leads to reduced photosynthesis, stomata functioning stunted growth and enhances leaf senescence and premature leaf fall in many plant species. Ozone directly affects all the functioning of plants as it produces free radicals which can damage cell membrane. Increase in concentration of NO_2 up to the level of 1900 µg m^{-3} leads to the necrosis in plants. While 989 µg m^{-3} concentrations lead to the chlorosis. Pollutants lead to the formation of Abscisic acid (ABA) which causes closure of stomata. Many of the lower plants like Lichens and moses are very sensitive to pollution. Lichens are said to be very good indicator of SO_2 pollution because they are affected by very low concentration of SO_2 in air. Sulphur and oxygen molecules react to form different types of compounds such as SO, SO_2, SO_3, *etc.* which are known as oxides of sulphur. With an increased of 50 µg/m^3 in concentration of sulphur dioxide in environment western European countries witnessed about 3% increase in daily mortality rate.

Nitric oxide, Nitrous oxide (N_2O), Nitrate (NO_3) and Nitrogen dioxide (NO_2) are the major types of the oxides of Nitrogen which variably affects plants, animals and environmental processes. NO_2 partially contributes to the health effects caused by the urban air pollution mixtures [39]. Oxides of nitrogen can inhibit the photosynthesis in majority of plants [34]. Nitric oxide (NO) which is a low molecular weight gas, effects the regulation of blood vessel dilation [40]. The normal functioning of biomolecules like proteins, lipids and even DNA is hampered by free radicals [41]. This oxidative state causes multifarious diseases such as cardiac arrest, atherosclerosis, chronic inflammatory diseases, Parkinson's, and Alzheimer's [37].

Carbon monoxide (CO) can show lethal effects in hemoglobic animals when its concentrations rises over about 35 ppm. However, it is also brought out in normal animal metabolic processes in low quantities which is significant in some regular biological functions. The photosynthetic activity of the crop plants, temperate trees and forests are influenced by the increasing concentration of CO_2 in the environment [35, 36]. High CO_2 concentration could increase of yield and water use efficiency by minimizing transpiration by 34% [42]. CO captures heme iron of hemoglobin and myoglobin. Thus, it affects several intracellular signaling pathways and functioning of cell [40]. Most of the gaseous pollutants inhaled impact the respiratory system and promote hematological disorder and cancer [37].

Crude oil is the main source of hydrocarbons found on earth, three major types of saturated hydrocarbons, unsaturated hydrocarbons and aromatic hydrocarbons which significantly affect the environment can be of different forms like gases, liquids, waxes or low melting solids or polymers like polyethylene.

Soil can concentrate PAH's from 1 to 10 μg / kg which can aggregate almost double in amount in vegetation from 10 to 20 μg / kg [43]. Five different types of lung cancer cases are recorded annually out of one million people in urban areas of central Copenhagen because of the presence of toxic air pollutants and polycyclic aromatic hydrocarbons [44].

Persistent Organic Pollutants

Dioxin

More than 90% of humans are exposed to dioxin by consuming contaminated food mainly meat and dairy products and fishes. Dioxin causes the alteration of metabolism by inducing many metabolic enzymes like glutathione-transferase, tyrosine kinase *etc* [45]. The interference of with hormones and can harm immune system, reproduction and development, and can also lead to cancer. Even benzene, basic component of dioxin can volatilize to air and can potentially cause severe respiratory diseases. Trinitrotoluene derived herbicides may consist of dioxin, which is having toxic properties.

Hazardous Pesticides

The use of synthetic insecticides like DDT which is banned by most of the countries is still prevalent worldwide. The use of organic pesticide has a cascading effect in the form of food chain. Organic chemicals, including pesticides, herbicides, fungicides *etc*. cause wide range of health problems.

Heavy Metals

The normal functioning of enzymes can be hampered by binding of Heavy metals [46, 47]. Metals like nickel can act as a carcinogenic agent when it binds to DNA and affect the tumor suppressor genes.

Lead is extensively used in batteries, construction, plumbing, paints, solders, alloys and leaded gasoline. In the late 19th century, as soon as the toxicity of the lead was recognized, its use was strictly restricted from many applications. Lead possesses neurotic properties and affects soft tissues and bones, causes blood disorders and damages the nervous system particularly in case of children. The content of lead in the atmosphere should not exceed 2ug/m3. Improperly disposed Batteries, soldering equipments, ammunition, hair colour, leaded gasoline from vehicle exhausts) mining, and coal burning are the main sources that cause lead content in the air atmosphere. Inhalation causes reduced haemoglobin formation and leads to anemia. Lead can damage RBC which leads to organ damage. Lead exposure negatively affects functioning of glutamate, dopamine and other molecules involved in functioning of nervous system [48]. Another severe effect of lead can be seen in fetus where disfunctioning of endocrine system affects the development of central nervous system [49]. Lead contaminated soil can cause damage the development of brain in young children. Some frequent disorders caused by lead contaminated soil, water and food are renal disease, Neurological damage, impairment of hand-eye co-ordination Lack of attention, bone weakening.

Cadmium can accumulate in plants through the use of cadmium contaminated soil or water used for irrigation which subsequently enters the food chain. Cadmium when accumulated in some animals can be transferred to human on consumption of contaminated meat. Itai-itai is a famous known disease, first identified when high levels of cadmium in rice was noticed from mining areas of Toyoma Region of Japan. Beside the above mentioned disease it also causes osteomalacia and proteinuria which results in softness of bones and secretion of excess proteins through urine. The many cases of lung cancer have been reported which advocates the carcinogenic property of cadmium.

Zinc is released from zinc smelters and refineries which is highly toxic for humans. Fluoride is a micronutrient whereof their appropriate level strengthens teeth. However, accumulation of fluoride in the bone over the years can cause "Skeletal fluorosis" A disease which causes stiffness and pain in the joints. Some other symptoms of fluoride are osteosclerosis and bone malformations.

Mercury is a part of many household, if not handled properly it can easily affect humans. Main cause of exposure for humans is through the consumption of contaminated seafood. It severely affects central nervous system, gastric system, brain development, low IQ level, deformed infants, touch and visual senses. Inhalation of 1mg/m3 of air for three months can lead to death. It also effects nervous system, livers, eyes are damaged, headache, fatigue anxiety, loss of appetite. Chronic exposure *i.e.* for one month or long can cause swelling of gums and renal and neurotic disorder. It also causes pink disease having symptoms like pain and pink discoloration of hands and feet. While in case of arsenic a long period exposure can lead to chronic arsenic poisoning known as arsenicosis. It also enhances the risk of miscarriage, hinders the functions of heart, liver, bone marrow and neurons.

Particulate Matter

Atmospheric aerosol particles, particulates, particulate matter (PM), or suspended particulate matter (SPM) can be used as synonyms to each other. Term aerosol is a broad term which refers to the particulate and the air mixture present in the atmosphere. Particulates directly impacts environmental processes as they affect radiation and sunlight coming from sun which adversely affect human health. The dust particles and particulates larger than 2.5 µm settles down easily but lesser size than 2 μm diameter is a major problem as they are very difficult to remove from atmosphere even after rain, whereas particles lesser than 1 μm in size can penetrate deep in lungs and alveoli causing serious health disorders like Cardiovascular problems, Irritation of eyes, defects in lung functioning, inefficient endocrine, reproductive and immune systems, DNA mutations, bronchitis, Asthma, cough, Headaches, fatigue and dizziness neurobehavioral changes even premature death in some cases. The minute particles like lead particles, asbestos, flash, volcanic emission, pesticide spray and metallic dust cause respiratory diseases. Byssinosis is a very common disease in India caused by cotton dust. The fungal spores, bacterial cells, pollen grains are also a part of particulates and can cause allergy, respiratory and other diseases.

Microscopic asbestos fibers are released into the air and dust when asbestos-containing products are crumbled. Health effects may not appear for decades. Inhalation is the common cause but, in some cases, it can be ingested in the skin. Inhalation of asbestos fibers can lead to lung tissues and retains for many years causing asbestos-related pleural abnormalities and lung carcinoma.

Other Pollutants

Photochemical Products

London has faced smog as a serious problem till the 20th century. Many of the major cities throughout the world are facing problem of smog. It has become a very common phenomenon in National Capital Region of India which has resulted mainly because of agricultural burnings in nearby states and vehicular pollution. In humans, smog has introduced severe respiratory diseases. Studies in Canada revealed increased asthma patients with increase in ozone concentration [38]. Plants are no more unaffected by the smog. Presence of smog can significantly injure foliage in some species of pines [50]. In humans, severe respiratory diseases are caused by photochemical smog. Olefins which is product of ethylene directly affects the growth of plants and can enhance the withering of sepals and petals in flowering plants. Aldehydes cause skin irritation and respiratory disorders. The photochemical smog enters with the inhaled air and effects respiratory system and causes asthma and bronchitis. Emphysema is a serious disease caused by smog pollutants which effects the lungs and cause severe breathlessness. The smog particles like fog, dust, mist *etc.* reduces the visibility, cause corrosion of metals, paints, leather, textiles and buildings. Aerosols decreases the total amount of sunlight reaching the earth therefore it reduces the photosynthetic activities thus it effects agricultural production also.

Acid Rain

Acid rain is having high levels of hydrogen ions which increase the acidity of rainwater. It shows destructive effects on flora, fauna and infrastructure. It damages the leaf cover of tree thus reduces the photosynthetic activity and affects its growth. Acid rain leads to soil acidification and exposes plants to toxic chemicals and affects mycorhiza functioning and nitrogen fixation by bacteria. Acid rain upsurges acidity of aquatic bodies thus directly harms organisms. Corals and mollusks are highly sensitive o acidity. It also affects zooplankton and phytoplankton which directly affects the productivity and the food chain system of water bodies. Higher acidity directly affects the fish mortality rate and species diversity in aquatic ecosystem. Weathering of limestone marble, stone buildings and statues causes a serious effect of monuments. It enhances corrosion of metals, fading of paints, degrades building materials. Acid rain significantly affects the environmental nutrient cycling [51]. Acid rain negatively affects breeding of many bird species by interacting with soil pH, and habitat fragmentation.

Effects of Water and Soil Contaminants

As the pollutants of soil and water are common, the effects produced by the water pollutants are similar in case of soil contaminants. The surface runoff from contaminated soil can easily pollute the water bodies. In addition to the already discussed different types of pollutants, water can be polluted by many types of disease-causing bacteria, viruses and other microbes which can potentially cause many diseases like Typhoid, Giardiasis, Ascariasis, and Amoebiasis. Polluted water of swimming pool and sea beaches can cause rashes, ear ache, respiratory infections, hepatitis, pink eye, encephalitis, diarrhea, stomach aches, gastro-enteritis, vomiting. Polluted water can cause severe fatal diseases like failure of many organs and can even mutilate DNA. Exposure to mercury through contaminated water can severely affect the development of fetus in the womb and can cause diseases like Parkinson, Alzheimer, Cardiac diseases. Too much amount of chemical like sodium chloride (ordinary salt) in water can alter the metabolism of microorganisms. Water contaminated from herbicides, pesticides may disrupt photosynthesis in aquatic plants or can even kill plants and animals important for food chain and ecosystem. Consumption of chemical contaminants by microbes from water can easily enter in food chain thus and can affect organisms at higher trophic level though they are not in direct contact of contaminants.

The contaminated soil can directly or indirectly affects human on the basis of contact. Sometimes excess flow of nitrogen, phosphate and potassium in water bodies from soil can lead to overgrowth of toxic algae pollution which is a kind of nutrient pollution and the phenomenon is known as eutrophication. This excess grown algae doesn't allow enough sunlight to penetrate in water body thus affecting the other flora and fauna and if eaten by other animals it may cause death. Most of the pesticides enter the water bodies from surface water drainage [52].

Chemical contamination or slight change in pH can cause decrease amphibians diversity specially frogs. Oil pollution caused by oil spills or accidents can negatively affect the marine and the nearby environment drastically. It causes severe effects on the productivity of nearby soil in contact and can disturb the diversity of microbes, planktons, fishes and birds. Persistent organic pollutants (POPs) may cause severe effect on fish life.

Effects of Radioactive Contaminants

When radioactive materials or byproducts of a nuclear reaction come in contact with human populations either by man-made or natural factors causes radioactive pollution. Its major factor is the human created radioactive waste. As the increase in the radioactive nuclear fuel in nuclear power plants, nuclear reactors and in some other experiments has resulted in the nuclear pollution because the byproducts are disposed of without any precautionary measures. An electron can be easily separated from an atom converting it in to ionized form by highly energetic radiations.

These ionized molecules form free radicals which affect the cells and other components and initiate the growth of cancerous cells. The important types of the radioactive emissions are Alpha, Beta and Gamma emission. Gamma rays are the most dangerous which can easily penetrate human body and can be stopped by thick sheets of lead. When soil is contaminated by radioactive substances it can results in chromosomal mutation, stunted plant's growth, weak seeds development, other biochemical processes and can cause death in plants [38]. When water gets contaminated with radioactive waste it can affect the entire aquatic food web. On humans, the scale of the adverse effects largely depends on the extent and period of exposure to radioactivity. Exposure of shorter and low duration can cause skin rashes, while long and low-intensity exposures include vomiting diarrhea, hair loss, *etc.* Irreparable damage to DNA molecules can occur if they are exposed to long-term exposure of radiation. Bone marrow and alimentary canal cells are more profound towards radioactivity as compared to bone cells and nervous cells because of their rapidly growing nature. Some common types of cancers caused by radiations are skin, lung and thyroid cancer. Genetic mutation caused by radiations can be passed on to the next generations and can carry their effects on for many generations. This can be observed in the tragic radiation outburst happened in Hiroshima and Nagasaki. The other type of the nonionizing radiations is UV radiations which can severely affect plants, animals, and microorganisms [53].

CONCLUSION

We humans individually are responsible for the deterioration of our environment. Our personal actions can worsen or improve the quality of environment. We need to make major changes in our life style to get a cleaner environment. For this we have to make firm resolution to adopt an environment friendly life style. With the help of solar energy, natural processes can indefinitely renew the topsoil, air, water, grasslands, forests and wildlife on which all forms of life depend, but these renewable resources should not be used at a faster rate than they are replenished. Our waste can be diluted, recycled and decomposed by natural processes until these processes are not overloaded. We must respect our natural resources and use them sustainably.

CONSENT FOR PUBLICATION

Not applicable.

CONFLICT OF INTEREST

The authors declare no conflict of interest, financial or otherwise.

ACKNOWLEDGEMENTS

None Declare

REFERENCES

[1]   Edelstein MR. Contaminated communities: coping with residential toxic exposure. Cambridge: Westview Press 2004.

[2]   Benner ML, Mohtar RH, Lee LS. Factors affecting air sparging remediation systems using field data and numerical simulations. J Hazard Mater 2002; 95(3): 305-29.
       [http://dx.doi.org/10.1016/S0304-3894(02)00144-9] [PMID: 12423944]

[3]   Darvishzadeh T, Priezjev NV. Effects of cross flow velocity and transmembrane pressure on microfiltration of oil-in-water emulsions. J Membr Sci 2012; 423: 1-31.

[4]   Lindell MK, Perry RW. Hazardous materials releases in the Northridge earthquake: implications for seismic risk assessment. Risk Anal 1997; 17(2): 147-56.
       [http://dx.doi.org/10.1111/j.1539-6924.1997.tb00854.x]

[5]   Young S, Balluz L, Malilay J. Natural and technologic hazardous material releases during and after natural disasters: a review. Sci Total Environ 2004; 322(1-3): 3-20.
       [http://dx.doi.org/10.1016/S0048-9697(03)00446-7] [PMID: 15081734]

[6]   Cruz AM, Krausmann E. Hazardous-materials releases from offshore oil and gas facilities and emergency response following Hurricanes Katrina and Rita. J Loss Prev Process Ind 2009; 22: 59-65.

[http://dx.doi.org/10.1016/j.jlp.2008.08.007]

[7]　Cozzani V, Campedel M, Renni E, Krausmann E. Industrial accidents triggered by flood events: analysis of past accidents. J Hazard Mater 2010; 175(1-3): 501-9.
[http://dx.doi.org/10.1016/j.jhazmat.2009.10.033] [PMID: 19913354]

[8]　Huang WY, Uri ND. The determination of an optimal policy for protecting groundwater quality. Water Resour Bull 1989; 25(4): 775-82.
[http://dx.doi.org/10.1111/j.1752-1688.1989.tb05392.x]

[9]　Duttweiler DW, Nicholson HE. Environmental problems and issues of agricultural non point source pollution.Agricultural Management and Water Quality. Ames: Iowa State University Press 1983; pp. 1-16.

[10]　Tudge C. Whatever happens to nitrogen. New Sci 1984; 101: 13-5.

[11]　Powers JF, Broadbent FE. Proper accounting for N in cropping systems.Nitrogen Management and Groundwater Protection. Amsterdam: Elsevier 1989.
[http://dx.doi.org/10.1016/B978-0-444-87393-4.50012-5]

[12]　Li H, Shi A, Zhang X. Particle size distribution and characteristics of heavy metals in road-deposited sediments from Beijing Olympic Park. J Environ Sci (China) 2015; 32: 228-37.
[http://dx.doi.org/10.1016/j.jes.2014.11.014] [PMID: 26040749]

[13]　Chadar CK. Solid Waste Pollution: A Hazard to Environment. 2017.

[14]　Hunter PR, Hughes S, Woodhouse S, *et al.* Health sequelae of human cryptosporidiosis in immunocompetent patients. Clin Infect Dis 2004; 39(4): 504-10.
[http://dx.doi.org/10.1086/422649] [PMID: 15356813]

[15]　Fraser D, Bilenko N, Deckelbaum RJ, Dagan R, El-On J, Naggan L. Giardia lamblia carriage in Israeli Bedouin infants: risk factors and consequences. Clin Infect Dis 2000; 30(3): 419-24.
[http://dx.doi.org/10.1086/313722] [PMID: 10722422]

[16]　Fields BS, Benson RF, Besser RE, Richard E, Besser RE. Legionella and Legionnaires' disease: 25 years of investigation. Clin Microbiol Rev 2002; 15(3): 506-26.
[http://dx.doi.org/10.1128/CMR.15.3.506-526.2002] [PMID: 12097254]

[17]　Villanueva CM, Cordier S, Font-Ribera L, Salas LA, Levallois P. Overview of Disinfection By-products and Associated Health Effects. Curr Environ Health Rep 2015; 2(1): 107-15.
[http://dx.doi.org/10.1007/s40572-014-0032-x] [PMID: 26231245]

[18]　Cooper RG, Harrison AP. The exposure to and health effects of antimony. Indian J Occup Environ Med 2009; 13(1): 3-10.
[http://dx.doi.org/10.4103/0019-5278.50716] [PMID: 20165605]

[19]　Tchounwou PB, Patlolla AK, Centeno JA. Carcinogenic and systemic health effects associated with arsenic exposure--a critical review. Toxicol Pathol 2003; 31(6): 575-88.
[PMID: 14585726]

[20]　Yang H, Riviera YZ, Jube S, Nasu M. Pietro Bertino, Chandra Goparaju,Guido Franzoso, Michael T. Lotze, Thomas Krausz, Harvey I. Pass, Marco E. Bianchi, and Michele Carbone. Programmed necrosis induced by asbestos in human mesothelial cells causes high-mobility group box 1 protein release and resultant inflammation. Proc Natl Acad Sci USA 2009; 12611-6.
[http://dx.doi.org/10.1073/pnas.1006542107]

[21]　Kravchenko J, Darrah TH, Miller RK, Lyerly HK, Vengosh A. A review of the health impacts of barium from natural and anthropogenic exposure. Environ Geochem Health 2014; 36(4): 797-814.
[http://dx.doi.org/10.1007/s10653-014-9622-7] [PMID: 24844320]

[22]　Day GA, Stefaniak AB, Weston A, Tinkle SS. Beryllium exposure: dermal and immunological considerations. Int Arch Occup Environ Health 2006; 79(2): 161-4.
[http://dx.doi.org/10.1007/s00420-005-0024-0] [PMID: 16231190]

[23]   Godt J, Scheidig F, Grosse-Siestrup C, *et al.* The toxicity of cadmium and resulting hazards for human health. J Occup Med Toxicol 2006; 1: 22.
[http://dx.doi.org/10.1186/1745-6673-1-22] [PMID: 16961932]

[24]   De Flora S. Threshold mechanisms and site specificity in chromium(VI) carcinogenesis. Carcinogenesis 2000; 21(4): 533-41.
[http://dx.doi.org/10.1093/carcin/21.4.533] [PMID: 10753182]

[25]   Banerjee KK, Bishayee A, Marimuthu P. Evaluation of cyanide exposure and its effect on thyroid function of workers in a cable industry. J Occup Environ Med 1997; 39(3): 258-60.
[http://dx.doi.org/10.1097/00043764-199703000-00016] [PMID: 9093978]

[26]   Kanduti D, Sterbenk P, Artnik B. Fluoride: a review of use and effects on health. Mater Sociomed 2016; 28(2): 133-7.
[http://dx.doi.org/10.5455/msm.2016.28.133-137] [PMID: 27147921]

[27]   Tong S, von Schirnding YE, Prapamontol T, Prapamontol T. Environmental lead exposure: a public health problem of global dimensions. Bull World Health Organ 2000; 78(9): 1068-77.
[PMID: 11019456]

[28]   Fernandes Azevedo B, Barros Furieri L, Peçanha FM, *et al.* Toxic effects of mercury on the cardiovascular and central nervous systems. J Biomed Biotechnol 2012; 2012: 949048.
[http://dx.doi.org/10.1155/2012/949048] [PMID: 22811600]

[29]   van Grinsven HJM, Ward MH, Benjamin N, de Kok TM. Does the evidence about health risks associated with nitrate ingestion warrant an increase of the nitrate standard for drinking water? Environ Health 2006; 5: 26.
[http://dx.doi.org/10.1186/1476-069X-5-26] [PMID: 16989661]

[30]   MacFarquhar JK, Broussard DL, Melstrom P, Hutchinson R, Amy Wolkin MPH. Acute selenium toxicity associated with a dietary supplement. Arch Intern Med 2010; 170(3): 256-61. Colleen Martin, Raymond F, Burk, JR. Dunn, DVM, Alice L. Green, DVM; Roberta Hammond, William Schaffner, Timothy F. Jones doi:10.1001/archinternmed. 2009. 495.

[31]   Hoffman RS. Thallium toxicity and the role of Prussian blue in therapy. Toxicol Rev 2003; 22(1): 29-40.
[http://dx.doi.org/10.2165/00139709-200322010-00004] [PMID: 14579545]

[32]   Chauhan V, Howland M, Kutzner B, McNamee JP, Bellier PV, Wilkins RC. Biological effects of alpha particle radiation exposure on human monocytic cells. Int J Hyg Environ Health 2012; 215(3): 339-44.
[http://dx.doi.org/10.1016/j.ijheh.2011.11.002] [PMID: 22153871]

[33]   Harrison J, Day P. Radiation doses and risks from internal emitters. J Radiol Prot 2008; 28(2): 137-59.
[http://dx.doi.org/10.1088/0952-4746/28/2/R01] [PMID: 18495991]

[34]   Hill CA, Bennett JH. Inhibition of apparent photosynthesis by nitrogen oxides. Atmos Environ 1970; 4: 341-8.
[http://dx.doi.org/10.1016/0004-6981(70)90078-8]

[35]   Cure JD. Crop response to carbon dioxide doubling: A literature survey. Agric For Meteorol 1986; 38: 127-45.
[http://dx.doi.org/10.1016/0168-1923(86)90054-7]

[36]   Eamus D, Jarvis PG. The direct effects of increase in the global atmospheric CO_2 concentration on natural and commercial temperate trees and forests. Adv Eco Res 1989; 19: 1-55.
[http://dx.doi.org/http://dx.doi.org/10.1016/S0065-2504(08)60156-7]

[37]   Kampa M, Castanas E. Human health effects of air pollution. Environ Pollut 2008; 151(2): 362-7.
[http://dx.doi.org/10.1016/j.envpol.2007.06.012] [PMID: 17646040]

[38]   Cody RP, Weisel CP. GlennBirnbaum Lioy PJ. The effect of ozone associated with summer time

photochemical smog on the frequency of asthma visits to hospital emergency departments. Environ Res 1992; 58(1–2): 184-94.
[http://dx.doi.org/10.1016/S0013-9351(05)80214-2] [PMID: 1511672]

[39]   Liebrich A, Rapp U. Regula. Epidemiological effects of oxides of nitrogen, especially NO_2.. Air Pollut Health 1999; pp. 561-84.

[40]   Ryter SW, Otterbein LE. Carbon monoxide in biology and medicine. BioEssays 2004; 26(3): 270-80.
[http://dx.doi.org/10.1002/bies.20005] [PMID: 14988928]

[41]   Valko M, Rhodes CJ, Moncol J, Izakovic M, Mazur M. Free radicals, metals and antioxidants in oxidative stress-induced cancer. Chem Biol Interact 2006; 10: 160(1): 1-40.
[http://dx.doi.org/10.1016/j.cbi.2005.12.009]

[42]   Kimball BA, Idso SB. Increasing atmospheric CO_2: effects on crop yield, water use and climate. Agric Water Manage 1983; 7: 55-72.
[http://dx.doi.org/10.1016/0378-3774(83)90075-6]

[43]   Edwards NT. Polycyclic Aromatic Hydrocarbons (PAH's) in the terrestrial environment—a review. J Environmental Quality Abstract 1983; 12(4): 427-44.
[http://dx.doi.org/10.2134/jeq1983.00472425001200040001x]

[44]   Nielsen T, Jørgensen HE, Larsen JC, Poulsen M. City air pollution of polycyclic aromatic hydrocarbons and other mutagens: occurrence, sources and health effects. Sci Total Environ 1996; 189-190(190): 41-9.
[http://dx.doi.org/10.1016/0048-9697(96)05189-3] [PMID: 8865676]

[45]   Birnbaum LS. The mechanism of dioxin toxicity: relationship to risk assessment. Environ Health Perspect 1994; 102 (Suppl. 9): 157-67.
[http://dx.doi.org/10.1289/ehp.94102s9157] [PMID: 7698077]

[46]   Rossi E, Taketani S, Garcia-Webb P. Lead and the terminal mitochondrial enzymes of haem biosynthesis. Biomed Chromatogr 1993; 7(1): 1-6.
[http://dx.doi.org/10.1002/bmc.1130070102] [PMID: 8431673]

[47]   Goering PL. Lead-protein interactions as a basis for lead toxicity. Neurotoxicology 1993; 14(2-3): 45-60.
[PMID: 8247411]

[48]   Lasley SM, Green MC, Gilbert ME. Rat hippocampal NMDA receptor binding as a function of chronic lead exposure level. Neurotoxicol Teratol 2001; 23(2): 185-9.
[http://dx.doi.org/10.1016/S0892-0362(01)00116-7] [PMID: 11348836]

[49]   Wang SL, Lin CY, Guo YL, Lin LY, Chou WL, Chang LW. Infant exposure to polychlorinated dibenzo-p-dioxins, dibenzofurans and biphenyls (PCDD/Fs, PCBs)--correlation between prenatal and postnatal exposure. Chemosphere 2004; 54(10): 1459-73.
[http://dx.doi.org/10.1016/j.chemosphere.2003.08.012] [PMID: 14659948]

[50]   Bytnerowicz A, Olszyk DM, Huttunen S, Takemoto B. Effects of photochemical smog on growth, injury, and gas exchange of pine seedlings. Can J Bot 1989; 67(7): 2175-81.
[http://dx.doi.org/10.1139/b89-276]

[51]   Jhonson DW, Turner J, Kelly JM. The effects of acid rain on forest nutrient status. Water Resour Res 1982; 8(3): 449-61.
[http://dx.doi.org/10.1029/WR018i003p00449]

[52]   Wauchop RD. The pesticide content of surface water draining from agricultural fields—a review. J Environ Qual Abs 1977; 7(4): 459-72.
[http://dx.doi.org/10.2134/jeq1978.00472425000700040001x]

[53]   Weihs AW, Schmalwieser SG. UV Effects on living organisms. Environ Toxicol 2012; 609-88.

CHAPTER 2

Environmental Pollutants and Risk of Cancer

Indranil Chattopadhyay*

Department of Life Sciences, Central University of Tamil Nadu, Thiruvarur-610101, India

Abstract: Cancer is characterized by cell proliferation, prevention or bypass of programmed cell death, genomic instability, angiogenesis, invasion and metastasis which are influenced by environmental pollutants. It is the major cause of death in world wide. Lifestyle factors such as diet, smoking and use of alcohol are responsible for the development of cancer in a large part of the population of developed countries. Existence of carcinogens or co-carcinogens in polluted air and drinking water, as well as in food, played a significant contribution in our country. Endocrine disrupters modify the risk of breast, endometrial and prostate cancer. Laryngeal, oropharyngeal, hypopharyngeal, sinonasal, nasopharyngeal, oral and lung cancer are positively associated with smoking and air pollutants. Tobacco smoking induces DNA adducts formation which is responsible for mutations at *K-RAS and TP53* gene in the lung and pancreatic adenocarcinomas. Tobacco smoking induces promoter hypermethylation of p16 and DAPK genes in Non-Small Cell Lung Cancer (NSCLCs). Aflatoxin B1 (AFB1) causes promoter hypermethylation of tumour-suppressor genes *RASSF1, MGMT, and p16* in human hepatocellular carcinoma (HCC) patients. Down-regulation of p15, p16, PRKG1, PARD3, and EPHA8 genes at mRNA level due to hypermethylation and increased expression of STAT3, IFNGR1 at mRNA level due to hypomethylation were reported in patients having benzene exposure. Methylation-induced transcriptional inactivation of tumor suppressor genes, including *p53*, *CDKN2A (p16INK4A)*, Ras association domain family member 1 (*RASSF1A*), and death-associated protein kinase (*DAPK*) were reported in arsenic exposed individuals. The genetic and epigenetic alterations respond to environmental carcinogens have significant contribution for biomarker development in assessment of health risk.

Keywords : Arsenic Contamination, Asbestos, Benezene, Biomarkers, Bisphenol A, Cadmium, Cancer, DNA Methylation, Dioxins, Endocrine Disruptors, Gene Mutations, Inflammation, miRNA Alterations, Pesticides, PAHs, Tobacco, Tumor Microenvironment, Tumorigenesis.

INTRODUCTION

Cancer is among the leading cause of death in world, with ~14 million new cases

---
* Corresponding author Indranil Chattopadhyay: Department of Life Sciences, Central University of Tamil Nadu, Thiruvarur-610101, India; Tel: +91-9489054283; E-mails: indranil@cutn.ac.in; indranil_ch@yahoo.com

Ashita Sharma, Manish Kumar, Satwinderjeet Kaur & Avinash Kaur Nagpal (Eds.)

and 8.2 million cancer-related deaths in 2012 [1]. Carcinogens having different chemical properties can cause cancer through genetic mutations or genomic damage and also through the alterations of tumor microenvironment that facilitates tumor progression. Tumorigenesis is characterized by uncontrolled cell growth, inhibition of apoptosis, prevention of growth suppression, immortality, inflammation, alteration of cellular metabolism, genomic instability, angiogenesis, and cellular invasion that result metastasis [2]. Genome instability is strongly associated with the acquisition of cancer hallmarks. Genome instability is characterized by translocations, inversions, deletion and insertion of nucleotides, gene amplification and alterations of chromosomes numbers (aneuploidy) and base substitutions [3].

Environmental contaminants are substances which create harmful effects on organisms by accidentally and deliberately introduced into the environment. Key environmental contaminants such as arsenic, asbestos fibers, cadmium compounds, indoor coal combustion emissions, nickel and tobacco smoke are associated with the lung cancer development. Exposures of benzene are associated with an increased risk of developing leukemia and occupational exposures to benzidine are associated with development of bladder cancer. Exposures of vinyl chloride enhances the risk of development of hepatic angiosarcoma, tumors of brain and lung, and hematopoietic tumors such as lymphoma and leukemia (https://www.cancer.gov/about-cancer/causes-prevention/risk/ substances).

DNA damage induced by environmental pollutants can cause loss of genomic instability. People are exposed to various chemicals and carcinogens over their lifetimes [4]. In this chapter, we explore the role of environmental pollutants in genome instability that increases incidence of cancer.

Role of Endocrine Disruptors in Risk of Cancer Development

Endocrine disruptors (EDs) are exogenous chemicals which inhibit normal mammary and female genitalia development during early exposure of life. EDs enhance the risk of breast and endometrial cancers at very low doses through the prevention of production of natural hormones, their release and/or transport. Genotype and phenotype variations in individual are important determining factors of the effect of genetic disruptors [5].

Dioxins and dioxin-like compounds enhance risk of breast cancer through modification of certain genes which are involved in activation of carcinogen and metabolism of steroid hormone. Dioxins in females act as anti-estrogenic EDs as it acts through the ligand-activated nuclear transcription factor aryl hydrocarbon receptor (AhR). Activation of AhR leads to a reduction of ERβ transcriptional activity [5].

Cadmium is extensively spread into the environment through industrial emission, combustion of fossil fuels and by the use of cadmium-containing fertilizers and sewage sludge. It has been recognized as a human carcinogen by the International Agency for Research on Cancer (IARC, Lyon, France). Cadmium functions as a ligand of estrogens receptor. Cadmium may induce cancer through alteration of mRNA expression, prevention of DNA damage repair, enhancement of oxidative stress and prevention of apoptosis. Cadmium exposure enhances the risk of breast cancer through upregulation of ERK1/2 and AKT kinases in human breast cancer-derived cells due to its estrogenic mimic property [5].

Exposure of six agricultural pesticides such as chlorpyrifos, fonofos, coumaphos, phorate, permethrin, and butylate increased risk prostate cancer development [6, 7]. These compounds have significant inhibition capacity of p450 enzyme such as CYP1A2 and CYP3A4; CYP1A2 and CYP3A4 enzymes. These enzymes are involved in steroid hormone metabolism in prostate. These compounds may enhance risk of prostate cancer development by inhibiting steroid hormone metabolism in prostate [8].

Synthetic polymer such as Bisphenol A [BPA, 2, 2,-bis (hydroxyphenyl) propane] is used in the production of polycarbonate plastics in reusable plastic containers, beverage can liners, baby bottles, dental sealants and food. Due to continuous environmental exposure, it is observed in body fluids.

Early life exposure to BPA may increase susceptibility to develop prostate cancer by inducing point mutation (AR-T877A) in androgen receptor (AR) [9, 10]. BPA exposure enhanced cell proliferation and over expression of progesterone receptor and decreased apoptosis in breast cancer. The cytochrome P450 1A1 (CYP1A1) is involved in the metabolism of steroid hormones and polycyclic aromatic hydrocarbons. The A2455G G which is a allele of CYP1A1 is a risk factor for breast cancer development among Caucasian population [10].

Role of Tobacco in Risk of Cancer Development

Out of 4,000 chemicals, 60 carcinogens have been identified in cigarette smoke and 16 carcinogens have been identified in unburned tobacco. Tobacco smoking enhances the risk of oral, oropharyngeal, hypopharyngeal, laryngeal, sinonasal, nasopharyngeal, and lung cancer. The use of smokeless tobacco and consumption of alcohol in combination with tobacco smoking have synergistic effect for the development of these cancers. Tobacco smoking also causes development of tumour in the stomach, liver, and pancreas, as well as urinary bladder, ureter and renal pelvis, and renal cell carcinoma. Cigarette smoking and smokeless tobacco are one of the most important causative agent for the development esophageal squamous cell carcinoma (ESCC) [11].

Tobacco-specific nitrosamines such as 4-(methylnitrosamino)-1-(3-pyrid-l)-1-butanone (NNK) and *N*-nitrosonornicotine (NNN) are strong carcinogens in unburned tobacco [12, 13]. Tobacco specific carcinogens such as polycyclic aromatic hydrocarbons (PAH) and NNK and their DNA adducts are important contributor for tobacco-induced lung cancer. Cigarette smoke contains lower amounts of strong carcinogens such as nitrosamines, aromatic amines, and polycyclic aromatic hydrocarbons (PAHs) having 1–200 ng per cigarette than the weak carcinogens such as acetaldehyde, catechol, and isoprene having 1mg per cigarette. Unburned tobacco such as oral snuff, chewing tobacco, tobacco of cigarette contains less carcinogens than smoke of cigarette because smokes of cigarette are formed during combustion (Tables **1**-**2**).

Table 1. Carcinogens in tobacco

Chemical class	Representative carcinogens
PAH	dibenz[a,h]anthracene, BaP
Nitrosamines	NNN, NNK
Aldehydes	Formaldehyde, acetaldehyde
Aromatic amines	4-Aminobiphenyl, 2-naphthylamine
Phenols	Catechol
Volatile hydrocarbons	Benzene, 1,3-butadiene
Inorganic compounds	Cadmium

Table 2. List of cancers induced by tobacco specific carcinogens

Cancer type	Involvement of Carcinogen
Lung	NNK (major) 1,3-butadiene, PAH, isoprene, ethylene oxide, ethyl carbamate, aldehydes, benzene, metals
Larynx	PAH
Nasal	NNK, NNN, other nitrosamines, aldehydes
Oral	PAH, NNK, NNN
Esophagus	NNN
Liver	NNK, other nitrosamines, furan
Pancreas	NNK, NNAL
Cervix	PAH, NNK
Bladder	4-Aminobiphenyl, other aromatic amines
Leukaemia	Benzene

Effects of Tobacco on Genes Mutation

Unburned tobacco has low levels of Levels of PAH. The levels of NNK and NNN are higher in smokeless tobacco products. Metabolic activation of tobacco carcinogens exert their carcinogenic effects through the formation of DNA adducts. If DNA adducts are unable to escape DNA repair mechanisms, they may lead to mutation in an oncogene or tumor suppressor gene. This induces activation of the oncogenes such as such as *RAS* and *MYC* or inactivation of the tumor suppressor genes such as *TP53* and *CDKN2A*. PAH-DNA adducts and mutations in the TP53 gene have been reported in lung tumors. TP53 gene mutations have also been reported in tumors of larynx. G–T and G–A mutations are induced by carcinogens of tobacco-smoke [12].

NNK and its major metabolite 4-(methylnitrosamino)-1-(3-pyridyl)-1-butanol (NNAL) are responsible for the development of pancreatic cancer. Metabolic activity of NNK and PAH are reported in the cervix. These compounds may enhance the risk of cervical cancer development in smokers along with human papilloma virus infection. Polymorphisms in genes of metabolic activation and detoxification, as well as DNA repair may contribute to this variability. Genetic variations and environmental factors can influence carcinogenic risk. Adducts of DNA and protein and urinary metabolites provide important information about dose and metabolism of carcinogen. They are used as biomarkers for carcinogen exposure. Other than ethylene oxide, formaldehyde and acetaldehyde, most carcinogens of tobacco products require metabolic activation before formation of DNA adducts. Metabolic activation is enhanced by cytochrome P450 enzymes (P450s). Cytochrome P450s (*CYP1A1*) and glutathione-*S*-transferases (*GSTM1*) are responsible for the metabolic activation and detoxification of carcinogens. Metabolic activation of carcinogens forms DNA adducts.

Levels of serum-albumin and carcinogen–haemoglobin adducts are used to determine levels of carcinogen exposure and activation. Due to abundant amount of haemoglobin in the blood and long lifetime of erythrocytes in humans, haemoglobin-adduct measurement is more advantageous to determine levels of carcinogen exposure and activation. Levels of 3- and 4-aminobiphenyl–haemo-globin adducts was significantly higher in women than in men in relation to the number of cigarettes smoked per day. Adducts formation with the amino-terminal end of valine of haemoglobin have also been reported in smokers having carcinogenic dose. Ethylation of valine in haemoglobin at the amino-terminal end is increased in smokers as compared to non-smokers. O4-Ethylthymidine levels are increased in lung DNA of smokers as compared to non-smokers. Urinary metabolites are useful to determine dose, metabolic activation and detoxification of carcinogens due it's readily accessibility and presence in sufficient quantities.

Three urinary carcinogen biomarkers are *t,t*-Muconic acid (a metabolite of benzene), 1-hydroxypyrene (an indicator of PAH uptake) and NNAL and its glucuronides (NNAL-Glucs) which are produced due to metabolic activity of tobacco specific NNK. NNAL and NNALGlucs have relatively long half-lives compared with other urinary metabolites. These are used to quantify levels of NNK consumption in smokers and tobacco chewers [14].

Molecular Mechanisms Involved in Tobacco Smoking Induced Lung and Pancreatic Adenocarcinoma

DNA adducts formation and *K-RAS* gene mutations are reported in ~30% of pulmonary and 50–90% of pancreatic adenocarcinomas. These genetic alterations are associated with nicotine-derived nitrosamine, nitrosamine 4-(methylnitrosamino)-1-(3-pyridyl)-1-butanone (NNK). Epigenetic effects of NNK by functioning as an agonist for β-adrenergic receptors on pulmonary and pancreatic cells are also reported. This causes the secretion of arachidonic acid (AA) followed by the enzymatic formation of mitogenic AA metabolites which induce cell proliferation. Over-expression of AA-metabolizing enzymes 5-, 12- or 15-lipoxygenases (LOXs) and cyclooxygenase-2 (COX2) have been reported in pulmonary and pancreatic adenocarcinoma.

In pulmonary and pancreatic epithelial cells, nitrosamine 4-(methylnitrosamino--1-(3-pyridyl)-1-butanone (NNK) binds to β1- or β2-adrenergic receptors which activates G-proteins coupled receptor signalling pathway. This event activates adenylyl cyclase and cyclic AMP (cAMP) that drive activation of protein kinase A (PKA).

PKA activates transcription factors that are involved in cell proliferation such as cAMP response element binding protein (CREB), cAMP response element modulator (CREM), activating transcription factor-1 (ATF-1), activator protein 1 (AP1) or nuclear factor κB (NF-κB).

PKA activates RAP-1 (small G protein) which activates serine/threonine kinase B-RAF that initiates mitogen activate protein kinase (MAPK) cascade. This MAPK induces transcriptional activation of genes involved in cell proliferation. PKA induces phospholipase-A2 (PLA2) to release arachidonic acid (AA) from cell-membrane phospholipids. Arachidonic acid and its metabolites such as prostaglandins and leukotrienes induce cell proliferation and malignant transformation.

NNK also binds to the β2-adrenergic receptors which activate the kinase c-SRC that lead to signal transducer and activator of transcription 3 (STAT3) pathways and the RAS-mediated MAPK pathway. C-SRC also activates K^+ channels which lead to production of AA metabolites (prostaglandins, leukotrienes). Nicotine binds to nicotinic acetylcholine receptors which cause cell-membrane depolarization by opening and influx of voltage-activated Ca^{2+} channels in the cell membrane. NNK increases anchorage-independent growth of human bronchial epithelial cells. Increased expression of EGFR, EGFR-associated phosphotyrosine kinase and cyclin D1 were reported in NNK-induced malignant transformation of pulmonary and pancreatic region. Inactivation of tumour-suppressor gene *CDKN2A* and *O6*-methylguanine-DNA methyltransferase by promoter hypermethylation is a key mechanism for tobacco induced carcinogenesis [15].

Role of Environmental Exposure of Asbestos for Cancer Risk

Asbestos induced lung cancer is one of the leading occupational cancers. Asbestos are naturally occurring fibrous silicates which are mainly in acoustical and thermal insulation. Chrysotile is the most commonly used asbestos. As per WHO report, 5% of the European population have experience residential asbestos exposure. The relative risk (RR) of mesothelioma and lung cancer with environmental exposure to asbestos has significant positive correlation [16].

Cancer Risk from Outdoor and Indoor Air Pollution

Air pollution which include different gaseous and particulate components (SO_2, smoke particles, volatile organic compounds, nitrogen oxides, benzo[a]pyrene, benzene, some metals, particles especially fine particles) cause cancers of head and neck, oesophagus, stomach, colon, rectum, larynx, female breast, bladder and prostate. Several studies reported that SO_2, particulates or fuel consumption induced stomach cancer. Occupational benzene exposure was related to development of leukaemia. Heating fuel, cooking fuel, fumes from frying oils are the main source of indoor air pollution. A case control study from the Northern Province of South Africa reported that women using wood or coal as main fuel significantly developed lung cancer [16].

Cancer Risk from Arsenic and Chlorine Contaminated Drinking Water

A high level of arsenic contaminated groundwater is found in areas of Bangladesh, India (West Bengal), China (Xinjiang, Shanxi), Thailand, Mongolia, Taiwan, Vietnam, Argentina, Bolivia, Chile, Mexico, and the USA (Arizona, California, Nevada). In terms of levels and populations, the most significant exposures were reported in the Gulf of Bengal, South America and Taiwan. Arsenic and arsenic trioxide are used in the production of pesticides, herbicides, and insecticides. Arsenic contaminated water enhanced risk of bladder, skin and lung cancers. Drinking water may contain chlorination by products such as trihalomethanes, chloroform, bromodichloromethane, chlorodibromomethane and trihalomethanes which are potent carcinogenic agents [16].

Cancer Risk from other Drinking Water Pollutants

Organic compounds such as nitrites & nitrates, radionuclides and asbestos released from agricultural land, industries and commercial activities, and from waste disposals are possible sources of cancer risk in humans. High nitrate levels in drinking water increased risk of cancer in stomach, bladder and non-Hodgkin lymphoma [16].

Molecular Alterations by Environmental Carcinogens

Exposure to environmental mutagens induce carcinogenic effects in humans through DNA adducts formation and persistent mutations in DNA. This increases error rates during DNA replication. Molecular characterizations of environmental carcinogens induced tumours provide information about biologic responses to environmental carcinogens. Distinct environmental carcinogens induce somatic mutations in the DNA of cancer cells because they induce base exchange. Next generation sequencing of cancer genomes has contributed significant awareness of mutational pattern in environmental mutagens induced tumours. Environmental carcinogens modify gene expression by altering structural changes in DNA, DNA methylation, histone modifications, and microRNAs alterations [17].

Effect of Environmental Carcinogens on TP53 Gene

Tumour suppressor gene TP53 is a carcinogen-specific biomarker. Differential mutational spectrum of TP53 in different tumours reflects different mode of action mechanisms of different carcinogens on TP53 in cancer cell. TP53 is frequently mutated gene in human cancers. 80% of P53 mutations are nonsynonymous substitutions which lead to amino acid substitutions in proteins and can modify the conformation of wild type TP53 protein. These alterations prevent P53-dependent pathways such as cell-cycle regulation, DNA repair, differentiation, genomic instability, and apoptosis. Transversion of G: C to T: A of the P53 gene which induces substitution of arginine by serine at codon 249 was reported in liver cancer patients from Africa and China having habit of using aflatoxin B1 contaminated food. Best known carcinogen in cigarette smoke benzo[a]pyrene (BP) induces G: C-T: A transversions in TP53 gene in lung carcinoma. The codons 248, 273, and 157 of TP53 gene are commonly mutated in lung cancer by benzo[a]pyrene (BP).

Effect of Environmental Carcinogens on RAS Oncogene

RAS oncogenes such as N-RAS, H-RAS, and K-RAS are located on chromosome

1, 11, and 12, respectively. All genes possess a guanosine triphosphatase activity. K-RAS is most frequently altered oncogene in cancers. Majority of K-RAS gene mutations have been reported in codons 12 and 13. Each carcinogen induces a specific point mutation within the RAS oncogene. In colorectal tumors, frequent G to A transitions mutations in RAS oncogene are reported, but in pancreatic adenocarcinomas and lung carcinomas, this mutation is very rare. 4-Amino-biphenyl induces specific mutations C to A transversions in codon-61of H-RAS gene. Benzene induced mutation glycine to valine (G to T) or to aspartic acid (G to A) in codon 12 of K-RAS; whereas benzene metabolites induced GC to AT transitions and GC to TA transversions. Different radon-induced mutations were found in codon 12 (G to A transitions and G to T transversions) and 13 (G to A transitions) of K-RAS gene in uranium miners workers who developed lung adenocarcinomas [18].

Effect of Environmental Carcinogens on Genes Transcription

Increased expression of cancer-related genes such as C-MYC, H-RAS, C-FOS and decreased protein expression of β4 integrin have been reported in arsenic exposed human small airway epithelial cells. Arsenic induced over expression of dipeptidyl-peptidase 4, fibroblast activation protein alpha, and interleukin receptor-associated kinase in human bronchial epithelial cell line (BEAS-2B). ADAM 28 having disintegrin and metalloproteinase domain interacts with integrins. ADAM 28 showed higher expression in patients with primary lung adenocarcinoma having asbestos exposure. Annexin 2 also showed over expression in asbestos exposed lung cancer patients. Cadmium chloride induces promoter hypermethylation and a decrease expression of four DNA repair genes such as *ERCC1, XRCC1, hMSH2 and hOGG1* at mRNA and protein levels in human bronchial epithelial cells. Over expression of genes involved in RAS/MAPK signalling pathways, calcium regulation, and receptor-associated kinases in cardiovascular and pulmonary function are associated with Dioxins mediated toxic effect. Increased mRNA expression of p53 and a decreased expression of p21 were observed in a vinyl chloride monomer (VCM)-exposed workers. Environmental carcinogens such as benzo[a]pyrene, dibenzo[a,l]pyrene, and coal tar induce increased expression of heat shock proteins (HSP) such as HSP 70 and HSP 27 in breast cancer MCF-7 cells. Signal transduction proteins (Nucleoside diphosphate kinase-A and Galectin-3), cytoskeletal proteins (actin cytoplasmic 1, tubulin alpha-1C chain, Myosin light chain alkali, cyclophilin B), DNA-associated proteins (Dihydrodrofolate reductase, Heterogeneous nuclear ribonucleoprotein, histone H2A type 1) and glycolytic and mitochondrial proteins (Alpha enolase and glyceraldehyde-3-phosfate dehydrogenase, ATP synthase) were increased in benzo[a]pyrene exposed MCF-7 cells. Down regulation of apoptosis inducing gene (BAX), cell cycle regulatory genes (p53 and p21) and up-regulation of proliferating cell nuclear antigen, cyclins, CDKs and phosphorylated pRB were reported in BPA-exposed T47D breast cancer cells [18]. Diesters of phthalic acid are commonly used in consumer products, food processing and medical applications. It enhances progression of ER-negative breast cancer through regulation of AhR/HDAC6/c-Myc signaling pathway. It also induces the expression of cyclin D and cdk-4 which are involved in cell proliferation of ovarian cancer [1].

Effect of Environmental Carcinogens on Genes Methylation

Methylation of promoter in oestrogen receptor has been reported in lung tumors from smokers of tobacco. Aberrant hypermethylation of TRIM36 could be served as a biomarker for PAHs induced Non-Small Cell Lung Cancer (NSCLCs) development. Cigarette smoking is significantly associated with the increase of promoter hypermethylation of p16 and DAPK genes in Non-Small Cell Lung Cancer (NSCLCs). Hypermethylation of *P16* and *hMLH1* has been reported in lung cancer patients having previous exposure of chromium [18]. Significant hypo-methylation of LINE-1, *TP53* gene, and *IL-6* gene and hypermethylation of *HIC1* have been reported in workers of industrial estate in Thailand in blood leukocyte DNA compared with rural residents [19]. BPDE exposed human bronchial epithelial cells showed increased expression of DNA methyltransferase proteins and decreased expression of CDH13 which is frequently down-regulated in lung cancer [20].

Promoter hypermethylation and reduced expression of the IFNγ gene was reported in two human adenocarcinoma cell lines exposed to low and non-cytotoxic doses (0.1 and 1 nM) of BaP. Reduced expression of tumor suppressor gene p16INK4α due to hypermethylation of CpG islands within this gene was observed in BaP exposed human bronchial epithelial cells [21, 22]. Promoter hypermethylation and reduced expression of DUSP22 was reported in BaP exposed human Jurkat T lymphocyte cells and normal human prostate cells [23]. Gene-specific H3K9 hyperacetylation was reported in BaP exposed MCF7 breast cancer cells [24]. Down regulation of H3K4me2 was reported in the promoter region of the estrogen receptor α gene (ER) in human breast cancer cell line exposed to BaP [25].

Zhang *et al.* reported that tumour-suppressor genes RASSF1, MGMT, and p16 showed promoter hypermethylation in human hepatocellular carcinoma (HCC) patients who were exposed to aflatoxin B1 (AFB1) [26, 27]. Promoter hypermethylation of glutathione S-transferase pi (GSTP1) gene and level of AFB1-DNA adducts showed significant association in human HCC tumour tissue exposed to AFB1 [28]. A significant association has been reported between increased risks of lung cancer and coke oven workers. Promoter methylation of the tumor suppressor genes p14ARK and p16INK4 was increased in peripheral blood mononuclear cells of coke oven workers relative to water pump workers [29].

Various types of leukemia (primarily acute myelogenous leukemia) have been reported in workers exposed to benzene [30]. Hypermethylation along with mRNA down-regulation of p15, p16, PRKG1, PARD3, and EPHA8 genes and hypomethylation with increased mRNA level of STAT3, IFNGR1 genes were reported in patients with benzene poisoning [31, 32]. Down regulation of acetylation at histone H4 and H3 and methylation at H3K4, and upregulation of methylation at H3K9 were reported in the promoter region of topoisomerase IIα (Topo IIα) in patients having benzene exposure [33]. Formaldehyde enhances the risk of nasopharyngeal cancer and leukemia. Down-regulation of DNA methyltransferase genes DNMT3a and DNMT3b and up-regulation of DNMT1 and MBD2 at both the mRNA and protein level were reported in formaldehyde exposed individuals [34]. High level of phosphorylation of histone H3 H3S10 and H3S28 within the promoter region of the proto-oncogenes FOS and JUN were reported in human pulmonary epithelial cells after exposure to formaldehyde [35 - 37]. Vinyl chloride exposure is associated with angiosarcoma of the liver, hepatocellular carcinoma (HCC), lung cancer, and malignant neoplasms of connective and soft tissues. Promoter methylation of p14ARF and p16INKa were reported in angiosarcoma patients who had confirmed chronic occupational exposure to vinyl chloride [38]. Transcriptional inactivation of tumor suppressor genes, including *p53, CDKN2A (p16INK4A)*, death-associated protein kinase (*DAPK*), and Ras association domain family member 1 (*RASSF1A*) due to methylation were reported in arsenic exposed individuals [39 - 41] (Tables **3**-**7**).

Table 3. Role of chemical disruptors in carcinogenesis.

Environmental chemical disruptors	Molecular target	Molecular Mechanisms
Benzo(a)pyrene, bisphenol A, DDT, aflatoxin	p53	• Down regulation of p53 expression 25CF; Induce MDM2 expression • Mutation of *p53*
Arsenic [As(III)], DDT, Radon, Butadiene	Retinoblastoma	• Loss of heterozygosity of *RB1* • Hyperphosphorylation of Rb through increased activity of cdk • Increased expression of cyclin D1 • Loss of *INK4a*

(Table 3) cont.....

Environmental chemical disruptors	Molecular target	Molecular Mechanisms
Cigarette smoke & polycyclic aromatic hydrocarbons	LKB1, Gap junctions (connexins)	• Down-regulation of connexins expression • *LKB1* mutation and reduced LKB1 expression

Table 4. Effect of selected chemicals on Carcinogenesis

Chemicals	Cellular alteration of Carcinogenesis
Bisphenol A	• Promotes angiogenesis, promotes genetic instability, inflammatory processes and inhibits programmed cell death • Promotes replicative immortality and activates sustained proliferative signaling
DDT	• Promotes angiogenesis, promotes genetic instability, inflammatory processes and inhibits apoptosis • Promotes replicative immortality and activates sustained cell proliferation • It cause immune evasion through preventing the function of natural killer cells

Table 5. Effect of chemical disruptors on genome stability

Chemical disruptor (IARC group)	Mode of Exposure in Human	Mode of action
Lead	Ingestion of contaminated crops and water	Binding of DNA repair enzymes resulting in inhibition of DNA repair
Acrylamide	Ingestion of heated, fried and baked food	Interaction with the sulfhydryl groups of DNA repair proteins/enzymes to make them non-functional
Bisphenol A	Discharge from plastics into food and water	Reduce level of histone acetylation by alteration of histone acetylases and deacetylases. Alteration of DNA methylation patterns by affecting DNMTs
Nickel	Oral intake of contaminant drinking water, contaminant of foods such as chocolate, nuts and grains	Alteration of enzymes that modulate posttranslational histone modification

Table 6. Carcinogenic chemicals induced DNA adducts

Chemical	Source	Type of Cancer
Aflatoxin B1	Fungal contaminated food	Liver
Benzidine	Dye manufacture	Bladder
Benzo[*a*]pyrene	Tobacco smoke	Lung
NNK and NNN	Tobacco	Lung and Head and Neck
Vinyl chloride	Polyvinylchloride manufacture	Angiosarcoma

Table 7. Environmental carcinogens induced mutations reported in p53 Gene.

Carcinogen	Cancer Tissue	P53 Mutation
Chromium (VI)	Lung	transitions of GC base-pairs and transversions of AT base-pair at Exons from 5 to 8
Radon-222	Lung	transversion of AGG (arg) →ATG (met)at Codon 249
Aflatoxin B1	Liver	G-T transversions at codon 249 resulting in substitution of arginine by serine
4-Aminobiphenyl	Bladder	Mutation at codons 175, 248, 280, and 285
Arsenic	Skin & Bladder	C→T at codon 149 T→C at codon 172 G→A at codon 175 G→A at codon 175 G→C at codon 273 G→T at codon 283 C→A at codon 284 T→A at codon 292
Benzidine	Bladder	C→ T transitions at Exon-5 of codons 151 and 152
Benzo[a]pyrene	Lung	G → T transversions at codon 157, 248, and 273

Epigenetic alterations are considered as early indicators for both genotoxic and non-genotoxic carcinogens induced tumorigenesis. Identification of those epigenetic events is essential to understand carcinogens induced tumorigenesis.

Alterations of miRNA Signatures by Environmental Carcinogens

Irreversible miRNA alterations induced by long-term exposures of environmental carcinogens are used as a predictive biomarker for cancer [42]. Alteration of miRNA expression by environmental carcinogens involves down regulation of tumour suppressor miRNA and up-regulating oncogenic miRNA. MiR-17 is over-expressed by cigarette smoke, silica dust, and radon. Up-regulation of miR-19 by arsenic, miR-21 by TPA and miR-155 by benzene, toluene, ethanol, and TPA are reported. MiR-126 expression was down regulated in alcohol induced hepatocellular carcinoma.

Cigarette smoke (CS) condensed significantly repressed expression of miR-487b which directly targets SUZ12, BMI1, WNT5A, MYC, and KRAS genes. MiR-218, miR-15a, miR-199b, and miR-125b were extensively down regulated in bronchial epithelial cells of smokers [43]. Tumour suppressive miRNAs such as miR-200b, miR-200c, and miR-205 were significantly down regulated in tobacco carcinogens induced bronchial epithelial cells and in primary lung tumours due to DNA methylation [44]. MiR-126 regulates the expression of CYP2A3 which is involved bio-activation pathway for NNK [45]. Down regulation of miR-125a by CS is correlated with the activation of the oncogene ERBB2 in lung [46, 47]. Down regulation of miR-466 by CS exposure is functionally associated with cell proliferation, apoptosis, and k-Ras activation [48]. Metabolically activated form of Benzo(a)pyrene such as benzo[a]pyrene-7,8-diol-9,10-epoxide (BPDE) enhances the down regulation of miR-506 [49] and miR-10a [50], and the up regulation of miR-106a [51], miR-494 [50], and miR-22 [52] in transformed malignant bronchial epithelial cell. MiR-506 inhibits cell proliferation by targeting N-Ras [49]. MiR-22 regulates PTEN expression by binding to 3'-untranslated region of PTEN which leads to resistance to cellular apoptosis [52].

Over-expression of miR-24, let-7d, let-7e, miR-96, miR-148b, miR-199b-5p, miR-331-3p and miR-374a and down regulation of miR-202, miR-671-5p, miR-939, miR-605, and miR-1224-5p were reported by asbestos exposure. MiR-103 was identified as a potential blood based biomarker in the mesothelioma patients and asbestos-exposed individuals [53].

Among smokers and non-smokers, radon is the most frequent cause of lung cancer (**www.epa.gov, United States Environmental Protection Agency, October 12, 2010**). Significant differential expression of miRNAs including miR-17, miR-18a, miR-33b, miR-125b, miR-483-3p, miR-494, miR-886-3p, miR-2115, miR-1246, and miR-3202 were reported in radon gas exposed bronchial BEAS2B cells. Genes involved in the regulation of cell proliferation, differentiation, and adhesion during malignant transformation is targeted by these miRNAs [54].

Up-regulation of miR-146, let-7f, let-7g, miR-20, miR-21, and miR-335 reported in cytotrophoblast cell lines with BPA exposure. Both BPA and dichloro-diphenyl-trichlorethane (DDT) alter expression of 10 miRNAs in breast carcinoma cells [55]. Expression of miR-16, miR-92, miR-99, and miR-193a are only altered by DDT in breast carcinoma cells [55].

Vinyl carbamate which is a metabolite of ethylcarbamate (EC) is found in alcoholic beverages and a variety of food products, including yogurt, bread, soy sauce, cheese, and apple cider vinegar [56, 57]. Up regulation of miR-21, miR-31, miR-130a, miR-146b, and miR-377 and down regulation of miR-1 and miR-143 by vinyl carbamate in lung tumours compared with normal lungs was reported by Melkamu *et al.* (2010) [58]. MiR-21 acts as an oncogenic miRNA by down regulating the expression of the tumour-suppressor genes PTEN [59], p53, transforming growth factor-beta [60], tropomyosin 1 [61], PDCD4 [62], and RECK [63]. Up regulation of miR-22, miR-34, miR-221, and miR-222 and down regulation of miR-210 in arsenic exposed human immortalized lymphoblast cell line TK-6 are reported by Marsit *et al.* (2006) [64]. MiR-22 is involved in suppression of NF- kB [65], inhibition of angiogenesis [66], and induces cellular senescence [67]. Oncogenic miRNA MiR-221 regulates cell proliferation and migration. MiR-222 and miR-221 is involved in apoptosis, cell growth, and cell-cycle progression [68].

miRNA alterations in response to environmental carcinogens increase the risk of carcinogenesis. Therefore, miRNAs are considered as a potential biomarker for environmental health risk assessment.

Carcinogens Induced Alterations in Tumor Microenvironment

Tumor microenvironment is defined as particular independent cell populations inside the tumor. Alterations in the tumor microenvironment induce carcino-genesis through inflammation, blood vessel formation, innate and adaptive immune response and release of cytokines. Tumor microenvironment is influ-enced by potential carcinogens. Understanding the role of the tumor micro-environment in carcinogenesis depends on three parameters. First, carcinogens may initiate field of cancerization development and induces development of premalignant lesions in the tumor microenvironment that develop inflammation through release of pro-tumorigenic cytokines and immunosuppressive cytokines [69]. Secondly, matrix metalloproteinases (MMPs) may be deregulated. Carcino-gens in the tumor microenvironment effect functional regulation of Matrix metalloproteinases (MMPs), stromal fibroblasts, infiltration of immune cells and release of specific chemokines. MMP-2, MMP-3, MMP-7, MMP- 9, MMP-28 and membrane type 1-MMP are involved in epithelial–mesenchymal transition [70]. Third, carcinogens induce inflammation by regulating macrophages, NK cells and T cells of immune function. Inflammatory cytokines, free radicals such as reactive oxygen species (ROS) and nitrogen species (RNS) regulate carcinogenesis process to induce tumor growth. ROS induces cell proliferation, angiogenesis through the production of VEGF and immune evasion [71 - 73]. Pro-tumorigenic cytokines such as IL-6 and TGF-β activate NF-κB and Wnt which induce cell proliferation and prevent apoptosis during chronic inflammation [74]. IL-6 regulates genes involved in cell cycle progression, cellular senescence and inhibition of apoptosis. Known and potential carcinogens effect on the tumor microenvironment through infiltration of regulatory T cells (Tregs), APCs and NK cells [75]. IL-6 also regulates non-canonical Notch signaling pathway through p53 in the cell. Cell proliferation, EMT, invasion and migration events are positively regulated by secretion of IL-6 and tumor associated dendritic cells of tumor microenvironment [76, 77]. Nickel exposures enhance levels of ROS and up-regulation of c-Myc. Nickel induces hypoxia inducible factor-1 which promotes transcription of genes that drive angiogenesis, glucose transport and glycolysis [78]. Nickel chloride can induce EMT through the down regulation of E-cadherin by promoter hypermethylation and ROS generation [79]. Nickel chloride induces DNA damage and regulates inflammation and cell proliferation through the alteration of miR-222 and its target genes CDKN1B and CDKN1C [80].

Asbestos fibers induce inflammation associated with tumorigenesis in lung cancer and mesothelioma. In Lung epithelial cells of asbestos fibers exposure macrophages activate the Nalp3 inflammasome which enhances the production of TGF-β, TNF-α and IL-1β [81, 82]. Oxidative stress induced by methylmercury promotes carcinogenesis. Urethane-induced lung tumors showed over expression of myeloid-derived suppressor cells (CD11b+, GR-1+) and Tregs [83]. BPA, diethyl stilbesterol, bis (2-ethylhexyl) phthalate and *p*-nonylphenol induces cytokines production from macrophages [84]. Low exposure levels of BPA activate the mammalian target of rapamycin (mTOR) pathway that promote cell survival through insulin and insulin-like growth factors [85, 86]. NNK in cigarette smoke acts on alveolar macrophages. Pulmonary metabolism of NNK generates ROS which activates NF-κB, cyclooxygenase-1 and prostaglandin E2 production. Level of TNF, macrophage inflammatory protein-1α, IL-12 and nitric oxide are decreased by NNK [87, 88]. Oxidative stress in the tumor microenvironment generated by environmental pollutants may induce tumorigenesis [89 - 92].

SUMMARY

Environmental contaminants such as arsenic, asbestos fibers, cadmium compounds, indoor coal combustion emissions, nickel and tobacco smoke influence host genomes to develop tumorigenesis such as lung, bladder, and other hematopoietic malignancies. Initiation, progression, growth and relapse of cancer may be mediated through known and possible carcinogens. The environmental pollutants such as heavy metals, BPA, and acrylamide promote genome instability by alteration of DNA damage response, epigenetic alterations, DNA repair mechanism, cell division, and telomere integrity.

The modern public health problem is associated with low dose effects of carcinogens or tumor-inducing chemicals in our daily life. In the 20th century, cancer incidence has increased worldwide along with the advancement of western technology, pollution and consumer products. Lifestyle factors including diet, smoking contributes the advancement of risk of cancer in a major population of developing countries like India. Carcinogens or co-carcinogens present in air and drinking water, as well as in food, played significant contribution for cancer development in our country. Several biomarker such as microsatellite mutations, DNA damage/repair assessment, miRNA alterations, DNA methylation in promotor or CpG islands regions, mtDNA alterations, and telomere length are used monitoring individuals exposed to environmental carcinogens. Therefore, the implementation of physical and chemical hygiene and a reduction of exposure to chemical mixtures and genomic instability inducing agents are necessary condition for an effective prevention of cancer. Understanding the role of low-dose exposure of environmental carcinogens could help us to assess cancer risk assessment and may prevent risk of cancer by reducing or eliminating exposures to synergistic mixtures of environmental carcinogens. Carcinogen-biomarker can potentially identify crucial targets for chemoprevention and data can be important for developing effective chemoprevention strategies for clinical trials.

CONSENT FOR PUBLICATION

Not applicable.

CONFLICT OF INTEREST

The author declare no conflict of interest, financial or otherwise.

ACKNOWLEDGEMENTS

None Declare

REFERENCES

[1]   Narayanan KB, Ali M, Barclay BJ, *et al.* Disruptive environmental chemicals and cellular mechanisms that confer resistance to cell death. Carcinogenesis 2015; 36 (Suppl. 1): S89-S110.
        [http://dx.doi.org/10.1093/carcin/bgv032] [PMID: 26106145]

[2]   Hanahan D, Weinberg RA. The hallmarks of cancer. Cell 2000; 100(1): 57-70.
        [http://dx.doi.org/10.1016/S0092-8674(00)81683-9] [PMID: 10647931]

[3]   Hanahan D, Weinberg RA. Hallmarks of cancer: the next generation. Cell 2011; 144(5): 646-74.
        [http://dx.doi.org/10.1016/j.cell.2011.02.013] [PMID: 21376230]

[4]   Langie SA, Koppen G, Desaulniers D, *et al.* Causes of genome instability: the effect of low dose chemical exposures in modern society. Carcinogenesis 2015; 36 (Suppl. 1): S61-88.
        [http://dx.doi.org/10.1093/carcin/bgv031] [PMID: 26106144]

[5]   Del Pup L, Mantovani A, Cavaliere C, *et al.* Carcinogenetic mechanisms of endocrine disruptors in female cancers (Review). Oncol Rep 2016; 36(2): 603-12. [Review].
        [http://dx.doi.org/10.3892/or.2016.4886] [PMID: 27349723]

[6]   Alavanja MC, Samanic C, Dosemeci M, *et al.* Use of agricultural pesticides and prostate cancer risk in the Agricultural Health Study cohort. Am J Epidemiol 2003; 157(9): 800-14.
        [http://dx.doi.org/10.1093/aje/kwg040] [PMID: 12727674]

[7]   Mahajan R, Bonner MR, Hoppin JA, Alavanja MC. Phorate exposure and incidence of cancer in the agricultural health study. Environ Health Perspect 2006; 114(8): 1205-9.
        [http://dx.doi.org/10.1289/ehp.8911] [PMID: 16882526]

[8]   Sterling KM Jr, Cutroneo KR. Constitutive and inducible expression of cytochromes P4501A (CYP1A1 and CYP1A2) in normal prostate and prostate cancer cells. J Cell Biochem 2004; 91(2): 423-9.

[http://dx.doi.org/10.1002/jcb.10753] [PMID: 14743400]

[9]　　Prins GS, Korach KS. The role of estrogens and estrogen receptors in normal prostate growth and disease. Steroids 2008; 73(3): 233-44.
[http://dx.doi.org/10.1016/j.steroids.2007.10.013] [PMID: 18093629]

[10]　Prins GS. Endocrine disruptors and prostate cancer risk. Endocr Relat Cancer 2008; 15(3): 649-56.
[http://dx.doi.org/10.1677/ERC-08-0043] [PMID: 18524946]

[11]　Wogan GN, Hecht SS, Felton JS, Conney AH, Loeb LA. Environmental and chemical carcinogenesis. Semin Cancer Biol 2004; 14(6): 473-86.
[http://dx.doi.org/10.1016/j.semcancer.2004.06.010] [PMID: 15489140]

[12]　Tobacco habits other than smoking: betel quid and area nut chewing and some related nitrosamines. Lyon, France: International Agency for Research on Cancer Monographs on the Evaluation of the Carcinogenic Risk of Chemicals to Humans 1985; Vol. 37: pp. 37-202.

[13]　Hecht SS, Hoffmann D. Tobacco-specific nitrosamines, an important group of carcinogens in tobacco and tobacco smoke. Carcinogenesis 1988; 9(6): 875-84.
[http://dx.doi.org/10.1093/carcin/9.6.875] [PMID: 3286030]

[14]　Hecht SS. Tobacco carcinogens, their biomarkers and tobacco-induced cancer. Nat Rev Cancer 2003; 3(10): 733-44.
[http://dx.doi.org/10.1038/nrc1190] [PMID: 14570033]

[15]　Schuller HM. Mechanisms of smoking-related lung and pancreatic adenocarcinoma development. Nat Rev Cancer 2002; 2(6): 455-63.
[http://dx.doi.org/10.1038/nrc824] [PMID: 12189387]

[16]　Boffetta P, Nyberg F. Contribution of environmental factors to cancer risk. Br Med Bull 2003; 68: 71-94.
[http://dx.doi.org/10.1093/bmp/ldg023] [PMID: 14757710]

[17]　Chappell G, Pogribny IP, Guyton KZ, Rusyn I. Epigenetic alterations induced by genotoxic occupational and environmental human chemical carcinogens: A systematic literature review. Mutat Res Rev Mutat Res 2016; 768: 27-45.
[http://dx.doi.org/10.1016/j.mrrev.2016.03.004] [PMID: 27234561]

[18]　Ceccaroli C, Pulliero A, Geretto M, Izzotti A. Molecular fingerprints of environmental carcinogens in human cancer. J Environ Sci Health C Environ Carcinog Ecotoxicol Rev 2015; 33(2): 188-228.
[http://dx.doi.org/10.1080/10590501.2015.1030491] [PMID: 26023758]

[19]　Peluso M, Bollati V, Munnia A, et al. DNA methylation differences in exposed workers and nearby residents of the Ma Ta Phut industrial estate, Rayong, Thailand. Int J Epidemiol 2012; 41(6): 1753-60.
[http://dx.doi.org/10.1093/ije/dys129] [PMID: 23064502]

[20]　Damiani LA, Yingling CM, Leng S, Romo PE, Nakamura J, Belinsky SA. Carcinogen-induced gene promoter hypermethylation is mediated by DNMT1 and causal for transformation of immortalized bronchial epithelial cells. Cancer Res 2008; 68(21): 9005-14.
[http://dx.doi.org/10.1158/0008-5472.CAN-08-1276] [PMID: 18974146]

[21]　Tang WY, Levin L, Talaska G, et al. Maternal exposure to polycyclic aromatic hydrocarbons and 5'-CpG methylation of interferon-γ in cord white blood cells. Environ Health Perspect 2012; 120(8): 1195-200.
[http://dx.doi.org/10.1289/ehp.1103744] [PMID: 22562770]

[22]　Yang P, Ma J, Zhang B, et al. CpG site-specific hypermethylation of p16INK4α in peripheral blood lymphocytes of PAH-exposed workers. Cancer Epidemiol Biomarkers Prev 2012; 21(1): 182-90.
[http://dx.doi.org/10.1158/1055-9965.EPI-11-0784] [PMID: 22028397]

[23]　Ouyang B, Baxter CS, Lam HM, et al. Hypomethylation of dual specificity phosphatase 22 promoter correlates with duration of service in firefighters and is inducible by low-dose benzo[a]pyrene. J Occup Environ Med 2012; 54(7): 774-80.

[http://dx.doi.org/10.1097/JOM.0b013e31825296bc] [PMID: 22796920]

[24] Sadikovic B, Andrews J, Carter D, Robinson J, Rodenhiser DI. Genome-wide H3K9 histone acetylation profiles are altered in benzopyrene-treated MCF7 breast cancer cells. J Biol Chem 2008; 283(7): 4051-60.
[http://dx.doi.org/10.1074/jbc.M707506200] [PMID: 18065415]

[25] Khanal T, Kim D, Johnson A, Choubey D, Kim K. Deregulation of NR2E3, an orphan nuclear receptor, by benzo(a)pyrene-induced oxidative stress is associated with histone modification status change of the estrogen receptor gene promoter. Toxicol Lett 2015; 237(3): 228-36.
[http://dx.doi.org/10.1016/j.toxlet.2015.06.1708] [PMID: 26149760]

[26] Zhang YJ, Ahsan H, Chen Y, *et al.* High frequency of promoter hypermethylation of RASSF1A and p16 and its relationship to aflatoxin B1-DNA adduct levels in human hepatocellular carcinoma. Mol Carcinog 2002; 35(2): 85-92.
[http://dx.doi.org/10.1002/mc.10076] [PMID: 12325038]

[27] Zhang YJ, Chen Y, Ahsan H, *et al.* Inactivation of the DNA repair gene O 6-methylguanine-DNA methyltransferase by promoter hypermethylation and its relationship to aflatoxin B1-DNA adducts and p53 mutation in hepatocellular carcinoma. Int J Cancer 2003; 103(4): 440-4.
[http://dx.doi.org/10.1002/ijc.10852] [PMID: 12478658]

[28] Zhang YJ, Chen Y, Ahsan H, *et al.* Silencing of glutathione S-transferase P1 by promoter hypermethylation and its relationship to environmental chemical carcinogens in hepatocellular carcinoma. Cancer Lett 2005; 221(2): 135-43.
[http://dx.doi.org/10.1016/j.canlet.2004.08.028] [PMID: 15808399]

[29] Zhang H, Li X, Ge L, Yang J, Sun J, Niu Q. Methylation of CpG island of p14(ARK), p15(INK4b) and p16(INK4a) genes in coke oven workers. Hum Exp Toxicol 2015; 34(2): 191-7.
[http://dx.doi.org/10.1177/0960327114533576] [PMID: 24837742]

[30] Infante PF. Benzene exposure and multiple myeloma: a detailed meta-analysis of benzene cohort studies. Ann N Y Acad Sci 2006; 1076: 90-109.
[http://dx.doi.org/10.1196/annals.1371.081] [PMID: 17119195]

[31] Yang J, Bai W, Niu P, Tian L, Gao A. Aberrant hypomethylated STAT3 was identified as a biomarker of chronic benzene poisoning through integrating DNA methylation and mRNA expression data. Exp Mol Pathol 2014; 96(3): 346-53.
[http://dx.doi.org/10.1016/j.yexmp.2014.02.013] [PMID: 24613686]

[32] Xing C, Wang QF, Li B, *et al.* Methylation and expression analysis of tumor suppressor genes p15 and p16 in benzene poisoning. Chem Biol Interact 2010; 184(1-2): 306-9.
[http://dx.doi.org/10.1016/j.cbi.2009.12.028] [PMID: 20044985]

[33] Yu K, Shi YF, Yang KY, *et al.* Decreased topoisomerase IIα expression and altered histone and regulatory factors of topoisomerase IIα promoter in patients with chronic benzene poisoning. Toxicol Lett 2011; 203(2): 111-7.
[http://dx.doi.org/10.1016/j.toxlet.2011.02.020] [PMID: 21382456]

[34] Liu Q, Yang L, Gong C, *et al.* Effects of long-term low-dose formaldehyde exposure on global genomic hypomethylation in 16HBE cells. Toxicol Lett 2011; 205(3): 235-40.
[http://dx.doi.org/10.1016/j.toxlet.2011.05.1039] [PMID: 21745553]

[35] Ibuki Y, Toyooka T, Zhao X, Yoshida I. Cigarette sidestream smoke induces histone H3 phosphorylation *via* JNK and PI3K/Akt pathways, leading to the expression of proto-oncogenes. Carcinogenesis 2014; 35(6): 1228-37.
[http://dx.doi.org/10.1093/carcin/bgt492] [PMID: 24398671]

[36] Yoshida I, Ibuki Y. Formaldehyde-induced histone H3 phosphorylation *via* JNK and the expression of proto-oncogenes. Mutat Res 2014; 770: 9-18.
[http://dx.doi.org/10.1016/j.mrfmmm.2014.09.003] [PMID: 25771866]

[37]   Lu K, Boysen G, Gao L, Collins LB, Swenberg JA. Formaldehyde-induced histone modifications *in vitro*. Chem Res Toxicol 2008; 21(8): 1586-93.
        [http://dx.doi.org/10.1021/tx8000576] [PMID: 18656964]

[38]   Weihrauch M, Markwarth A, Lehnert G, Wittekind C, Wrbitzky R, Tannapfel A. Abnormalities of the ARF-p53 pathway in primary angiosarcomas of the liver. Hum Pathol 2002; 33(9): 884-92.
        [http://dx.doi.org/10.1053/hupa.2002.126880] [PMID: 12378512]

[39]   Huang YC, Hung WC, Chen WT, Yu HS, Chai CY. Sodium arsenite-induced DAPK promoter hypermethylation and autophagy via ERK1/2 phosphorylation in human uroepithelial cells. Chem Biol Interact 2009; 181(2): 254-62.
        [http://dx.doi.org/10.1016/j.cbi.2009.06.020] [PMID: 19577553]

[40]   Mass MJ, Wang L. Arsenic alters cytosine methylation patterns of the promoter of the tumor suppressor gene p53 in human lung cells: a model for a mechanism of carcinogenesis. Mutat Res 1997; 386(3): 263-77.
        [http://dx.doi.org/10.1016/S1383-5742(97)00008-2] [PMID: 9219564]

[41]   Cui X, Wakai T, Shirai Y, Hatakeyama K, Hirano S. Chronic oral exposure to inorganic arsenate interferes with methylation status of p16INK4a and RASSF1A and induces lung cancer in A/J mice. Toxicol Sci 2006; 91(2): 372-81.
        [http://dx.doi.org/10.1093/toxsci/kfj159] [PMID: 16543296]

[42]   Izzotti A, Cartiglia C, Longobardi M, Larghero P, De Flora S. Dose responsiveness and persistence of microRNA alterations induced by cigarette smoke in mice. Mutat Res Fund Mech 2011; 717: 9-16.
        [http://dx.doi.org/10.1016/j.mrfmmm.2010.12.008] [PMID: 21185844]

[43]   Izzotti A, Pulliero A. The effects of environmental chemical carcinogens on the microRNA machinery. Int J Hyg Environ Health 2014; 217(6): 601-27.
        [http://dx.doi.org/10.1016/j.ijheh.2014.01.001] [PMID: 24560354]

[44]   Tellez CS, Juri DE, Do K, *et al.* EMT and stem cell-like properties associated with miR-205 and miR-200 epigenetic silencing are early manifestations during carcinogen-induced transformation of human lung epithelial cells. Cancer Res 2011; 71(8): 3087-97.
        [http://dx.doi.org/10.1158/0008-5472.CAN-10-3035] [PMID: 21363915]

[45]   Jalas JR, Hecht SS, Murphy SE. Cytochrome P450 enzymes as catalysts of metabolism of 4-(methylnitrosamino)-1-(3-pyridyl)-1-butanone, a tobacco specific carcinogen. Chem Res Toxicol 2005; 18(2): 95-110.
        [http://dx.doi.org/10.1021/tx049847p] [PMID: 15720112]

[46]   Izzotti A, Calin GA, Arrigo P, Steele VE, Croce CM, De Flora S. Downregulation of microRNA expression in the lungs of rats exposed to cigarette smoke. FASEB J 2009; 23(3): 806-12. a
        [http://dx.doi.org/10.1096/fj.08-121384] [PMID: 18952709]

[47]   Izzotti A, Calin GA, Steele VE, Croce CM, De Flora S. Relationships of microRNA expression in mouse lung with age and exposure to cigarette smoke and light. FASEB J 2009; 23(9): 3243-50. b
        [http://dx.doi.org/10.1096/fj.09-135251] [PMID: 19465468]

[48]   Izzotti A, Cartiglia C, Steele VE, De Flora S. MicroRNAs as targets for dietary and pharmacological inhibitors of mutagenesis and carcinogenesis. Mutat Res 2012; 751(2): 287-303. a
        [http://dx.doi.org/10.1016/j.mrrev.2012.05.004] [PMID: 22683846]

[49]   Zhao Y, Liu H, Li Y, *et al.* The role of miR-506 in transformed 16HBE cells induced by anti-benzo[a]pyrene-trans-7,8-dihydrodiol-9,10-epoxide. Toxicol Lett 2011; 205(3): 320-6. a
        [http://dx.doi.org/10.1016/j.toxlet.2011.06.022] [PMID: 21726609]

[50]   Shen YL, Jiang YG, Greenlee AR, Zhou LL, Liu LH. MicroRNA expression profiles and miR-10a target in anti-benzo[a] pyrene-7, 8-diol-9, 10-epoxide-transformed human 16HBE cells. Biomed Environ Sci 2009; 22(1): 14-21.
        [http://dx.doi.org/10.1016/S0895-3988(09)60016-7] [PMID: 19462682]

[51]   Jiang Y, Wu Y, Greenlee AR, *et al.* miR-106a-mediated malignant transformation of cells induced by anti-benzo[a]pyrene-trans-7,8-diol-9,10-epoxide. Toxicol Sci 2011; 119(1): 50-60.
[http://dx.doi.org/10.1093/toxsci/kfq306] [PMID: 20889678]

[52]   Liu L, Jiang Y, Zhang H, Greenlee AR, Yu R, Yang Q. miR-22 functions as a micro-oncogene in transformed human bronchial epithelial cells induced by anti-benzo[a]pyrene-7,8-diol-9,10-epoxide. Toxicol In Vitro 2010; 24(4): 1168-75. b
[http://dx.doi.org/10.1016/j.tiv.2010.02.016] [PMID: 20170724]

[53]   Weber DG, Johnen G, Bryk O, Jöckel KH, Brüning T. Identification of miRNA-103 in the cellular fraction of human peripheral blood as a potential biomarker for malignant mesothelioma--a pilot study. PLoS One 2012; 7(1): e30221.
[http://dx.doi.org/10.1371/journal.pone.0030221] [PMID: 22253921]

[54]   Cui FM, Li JX, Chen Q, *et al.* Radon-induced alterations in micro-RNA expression profiles in transformed BEAS2B cells. J Toxicol Environ Health A 2013; 76(2): 107-19.
[http://dx.doi.org/10.1080/15287394.2013.738176] [PMID: 23294299]

[55]   Tilghman SL, Bratton MR, Segar HC, *et al.* Endocrine disruptor regulation of microRNA expression in breast carcinoma cells. PLoS One 2012; 7(3): e32754.
[http://dx.doi.org/10.1371/journal.pone.0032754] [PMID: 22403704]

[56]   Zimmerli B, Schlatter J. Ethyl carbamate: analytical methodology, occurrence, formation, biological activity and risk assessment. Mutat Res 1991; 259(3-4): 325-50.
[http://dx.doi.org/10.1016/0165-1218(91)90126-7] [PMID: 2017216]

[57]   Battaglia R, Conacher HB, Page BD. Ethyl carbamate (urethane) in alcoholic beverages and foods: a review. Food Addit Contam 1990; 7(4): 477-96.
[http://dx.doi.org/10.1080/02652039009373910] [PMID: 2203651]

[58]   Melkamu T, Zhang X, Tan J, Zeng Y, Kassie F. Alteration of microRNA expression in vinyl carbamate-induced mouse lung tumors and modulation by the chemopreventive agent indole---carbinol. Carcinogenesis 2010; 31(2): 252-8.
[http://dx.doi.org/10.1093/carcin/bgp208] [PMID: 19748927]

[59]   Meng F, Henson R, Wehbe-Janek H, Ghoshal K, Jacob ST, Patel T. MicroRNA-21 regulates expression of the PTEN tumor suppressor gene in human hepatocellular cancer. Gastroenterology 2007; 133(2): 647-58.
[http://dx.doi.org/10.1053/j.gastro.2007.05.022] [PMID: 17681183]

[60]   Papagiannakopoulos T, Shapiro A, Kosik KS. MicroRNA-21 targets a network of key tumor-suppressive pathways in glioblastoma cells. Cancer Res 2008; 68(19): 8164-72.
[http://dx.doi.org/10.1158/0008-5472.CAN-08-1305] [PMID: 18829576]

[61]   Zhu S, Si ML, Wu H, Mo YY. MicroRNA-21 targets the tumor suppressor gene tropomyosin 1 (TPM1). J Biol Chem 2007; 282(19): 14328-36.
[http://dx.doi.org/10.1074/jbc.M611393200] [PMID: 17363372]

[62]   Zhu S, Wu H, Wu F, Nie D, Sheng S, Mo YY. MicroRNA-21 targets tumor suppressor genes in invasion and metastasis. Cell Res 2008; 18(3): 350-9. b
[http://dx.doi.org/10.1038/cr.2008.24] [PMID: 18270520]

[63]   Gabriely G, Wurdinger T, Kesari S, *et al.* MicroRNA 21 promotes glioma invasion by targeting matrix metalloproteinase regulators. Mol Cell Biol 2008; 28(17): 5369-80.
[http://dx.doi.org/10.1128/MCB.00479-08] [PMID: 18591254]

[64]   Marsit CJ, Eddy K, Kelsey KT. MicroRNA responses to cellular stress. Cancer Res 2006; 66(22): 10843-8.
[http://dx.doi.org/10.1158/0008-5472.CAN-06-1894] [PMID: 17108120]

[65]   Takata A, Otsuka M, Kojima K, *et al.* MicroRNA-22 and microRNA-140 suppress NF-κB activity by regulating the expression of NF-κB coactivators. Biochem Biophys Res Commun 2011; 411(4): 826-

31.
[http://dx.doi.org/10.1016/j.bbrc.2011.07.048] [PMID: 21798241]

[66] Yamakuchi M, Yagi S, Ito T, Lowenstein CJ. MicroRNA-22 regulates hypoxia signaling in colon cancer cells. PLoS One 2011; 6(5): e20291.
[http://dx.doi.org/10.1371/journal.pone.0020291] [PMID: 21629773]

[67] Xu D, Takeshita F, Hino Y, *et al.* miR-22 represses cancer progression by inducing cellular senescence. J Cell Biol 2011; 193(2): 409-24.
[http://dx.doi.org/10.1083/jcb.201010100] [PMID: 21502362]

[68] Rao X, Di Leva G, Li M, *et al.* MicroRNA-221/222 confers breast cancer fulvestrant resistance by regulating multiple signaling pathways. Oncogene 2011; 30(9): 1082-97.
[http://dx.doi.org/10.1038/onc.2010.487] [PMID: 21057537]

[69] Casey SC, Vaccari M, Al-Mulla F, *et al.* The effect of environmental chemicals on the tumor microenvironment. Carcinogenesis 2015; 36 (Suppl. 1): S160-83.
[http://dx.doi.org/10.1093/carcin/bgv035] [PMID: 26106136]

[70] Orlichenko LS, Radisky DC. Matrix metalloproteinases stimulate epithelial-mesenchymal transition during tumor development. Clin Exp Metastasis 2008; 25(6): 593-600.
[http://dx.doi.org/10.1007/s10585-008-9143-9] [PMID: 18286378]

[71] Ralph SJ, Rodríguez-Enríquez S, Neuzil J, Saavedra E, Moreno-Sánchez R. The causes of cancer revisited: "mitochondrial malignancy" and ROS-induced oncogenic transformation - why mitochondria are targets for cancer therapy. Mol Aspects Med 2010; 31(2): 145-70.
[http://dx.doi.org/10.1016/j.mam.2010.02.008] [PMID: 20206201]

[72] Ushio-Fukai M, Nakamura Y. Reactive oxygen species and angiogenesis: NADPH oxidase as target for cancer therapy. Cancer Lett 2008; 266(1): 37-52.
[http://dx.doi.org/10.1016/j.canlet.2008.02.044] [PMID: 18406051]

[73] Zhou F, Shen Q, Claret FX. Novel roles of reactive oxygen species in the pathogenesis of acute myeloid leukemia. J Leukoc Biol 2013; 94(3): 423-9.
[http://dx.doi.org/10.1189/jlb.0113006] [PMID: 23715741]

[74] Grivennikov SI, Greten FR, Karin M. Immunity, inflammation, and cancer. Cell 2010; 140(6): 883-99.
[http://dx.doi.org/10.1016/j.cell.2010.01.025] [PMID: 20303878]

[75] Yamagiwa Y, Meng F, Patel T. Interleukin-6 decreases senescence and increases telomerase activity in malignant human cholangiocytes. Life Sci 2006; 78(21): 2494-502.
[http://dx.doi.org/10.1016/j.lfs.2005.10.015] [PMID: 16336976]

[76] Touboul C, Lis R, Al Farsi H, *et al.* Mesenchymal stem cells enhance ovarian cancer cell infiltration through IL6 secretion in an amniochorionic membrane based 3D model. J Transl Med 2013; 11: 28.
[http://dx.doi.org/10.1186/1479-5876-11-28] [PMID: 23369187]

[77] Na YR, Lee JS, Lee SJ, Seok SH. Interleukin-6-induced Twist and N-cadherin enhance melanoma cell metastasis. Melanoma Res 2013; 23(6): 434-43.
[http://dx.doi.org/10.1097/CMR.0000000000000021] [PMID: 24051540]

[78] Li J, Davidson G, Huang Y, *et al.* Nickel compounds act through phosphatidylinositol-3-kinase/At-dependent, p70(S6k)-independent pathway to induce hypoxia inducible factor transactivation and Cap43 expression in mouse epidermal Cl41 cells. Cancer Res 2004; 64(1): 94-101.
[http://dx.doi.org/10.1158/0008-5472.CAN-03-0737] [PMID: 14729612]

[79] Wu CH, Tang SC, Wang PH, Lee H, Ko JL. Nickel-induced epithelial-mesenchymal transition by reactive oxygen species generation and E-cadherin promoter hypermethylation. J Biol Chem 2012; 287(30): 25292-302.
[http://dx.doi.org/10.1074/jbc.M111.291195] [PMID: 22648416]

[80] Zhang J, Zhou Y, Ma L, *et al.* The alteration of miR-222 and its target genes in nickel-induced tumor. Biol Trace Elem Res 2013; 152(2): 267-74.

[http://dx.doi.org/10.1007/s12011-013-9619-6] [PMID: 23447020]

[81]   He Z, Li D, Ma J, *et al*. TRIM36 hypermethylation is involved in polycyclic aromatic hydrocarbons-induced cell transformation. Environ Pollut 2017; 225: 93-103.
[http://dx.doi.org/10.1016/j.envpol.2017.03.001] [PMID: 28359976]

[82]   Jin Y, Xu P, Liu X, *et al*. Cigarette Smoking, BPDE-DNA Adducts, and Aberrant Promoter Methylations of Tumor Suppressor Genes (TSGs) in NSCLC from Chinese Population. Cancer Invest 2016; 34(4): 173-80.
[http://dx.doi.org/10.3109/07357907.2016.1156689] [PMID: 27042875]

[83]   Exon JH. A review of the toxicology of acrylamide. J Toxicol Environ Health B Crit Rev 2006; 9(5): 397-412.
[http://dx.doi.org/10.1080/10937400600681430] [PMID: 17492525]

[84]   Avissar-Whiting M, Veiga KR, Uhl KM, *et al*. Bisphenol A exposure leads to specific microRNA alterations in placental cells. Reprod Toxicol 2010; 29(4): 401-6.
[http://dx.doi.org/10.1016/j.reprotox.2010.04.004] [PMID: 20417706]

[85]   Arita A, Niu J, Qu Q, *et al*. Global levels of histone modifications in peripheral blood mononuclear cells of subjects with exposure to nickel. Environ Health Perspect 2012; 120(2): 198-203.
[http://dx.doi.org/10.1289/ehp.1104140] [PMID: 22024396]

[86]   Dostert C, Pétrilli V, Van Bruggen R, Steele C, Mossman BT, Tschopp J. Innate immune activation through Nalp3 inflammasome sensing of asbestos and silica. Science 2008; 320(5876): 674-7.
[http://dx.doi.org/10.1126/science.1156995] [PMID: 18403674]

[87]   Liu G, Cheresh P, Kamp DW. Molecular basis of asbestos-induced lung disease. Annu Rev Pathol 2013; 8: 161-87.
[http://dx.doi.org/10.1146/annurev-pathol-020712-163942] [PMID: 23347351]

[88]   Rosin FC, Pedregosa JF, de Almeida JS, Bueno V. Identification of myeloid-derived suppressor cells and T regulatory cells in lung microenvironment after Urethane-induced lung tumor. Int Immunopharmacol 2011; 11(7): 873-8.
[http://dx.doi.org/10.1016/j.intimp.2010.12.025] [PMID: 21238620]

[89]   Yamashita U, Sugiura T, Yoshida Y, Kuroda E. Effect of endocrine disrupters on macrophage functions *in vitro*. J UOEH 2005; 27(1): 1-10.
[http://dx.doi.org/10.7888/juoeh.27.1_1] [PMID: 15794588]

[90]   Reichert WL, Myers MS, Peck-Miller K, *et al*. Molecular epizootiology of genotoxic events in marine fish: linking contaminant exposure, DNA damage, and tissue-level alterations. Mutat Res 1998; 411(3): 215-25.
[http://dx.doi.org/10.1016/S1383-5742(98)00014-3] [PMID: 9804956]

[91]   Sceneay J, Chow MT, Chen A, *et al*. Primary tumor hypoxia recruits CD11b+/Ly6Cmed/Ly6G+ immune suppressor cells and compromises NK cell cytotoxicity in the premetastatic niche. Cancer Res 2012; 72(16): 3906-11.
[http://dx.doi.org/10.1158/0008-5472.CAN-11-3873] [PMID: 22751463]

[92]   Therriault MJ, Proulx LI, Castonguay A, Bissonnette EY. Immunomodulatory effects of the tobacco-specific carcinogen, NNK, on alveolar macrophages. Clin Exp Immunol 2003; 132(2): 232-8.
[http://dx.doi.org/10.1046/j.1365-2249.2003.02142.x] [PMID: 12699410]

CHAPTER 3

Toxicity of Environmental Contaminants: Use of Battery of Standard Biological Assays

Tajinder Kaur[1], **Sneh Rajput**[2], **Gurpreet Kaur**[1], **Renu Bhardwaj**[2] and **Saroj Arora**[2,*]

[1] *Khalsa College, Amritsar, India*

[2] *Department of Botanical and Environmental Sciences, Guru Nanak Dev University, Amritsar, India*

Abstract: Environmental pollution had increased greatly due to fast urbanization and industrialization resulting in release of toxic compounds into the surrounding. Traditionally, physical and chemical parameters were used to detect these toxic compounds in the environment. Besides these it is also essential to study new assays to understand biological responses of undesirable compounds on living systems. Combination of all the methods will provide new prospects for integrated strategies for environmental and risk assessment. Bioanalytical techniques have become a rapidly emerging branch of environmental sciences. Therefore, the present communication deals with emerging environmental pollutants and bioanalytical tools for *in situ* environmental monitoring.

Keywords: Animal Bioassay, Bacterial Bioassays, Bioassays, Biomarkers, Carcinogenicity, Chemical Analysis, Ecological Risk Assessment, Environmental Pollution, Environmental Pollutants, Heavy Metals, Mutagencity, Plant Bio-assays, Risk Assessment, Toxicity, Xenobiotics.

INTRODUCTION

The environment is an extremely complex system that is divided into biotic and abiotic components that exchange the matter and energy constantly. There should be balance between these processes in nature. However, this balance may be disrupted by the release of various chemicals into the environment. Thus, any alteration or deterioration in the physical, chemical, and biological quality of the environment due to the presence of various pollutants leads to environmental pollution [1]. The increase in mutagenic pollutants as a result of anthropogenic activities in the recent years is a general and serious problem. Under favorable

* **Corresponding author Prof. Saroj Arora:** Department of Botanical and Environmental Sciences, Guru Nanak Dev University, Amritsar, India; Tel: +91-9417285485; E-mail: dr.sarojarora@gmail.com

Ashita Sharma, Manish Kumar, Satwinderjeet Kaur & Avinash Kaur Nagpal (Eds.)

conditions, these environmental contaminants may be transmitted over long distances by different environmental components such as water and air (as well as particulate matter and aerosols) or by living organisms [2 - 5]. Depending on their physicochemical properties, chemicals undergo a number of processes in the environment. Hydrophobic substances build up in soil or/and sediment, hydrophilic substances remain dissolved in water, and volatile compounds pollute the air. These pollutants may also get partially bioaccumulated by living organisms [6]. When present in environment, they have the ability to cause certain diseases including cancer and may also negatively affect the genetic makeup of future generations. Thus, evaluation of mutagens present in environment is important. However, the concentration of mutagenic compounds in environment is usually low and, hence, their detection is not an easy procedure. Moreover, mutagens are only a fraction of contaminating chemicals in natural environment [7].

Traditional tests employing physicochemical analysis are helpful in detecting and measuring the concentrations of chemicals in the environment. However, physicochemical assays have certain drawbacks that reduce their use in evaluating complex environmental contaminants. Environmental samples can be collected and analyzed *ex situ* using certain bioanalytical tools such as biosensors or bioassays. Bioanalytical techniques have become a rapidly emerging branch of environmental sciences. A wide range of tools is used for bioanalytical studies at present with bioassays being one of them. Biological environmental monitoring can be performed using a number of techniques and methods and with the appropriate tools. By conducting research with bioassays, not only one can find whether a sample contains toxic compounds, but can also perform a qualitative and/or quantitative assessment by using specially selected organisms or a group of organisms. Assays based on living organisms can provide a counter balance to classical chemical analysis [5]. The aim of this review is to provide evidence that bioassays can be a complementary alternative to classical methods of analysis and will ultimately provide sufficient information to estimate the risk to human health and environmental quality due to anthropogenic activities.

Sources of Environmental Pollutants

Environmental contamination due to toxic compounds has been linked with society ever since the beginning of industrialization [8]. These contaminants enter ecosystems by many pathways including natural and human activities and occur in concentrations higher than the natural limits. They are commonly divided into biodegradable and non-biodegradable categories. Effluents from sewage and organic matter decomposed readily under normal conditions are classified as biodegradable pollutants while non-biodegradable materials are not degraded by microorganisms, *e.g.,* plastics, heavy metals, and detergents [9]. Major sources of organic pollutants are petroleum refining, steel mining and coal conversion, organic chemical and synthetic industries, textile processing, and pulp and paper milling [10 - 12]. Use of gasoline, pesticides, and fertilizers, aerosol sprays also lead to the discharge of pollutants directly into the environment. The routes of environment contamination include uncontrolled human activities like industry, agriculture, urbanization and automobiles, accidental spillage, and illegal dumping [9]. Over the last few decades, these activities have resulted in pollution of air with carbon dioxide and greenhouse gases, water with chemicals and toxic metals, pesticides, oil spills and soil by the use non-biodegradable and hazardous materials (metals, pharmaceutical products, hormones *etc.*), before improper facilities for waste disposal. These substances are dispersed over larger distances by air and water flow. These are persistent in nature even at low concentrations resulting in acute and toxic effects on human health. They are not only persistent in nature but also bio-accumulates in food chain.

Metal pollution is a widespread problem as elevated levels of metal ions are usually found in developed countries. Different environmental problems caused by heavy metal pollution have been reported worldwide [13, 14]. Metals contaminate environment through different industrial operations like electroplating, mining, tanning, smelting and refining. The increased concentration of even trace metals can affect both plant and animal system. Long term exposure to metals results in many diseases like Parkinson, Dystrophy, Allergies and Cancer. Pesticides are used to control pests and increase the crop productivity. Most of the pesticides are persistent in nature due to restricted degradation by biological, physical, chemical and microbiological processes [15, 16].The pollutants present in environment affects human health either by direct exposure resulting in acute or chronic toxicity or indirectly by production of reactive oxygen species and free radicals resulting in oxidative stress by altering structural and chemical composition of molecules (DNA, Proteins and Lipids) and negative gene- environment interaction resulting in mutation/ DNA damage in living organisms. Health effects of various metal pollutants are given in Table **1**.

Table 1. Potential side effect of metals and other contaminants present in environment.

Metal	Potential side effects
Arsenic	Lung, bladder, liver, skin and kidney cancers. Arsenic toxicity may also result in neurological, reproductive, cardiovascular, hepatic, hematological, respiratory and diabetic effects in humans
Cadmium	Causes diarrhea, problem in reproductive system, gastrointestinal diseases, muscular cramps and joint pain, lung damage, nephrotoxicity, immunotoxicity, renal osteomalacia, tumors and cancer

(Table 1) cont.....

Metal	Potential side effects
Chromium	Skin diseases, allergic reactions, damage to kidney, liver, reproductive and respiratory system and cancer
Copper	Brain, liver, and kidney damage, hypertension and neurological complications
Iron	Cancer, diabetes, liver and heart diseases as well as neurodegenerative diseases
Mercury	Causes mental disturbance and impairment of hearing, speech, movement and vision
Manganese	Neurotoxic effects in children
Lead	Cause lead poisoning damaging the digestive, reproductive, renal, immune, nervous, muscular and skeletal system of living beings
Zinc	Cause hematological disorders and deteriorates human metabolism

Bioanalytical Tools and Ecological Risk Assessment

The chemical agents are genotoxic and mutagenic in nature and can cause genetic damage and several health problems to human beings. Environmental pollutants like industrial and agricultural waste, heavy metals and pharmaceutical drugs may disturb reproductive ability of the aquatic organisms by affecting the gamete development and viability. These toxic agents may also cause alterations which can be inheritable and affect future generations [17]. Hence, it is necessary to identify these compounds that react with DNA in order to assess the environmental quality. The presence of toxic chemicals or biological pollutants resulting in environmental hazards and risks could be evaluated through the application of various bioanalytical tools (*e.g.* bioassays, biomarkers and biosensors). Traditionally, analytical chemistry was used to evaluate the presence of compounds in the environment [18, 19]. However, presence of complex mixtures of known and unknown pollutants may cause toxic effects in the environment. Therefore, it is essential to investigate and to relate new strategies combining both chemical and biological analysis for determining the ecological risk from pollutants [20, 21].

Several genotoxic and mutagenic assays have been developed in a wide range of organisms. To quantify the effect of pollution that can cause genetic damage in living organisms, a number of short-term bioassays are used all over the world. These mutagenic/genotoxic bioassays are divided into categories *i.e.* bacterial, animal and plant assay. Based on the biological system employed and their genetic endpoint detected, test systems can be divided into different groups. Bioassays employing prokaryotic organisms facilitate the detection of compounds that stimulate gene mutation and DNA damage while use of eukaryotes allows the detection of a greater extent of damage, varying from gene mutations to chromosome damage and aneuploidies [16]. Biomarkers can be used as sensitive indicators of toxicants that have entered the organisms and have been distributed between tissues [22]. They generally involve measurements in body fluids, cells or tissues signifying biochemical or cellular modifications due to the presence of toxicants, or of host response. Table **2** shows different bioassays used for evaluating environmental quality.

Bioassay means the method for potential assessment of physical, chemical or biological agents under a standard set of conditions by measuring and comparing the extent of the response of the test with that of standard on a selected and appropriate biological system [23, 24]. Thus, Biological assays are a tool for the estimation of structure, composition, or potency of a material by means of the reaction that undertake into living matter. Moreover, it is possible to measure the individual pollutant both qualitatively and quantitatively. Although, these bioassays are more difficult to maintain repeatability and reproducibility, they give in hand information regarding the effect of particular pollutant on living organisms (Fig. **1**).

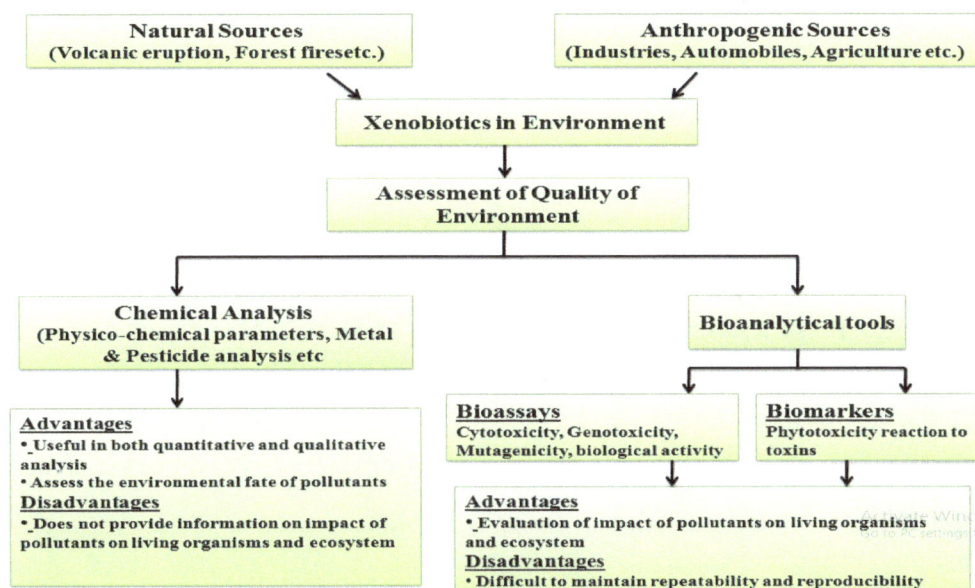

Fig. (1). Bioanalytical tools for evaluating environmental quality.

Plant Based Assays

Plant based bioassays are usually more sensitive, less expensive and have wider applications. Therefore, higher plants are used as excellent genetic models to evaluate the toxicity of environmental pollutants. However, this characteristic of higher plants is not only due to the sensitivity to identify mutagens present in the environment, but also to the possibility of evaluating several genetic endpoints

[25]. Presently, among the higher plants used to assess environmental contamination, *Allium cepa, Vicia faba, Zea mays, Tradescantia, Nicotiana tabacum, Crepis capillaris and Hordeum vulgare* have been used widely [25 - 30]. *Allium cepa* root chromosomal aberration assay is widely used for the detection of various mutagenic and genotoxic compounds among various plant-based bioassays. It can detect a wide range of compounds like, pesticides and herbicides, metals and dyes and sludge from solid waste treatment sewage plants, *etc* [25]. It is easily available, sensitive to a range of pollutants, mitotic phase can be observed clearly, has a stable karyotype *etc*. Another plant based bioassay is *Zea mays* assay. It is also used to evaluate various chemicals, pesticides, wastewater, contaminated soil, *etc* [31]. Another commonly used bioassay for the detection of mutagenic effects is *Tradescantia* micronucleus test. This assay is based on the formation of micronuclei due to the breakage of chromosome in meiotic pollen mother cells of *Tradescantia* sp. This assay determines the clastogenic effects of mutagens [32].

Lemma species are also very sensitive to inorganic and organic compounds. Therefore, these species are also used as test organisms for evaluating the adverse effects of various pharmaceutical drugs, herbicides and several heavy metals on aquatic plants. They are also used for testing the quality of water or testing of chemicals [33 - 35]. Now-a-days several kits are available commercially which can be stored for several months. Monocots and dicots are broadly used in the conventional phytotoxicity assays. Commonly used plants for the Phytotoxic kit microbiotest are: *Sorghum saccharatum* (monocotyl) and *Lepidiumsativum* (dicotyls). These test species are frequently used in phytotoxicity analysis due to their rapid germination and growth of roots and shoots [36].

Animal Based Bioassay

The use of animal-based bioassay helps to study the mechanism of carcinogenesis. The procedures for animal bioassays have been changed over time. According to the various reports published by the US Public Health Service the cancer bioassays using animal model organism has evolved in several stages. From the year 1940s to 1960s, animal bioassays were mainly used to appraise the safety of foods, drugs, and cosmetics. In the late 1960s and 1970s, these bioassays were used to evaluate the chemicals related to occupational or environmental exposures. Nowadays, animal bioassays are accentuating a group of tests which include multiple *in vitro* and both long-term and short-term animal tests. Animal based bioassays are used for evaluating the carcinogenic activity of compounds to assess the risk associated with these carcinogenic agents and to examine the mechanisms of carcinogenesis.

Aquatic organisms have adaptations to survive in the aquatic environment. The aquatic ecosystem serves as the final reservoir to a variety of pollutants present in the biosphere [37]. The quality of water is determined by physico-chemical, biological and hydro-morphological parameters. Therefore, it is crucial to investigate the pollution level of the aquatic ecosystems. A number of assays are available to assess the quality of water samples. These are used for the determination of effects of various pollutants like PAH, VOC, PCB and other toxic metals on the susceptible flora and fauna. Various species of organisms with different trophic levels can be used for this purpose such as fish, phytoplankton, rotifers, crustacean, amphipoda, worms *etc*. Quantitative assessment of water is done with the help of classical methods. These methods have limitations as they cannot be used for the detection of new chemicals formed by biotransformation in the water and possible interactions of chemicals within the ecosystem. As the amount of toxicants is increasing continuously, testing methods will continue to modify with increase in knowledge and understanding.

Bacterial Bioassays

To study the mutagenicity and genotoxicity of various compounds, a number of bacterial assays is standardized to detect the mutagenic potential of water samples *viz*. *Ames salmonella typhimurium* reverse assay, *Vibrio harveyi* bioassay, *E. coli* survival assay, UmuC assay, SOS chromotest. Among the various bacterial genotoxicity tests, Ames test (*Salmonella* assay) is widely used short term *in vitro* assay for identifying substances that can produce genetic damage and cause gene mutations [38]. Ames test employs the mutant strain of *Salmonella typhimurium*, which is characterised by gene mutations preventing the synthesis of an amino acid known as L-Histidine which is necessary for bacterial growth. When these mutant strains are exposed to the mutagens, these result in the reversal of mutation, and bacteria begins to synthesize L-Histidine and grow on medium which is deficient in this amino acid. Moreover, these bacteria have rfa mutation which allows the better penetration of mutagens through cell wall. The Ames Fluctuation Test is the liquid microplate version of the classical Ames test that was developed by Ames *et al*. [39]. The Ames fluctuation test is highly sensitive for aqueous samples containing low levels of mutagens [40]. *Vibrio harveyi* mutagenicity assay is a novel rapid microbiological bioluminescence assay for evaluation of mutagens in marine and other waters [41]. The assay is based on the detection of colonies of neomycin-resistant mutants which appears frequently after coming in contact with mutagens. Depending on the nature of the mutagen, the assay has shown to have sensitivity equal to or higher than Ames assay [7, 42]. To check the genotoxic potential of various chemical compounds UmuC-test was developed in 1985. This test is based on the ability of chemical compound to damage DNA and induce the expression of Umu operon. The genotoxicity of the sample can be measured by colorimetric determination of β-galactosidase activity. An alternative of Ames test is SOS chromotest. It is short-term, rapid and economical assay to study the genotoxic potential of test sample. The principal of this assay is the manifestation of SOS reaction, which results in the response of *E. coli* to genotoxic compounds.

Besides, the use of assays which involve plant, animal and bacteria as test organisms, DNA is also an important target of environmental stress among the various components of living cells [43]. DNA damage may be a useful parameter in screening the chemicals for their genotoxic properties [44]. DNA strand breaks have been proven as an effective biomarker of genotoxicity in environmental biomonitoring because they are common lesions produced by a wide range of agents and diverse mechanisms [43]. Furthermore, DNA damage caused by mutagenic/ genotoxic components lead to the alteration at molecular and tissue level in the organism which can be used as valuable biomarkers. Lesions in DNA strand is a sensitive biomarker for genotoxicity evaluation. DNA breaks can be studied by Gel Electrophoresis techniques like Comet assay, DNA nicking assays *etc*. Pollutants causing alteration in growth, biomolecules, enzymes and genetic material of plants are also very useful biomarkers. Comet assay (or single gel electrophoresis (SCGE) is a sensitive method used for measuring the DNA damage in individual cells [45]. The assay can be used as an important tool for monitoring the genotoxicity of waters and assessing the quality of water contaminated with effluents containing heavy metals [46, 47].

Table 2. Different bioassays used for evaluating environmental quality.

Endpoint	Bioassay name	Principle	Advantages	References
Mutagenicity	Ames fluctuation test	Mutations in DNA of bacterial cells (*Salmonella*)	Greater sensitivity, well adapted for the detection of genotoxicity compounds in water samples and potential hazards of weakly acting mutagens	[48]
	Vibrio harveyi Bioluminescene mutagenicity assay	Increase in level of luminescence when *Vibrio harveyi* A16 strain (a luxE mutant) comes in contact with chemical mutagens	Rapid, sensitive and useful bioindicator of mutagenic pollution in natural water samples	[41]
	UmuChromotest	UmuC gene induces overall SOS response	Capacity to detect various carcinogens, mutagenic and genotoxic compounds	[49]

(Table 2) cont.....

Endpoint	Bioassay name	Principle	Advantages	References
Mutagenicity	Ames test	Mutagenicity of the compounds leads to the formation of bacterial colonies	Sensitive to a varied pollutants; detection of carcinogens is possible, Capable to premetabolize the tested chemicals	[49]
	Comet assay	Intact DNA is compact, damaged DNA relaxes and forms tails when subjected to electrophoresis	Rapid, high sensitivity and relatively simple method for detecting DNA damage at the level of individual cells	[50]
	MICROTOX assay	Toxic or mutagenic compounds decreases the bioluminescence in marine bioluminescent bacteria	Reproducible; reliable; user friendly	[51]
Mutagenicity	UmuC test	UmuC gene induces overall SOS response	Hasty and user-friendly; high reproducibility; economical	[52]
	Umu-ChromoTest	UmuC gene induces overall SOS response	Hasty and user-friendly; high reproducibility; economical	[53]
	Chlorophyll content			
	Protein content			
	Antioxidant enzymes (SOD, POD, CAT, APX, GR)			
	Tradescantia micronucleus assay	Presence of micronuclei	High sensitivity towards various mutagenic and genotoxic compounds; economical; easy to grow and handle; can be used to determine the clastogenic effect and aneugenic agents	Sisenando *et al.*, 2011
Genotoxicity	SOS-Chromotest	lexA and recA genes induces overall SOS response	Fast and instant test; comparable with Ames test	Traczewska, 2011
	CFP-receptor yeast assay	Genotoxic compounds disrupt the endocrine system	Lucrative in comparison to various mammalian cells assays; highly sensitivity; manageable	[54]

(Table 2) cont.....

Endpoint	Bioassay name	Principle	Advantages	References
Genotoxicity	Mutatox	Toxic or mutagenic compounds increases the emission of bioluminescence in marine bioluminescent bacteria	Short incubation time; does not require sterilized conditions; does not require the maintenance of culture	[55]
	Zea mays chromosomal aberration assay	Gentoxic compounds brings changes in the phenotype of test organism	Cost effective and easily managed; test organism emits volatile organic compounds under stress conditions which can be easily determined	[31]
	Vicia faba chromosomal aberration assay	Gentoxic compounds causes aberration in meiotic chromosomes	Sensitive to a range of mutagenic compounds; point mutation and chromosomal aberration can be easily identified; easily available material; low-cost; easy to grow and handle; cell division take place at a higher rate; chromosomes readily available to mutagenic agent	[28]
Cytotoxicity	*Artemia franciscana* Microbiotest	Toxic compound caused mortality in test organism	Economical; high sensitivity to biotoxin produced by water	[56]
	Tetrahymena thermophile Microbiotest	Inhibition in the growth of test organism by toxic compounds	Multigenerational growth inhibition test based on optical density measurements; user-friendly and low cost; commercially available	[57]
	Ceriodaphniadubia Microbiotest	Toxic compounds cause inhibition in the growth, reproduction as well as mortality in the organism	Economical, maintenance not required; opportunity to study a range of liquid and solid samples	[58]
	Thamnocephalus platyurus Microbiotest	Reduction in the food intake capacity of the test organism	Fast; manageable and economical; higher sensitivity to biotoxin than bioluminescent bacteria; nominal equipment requirements	[59]

(Table 2) cont.....

Endpoint	Bioassay name	Principle	Advantages	References
Cytotoxicity	*Heterocyprisin congruens* Microbiotest	Growth inhibition and mortality	Range of samples can be detected; high sensitivity of the organism to ecotoxins and metals; minimal equipment required;easy to raise and maintain under laboratory conditions;commercially accessible	[60]
	Daphnia pulex Microbiotest	Immobility or mortality in test organism	Easy and lucrative; high reproducibility	[61]
	Brachinious calyciflorus Microbiotest	Reduced reproductive ability, mortality	Specially designed for acute and short-chronic tests; economical and easy; commercially accessible	[62]
	Daphnia magna Microbiotest	Immobility or mortality in test organism, Reduced reproductive ability, mortality	Easily managed and economical; commercially available	[63]
	Drosophila melanogaster bioassay	Chromosomal aberration, Gene mutation, Disruption of endocrine system	Completely sequenced DNA, extensively used as a model organism; easy to breed and maintain under laboratory conditions; nonspecialized equipment required; breeding cost is low; short generation time and high fertility;do not require the presence of S-9 fraction as the test organism have the capability to metabolize chemical compounds; due to morphological characteristics it is easy to distinguish females from males	[17]
	Folsomia candida Bioassay	Reduced reproduction	Easily available in the upper soil layer; short life and reproducing cycle	[64]
	Tubifex tubifex Bioassay	Reduced reproductive ability, mortality	Hasty; simple; low-cost; user friendly	[65]
	Apis mellifera Bioassay	Mortality in the test organism	Economical and easy to maintain under laboratory conditions; maintaining biocenosis	[66]

(Table 2) cont.....

Endpoint	Bioassay name	Principle	Advantages	References
Cytotoxicity	*Hyalella azteca* Bioassay	Reduced reproductive ability, mortality	High sensitivity to various xenobiotics; easy to rear and can be maintained under the laboratory conditions	[67]
	Daphnia pulex Bioassay	Immobility or mortality in test organism, Reduced reproductive ability, mortality	High sensitivity towards various chemicals	[68]
	Ceriodaphnia dubia Bioassay	Growth inhibition and mortality	Easy to breed and maintain under laboratory conditions; no special equipment is required	[68]
	Oryzias latipes Bioassay	Disruption of endocrine system, Growth inhibition and mortality	Short gestation period; high fertility; easy to culture in the laboratory; can be modified genetically	[69]
	Daphnia magna Bioassay	Immobility or mortality in test organism, Reduced reproductive ability, mortality	Highly sensitive to noxious chemicals and inhabits a dominant place in the food chain	[70]
	Brachydanio rerio Bioassay	Mortality, behavioral changes	Cost effective; short gestation period; familiar sexual behavior	[71]
	Eisenia fetida Bioassay	Reduced reproductive ability, mortality	High sensitivity towards pesticides; easily available; key indicator of ecotoxicity in soil	[72]

Battery of Bioassays

Biological assays can provide a counterbalance to chemical analysis for the assessment of environmental pollutants. For analyzing complex samples or probing the risk of environmental exposure to substances, a battery of bioassays can be an effective tool [5]. Source of contamination and the nature of pollutant(s) cannot be detected by the use of single bioassay. Instead of single bioassay, the use of two or more genotoxic bioassays is well justified as different bioassays respond in different ways to an array of different mutagens. Bioanalytical methods such as *in vitro* bioassays may be ideal screening tools that can detect a wide range of contaminants based on their biological effect. Use of different bioassays cannot be proposed to replace chemical analysis but can supplement conventional methods to provide risk assessment associated with chemical mutagens and to evaluate toxicity of pollutants present in environment [73].

A significant aspect of bioassays is the idealistic representation of an ecosystem by means of a single organism. However, due to different sensitivities to the same pollutant, their impact cannot be predicted on other species [74]. Bioassays can predict hazardous biological effects (*e.g.,* toxicity, cytotoxicity or genotoxicity) of pollutants by evaluating the effect of compounds or a mixture on living organisms or their constituent parts. The combination of these methods with the chemical analysis to identify the compounds recommends more integrated risk assessment [8]. Therefore, the application of a battery of bioassays employing different organisms is recommended to evaluate the quality of environment.

SUMMARY

Increased environmental pollution due to toxic chemicals has resulted in the need for sensitive assays to be used in risk assessment of polluted sites. Genotoxic compounds present in environment causes mutations and DNA damage in humans. However, the assessment of risk associated with chemical pollutants occurring at low concentrations is difficult by using chemical analysis only. Therefore, an alternative methodology is required to evaluate the toxicity of cytotoxic and genotoxic contaminants in environmental samples providing data useful in risk assessment. Environmental quality monitoring using *in vitro* bioanalytical tools is commonly accepted as an ideal tool for screening wide range of contaminants present in environment. Bioanalytical tools provide qualitative information on the toxicity of environmental samples and effect of particular pollutant on living organisms. Various bioassays based on plant, animal and bacterial systems are widely used now days for detection of various genotoxic and mutagenic compounds as an alternative to classical methods of analysis. Moreover, combination of different bioassays gives more certain results to calculate the hazards to human health associated with environmental contaminants.

CONSENT FOR PUBLICATION

Not applicable.

CONFLICT OF INTEREST

The authors declare no conflict of interest, financial or otherwise.

ACKNOWLEDGEMENTS

None Declared

REFERENCES

[1]   Moschella PS, Abbiati M, Åberg P, *et al*. Low-crested coastal defence structures as artificial habitats for marine life: using ecological criteria in design. Coast Eng J 2005; 52(10): 1053-71.

[http://dx.doi.org/10.1016/j.coastaleng.2005.09.014]

[2]   Walker K, Vallero DA, Lewis RG. Factors influencing the distribution of lindane and other hexachlorocyclohexanes in the environment. Environ Sci Technol 1999; 33(24): 4373-8. [http://dx.doi.org/10.1021/es9906 47n]

[3]   Oke TR. Boundary layer climates. Routledge 2002. [http://dx.doi.org/10.4324/9780203407219]

[4]   Hung H, Kallenborn R, Breivik K, *et al.* Atmospheric monitoring of organic pollutants in the Arctic under the Arctic Monitoring and Assessment Programme (AMAP): 1993-2006. Sci Total Environ 2010; 408(15): 2854-73. [http://dx.doi.org/10.1016/j.scitotenv.2009.10.044] [PMID: 20004462]

[5]   Wieczerzak M, Namieśnik J, Kudłak B. Bioassays as one of the Green Chemistry tools for assessing environmental quality: A review. Environ Int 2016; 94: 341-61. [http://dx.doi.org/10.1016/j.envint.2016.05.017] [PMID: 27472199]

[6]   Zenker A, Cicero MR, Prestinaci F, Bottoni P, Carere M. Bioaccumulation and biomagnification potential of pharmaceuticals with a focus to the aquatic environment. J Environ Manage 2014; 133: 378-87. [http://dx.doi.org/10.1016/j.jenvman.2013.12.017] [PMID: 24419205]

[7]   Czyż A, Szpilewska H, Dutkiewicz R, Kowalska W, Biniewska-Godlewska A, Wegrzyn G. Comparison of the Ames test and a newly developed assay for detection of mutagenic pollution of marine environments. Mutat Res 2002; 519(1-2): 67-74. [http://dx.doi.org/10.1016/S1383-5718(02)00112-2] [PMID: 12160892]

[8]   Blasco C, Picó Y. Prospects for combining chemical and biological methods for integrated environmental assessment. Trends Analyt Chem 2009; 28(6): 745-57. [http://dx.doi.org/10.1016/j.trac.2009.04.010]

[9]   Wasi S, Tabrez S, Ahmad M. Toxicological effects of major environmental pollutants: an overview. Environ Monit Assess 2013; 185(3): 2585-93. [http://dx.doi.org/10.1007/s10661-012-2732-8] [PMID: 22763655]

[10]   Tabrez S, Ahmad M. Effect of wastewater intake on antioxidant and marker enzymes of tissue damage in rat tissues: implications for the use of biochemical markers. Food Chem Toxicol 2009; 47(10): 2465-78. [http://dx.doi.org/10.1016/j.fct.2009.07.004] [PMID: 19596398]

[11]   Tabrez S, Ahmad M. Oxidative stress-mediated genotoxicity of wastewaters collected from two different stations in northern India. Mutat Res 2011; 726(1): 15-20. [http://dx.doi.org/10.1016/j.mrgentox.2011.07.012] [PMID: 21855648]

[12]   Gupta AK, Ahmad M. Assessment of cytotoxic and genotoxic potential of refinery waste effluent using plant, animal and bacterial systems. J Hazard Mater 2012; 201-202: 92-9. [http://dx.doi.org/10.1016/j.jhazmat.2011.11.044] [PMID: 22169142]

[13]   Tyagi P, Buddhi D, Choudhary R, Sawhney RL. Degradation of ground water quality due to heavy metals in industrial areas of India-A review. IJEP 2000; 20(3): 174-81.

[14]   Mena S, Ortega A, Estrela JM. Oxidative stress in environmental-induced carcinogenesis. Mutat Res 2009; 674(1-2): 36-44. [http://dx.doi.org/10.1016/j.mrgentox.2008.09.017] [PMID: 18977455]

[15]   Darko G, Acquaah SO. Levels of organochlorine pesticides residues in meat. IJEST 2007; 4(4): 521-4.

[16]   Afful S, Anim AK, Serfor-Armah Y. Spectrum of organochlorine pesticide residues in fish samples from the Densu Basin. Res J Environ Earth Sci 2010; 2(3): 133-8.

[17]   Adams MD, Celniker SE, Holt RA, *et al.* The genome sequence of *Drosophila melanogaster*. Science 2000; 287(5461): 2185-95.

[http://dx.doi.org/10.1126/science.287.5461.2185] [PMID: 10731132]

[18]   Kosjek T, Heath E, Petrović M, Barceló D. Mass spectrometry for identifying pharmaceutical biotransformation products in the environment. Trends Analyt Chem 2007; 26(11): 1076-85.
       [http://dx.doi.org/10.1016/j.trac.2007.10.005]

[19]   La Farre M, Pérez S, Kantiani L, Barceló D. Fate and toxicity of emerging pollutants, their metabolites and transformation products in the aquatic environment. Trends Analyt Chem 2008; 27(11): 991-1007.
       [http://dx.doi.org/10.1016/j.trac.2008.09.010]

[20]   Brack W, Schmitt-Jansen M, Machala M, *et al.* How to confirm identified toxicants in effect-directed analysis. Anal Bioanal Chem 2008; 390(8): 1959-73.
       [http://dx.doi.org/10.1007/s00216-007-1808-8] [PMID: 18224304]

[21]   Fernandez MP, Noguerol TN, Lacorte S, Buchanan I, Piña B. Toxicity identification fractionation of environmental estrogens in waste water and sludge using gas and liquid chromatography coupled to mass spectrometry and recombinant yeast assay. Anal Bioanal Chem 2009; 393(3): 957-68.
       [http://dx.doi.org/10.1007/s00216-008-2516-8] [PMID: 19057898]

[22]   McCarthy JF, Shugart LR. Biomarkers of environmental contamination

[23]   Agatonovic-Kustrin S, Morton DW, Yusof AP. Thin-layer chromatography-bioassay as powerful tool for rapid identification of bioactive components in botanical extracts. Modern Chemistry & Applications 2015.

[24]   Hemanta MR, Mane VK, Bhagwat A. Analysis of Traditional Food Additive Kolakhar for its Physico-Chemical Parameters and Antimicrobial Activity. J Food Process Technol 2014; 5(387): 2.

[25]   Leme DM, Marin-Morales MA. *Allium cepa* test in environmental monitoring: a review on its application. Mutat Res 2009; 682(1): 71-81.
       [http://dx.doi.org/10.1016/j.mrrev.2009.06.002] [PMID: 19577002]

[26]   Iqbal M, Bhatti IA. Re-utilization option of industrial wastewater treated by advanced oxidation process. Pak J Agric Sci 2014; 51(4)

[27]   Iqbal M, Bhatti IA. Gamma radiation/H_2O_2 treatment of a nonylphenol ethoxylates: Degradation, cytotoxicity, and mutagenicity evaluation. J Hazard Mater 2015; 299: 351-60.
       [http://dx.doi.org/10.1016/j.jhazmat.2015.06.045] [PMID: 26143198]

[28]   Iqbal M. *Vicia faba* bioassay for environmental toxicity monitoring: A review. Chemosphere 2016; 144: 785-802.
       [http://dx.doi.org/10.1016/j.chemosphere.2015.09.048] [PMID: 26414739]

[29]   Iqbal M, Nisar J. Cytotoxicity and mutagenicity evaluation of gamma radiation and hydrogen peroxide treated textile effluents using bioassays. JECE 2015; 3(3): 1912-7.

[30]   Qureshi K, Ahmad MZ, Bhatti IA, Iqbal M, Khan A. Cytotoxicity reduction of wastewater treated by advanced oxidation process. Chem Int 2015; 1(5)

[31]   Schmelz EA, Alborn HT, Tumlinson JH. The influence of intact-plant and excised-leaf bioassay designs on volicitin- and jasmonic acid-induced sesquiterpene volatile release in Zea mays. Planta 2001; 214(2): 171-9.
       [http://dx.doi.org/10.1007/s004250100603] [PMID: 11800380]

[32]   Sisenando HA, de Medeiros SR, Saldiva PH, Artaxo P, Hacon SS. Genotoxic potential generated by biomass burning in the Brazilian Legal Amazon by Tradescantia micronucleus bioassay: a toxicity assessment study. Environ Health 2011; 10(1): 41.
       [http://dx.doi.org/10.1186/1476-069X-10-41] [PMID: 21575274]

[33]   Standard Guide for conducting static toxicity tests with Lemna gibba G3 E 1415 – 91. West Conshohocken, PA, USA: ASTM International 2004.

[34]   Water quality – determination of toxic effect of water constituents and waste water to duckweed (Lemna minor) – duckweed growth inhibition test. Geneva, Switzerland: International Standard ISO

20079 2005.

[35]   Lemna sp Growth inhibition test Guideline 221. Paris, France: Organisation for Economic Co-operation and Development 2006.

[36]   Czerniawska-Kusza I, Ciesielczuk T, Kusza G, Cichoń A. Comparison of the Phytotoxkit microbiotest and chemical variables for toxicity evaluation of sediments. Environ Toxicol 2006; 21(4): 367-72.
       [http://dx.doi.org/10.1002/tox.20189] [PMID: 16841321]

[37]   Fleeger JW, Carman KR, Nisbet RM. Indirect effects of contaminants in aquatic ecosystems. Sci Total Environ 2003; 317(1-3): 207-33.
       [http://dx.doi.org/10.1016/S0048-9697(03)00141-4] [PMID: 14630423]

[38]   Mortelmans K, Zeiger E. The Ames Salmonella/microsome mutagenicity assay. Mutat Res/Fund and Mol Mech of Mutagen 2000; 455(1): 29-60.

[39]   Ames BN, Lee FD, Durston WE. An improved bacterial test system for the detection and classification of mutagens and carcinogens. Proc Natl Acad Sci USA 1973; 70(3): 782-6.
       [http://dx.doi.org/10.1073/pnas.70.3.782] [PMID: 4577135]

[40]   Siddiqui AH, Ahmad M. The Salmonella mutagenicity of industrial, surface and ground water samples of Aligarh region of India. Mutat Res 2003; 541(1-2): 21-9.
       [http://dx.doi.org/10.1016/S1383-5718(03)00176-1] [PMID: 14568291]

[41]   Podgórska B, Węgrzyn G. A modified Vibrio harveyi mutagenicity assay based on bioluminescence induction. Lett Appl Microbiol 2006; 42(6): 578-82.
       [PMID: 16706895]

[42]   Czyz A, Wróbel B, Węgrzyn G. Vibrio harveyi bioluminescence plays a role in stimulation of DNA repair. Microbiology 2000; 146(Pt 2): 283-8.
       [http://dx.doi.org/10.1099/00221287-146-2-283] [PMID: 10708366]

[43]   Gupta AK, Ahmad I, Ahmad M. Genotoxicity of refinery waste assessed by some DNA damage tests. Ecotoxicol Environ Saf 2015; 114: 250-6.
       [http://dx.doi.org/10.1016/j.ecoenv.2014.03.032] [PMID: 24836934]

[44]   Siddiqui AH, Tabrez S, Ahmad M. Short-term *in vitro* and *in vivo* genotoxicity testing systems for some water bodies of Northern India. Environ Monit Assess 2011; 180(1-4): 87-95.
       [http://dx.doi.org/10.1007/s10661-010-1774-z] [PMID: 21116844]

[45]   Lah B, Zinko B, Tisler T, Marinsek-Logara R. Genotoxicity detection in drinking water by Ames test, Zimmermann test and Comet assay. Acta Chim Slov 2005; 52(3): 341.

[46]   Matsumoto ST, Mantovani MS, Mallaguti MI, Marin-Morales MA. Investigation of the genotoxic potential of the waters of a river receiving tannery effluents by means of the *in vitro* comet assay. Cytologia (Tokyo) 2003; 68(4): 395-401.
       [http://dx.doi.org/10.1508/cytologia.68.395]

[47]   Tamie MS, Rigonato J, Mantovani MS, Marin-Morales MA. Evaluation of the genotoxic potential due to the action of an effluent contaminated with chromium, by the comet assay in CHO-K1 cultures. Caryologia 2005; 58(1): 40-6.
       [http://dx.doi.org/10.1080/00087114.2005.10589430]

[48]   Legault R, Blaise C, Rokosh D, Chong-Kit R. Comparative assessment of the SOS chromotest kit and the Mutatox test with the Salmonella plate incorporation (Ames test) and fluctuation tests for screening genotoxic agents. Environ Toxicol 1994; 9(1): 45-57.

[49]   Traczewska TM. Biologicznemetodyocenyskażeniaśrodowiska. Oficyna Wydawnicza Politechniki Wrocławskiej 2011.

[50]   Singh NP, McCoy MT, Tice RR, Schneider EL. A simple technique for quantitation of low levels of DNA damage in individual cells. Exp Cell Res 1988; 175(1): 184-91.
       [http://dx.doi.org/10.1016/0014-4827(88)90265-0] [PMID: 3345800]

[51]   De Zwart D, Slooff W. The Microtox as an alternative assay in the acute toxicity assessment of water pollutants. Aquat Toxicol 1983; 4(2): 129-38.
[http://dx.doi.org/10.1016/0166-445X(83)90050-4]

[52]   Giuliani F, Koller T, Würgler FE, Widmer RM. Detection of genotoxic activity in native hospital waste water by the umuC test. Mutat Res 1996; 368(1): 49-57.
[http://dx.doi.org/10.1016/S0165-1218(96)90039-7] [PMID: 8637510]

[53]   QU JH, WU W. Mutation Effect of Alkyl-hydroxybenzene with SOS/UmuChromotest [J]. J Zhejiang Ocean Univ 2003; 4: 007.

[54]   Beck V, Pfitscher A, Jungbauer A. GFP-reporter for a high throughput assay to monitor estrogenic compounds. J Biochem Biophys Methods 2005; 64(1): 19-37.
[http://dx.doi.org/10.1016/j.jbbm.2005.05.001] [PMID: 15992933]

[55]   Klamer HJ, Villerius LA, Roelsma J, de Maagd P, Opperhuizen A. Genotoxicity testing using the Mutatox™ assay: evaluation of benzo [a] pyrene as a positive control. Environ Toxicol Chem 1997; 16(5): 857-61.
[http://dx.doi.org/10.1002/etc.5620160504]

[56]   Velasco Martínez JAE. Estandarizacióndelbioensayo con Artemiafranciscana,(Flössner, 1972) y el efectoecotoxicológico del Sulfato de cobre (II) pentahidratado 1972.

[57]   Dayeh VR, Lynn DH, Bols NC. Cytotoxicity of metals common in mining effluent to rainbow trout cell lines and to the ciliated protozoan, Tetrahymena thermophila. Toxicol In Vitro 2005; 19(3): 399-410.
[http://dx.doi.org/10.1016/j.tiv.2004.12.001] [PMID: 15713547]

[58]   Rojíčková-Padrtová R, Maršálek B, Holoubek I. Evaluation of alternative and standard toxicity assays for screening of environmental samples: selection of an optimal test battery. Chemosphere 1998; 37(3): 495-507.
[http://dx.doi.org/10.1016/S0045-6535(98)00065-4]

[59]   Nałecz-Jawecki G, Persoone G. Toxicity of selected pharmaceuticals to the anostracan crustacean Thamnocephalus platyurus: comparison of sublethal and lethal effect levels with the 1h Rapidtoxkit and the 24h Thamnotoxkit microbiotests. Environ Sci Pollut Res Int 2006; 13(1): 22-7.
[http://dx.doi.org/10.1065/espr2006.01.005] [PMID: 16417128]

[60]   Kudłak B, Wolska L, Namieśnik J. Determination of EC50 toxicity data of selected heavy metals toward Heterocypris incongruens and their comparison to "direct-contact" and microbiotests. Environ Monit Assess 2011; 174(1-4): 509-16.
[http://dx.doi.org/10.1007/s10661-010-1474-8] [PMID: 20431939]

[61]   Kyselkova I. MARŠÁLEK B. Use of Daphnia pulex, Artemiasalina and Tubifextubifex for cyanobacterialmicrocystins toxicity detection. Biologia 2000; 55: 637-43.

[62]   Belgis C, Guido P. Cyst-based toxicity tests. XI. Influence of the type of food on the intrinsic growth rate of the rotifer Brachionus calyciflorus in short-chronic toxicity tests. Chemosphere 2003; 50(3): 365-72.
[http://dx.doi.org/10.1016/S0045-6535(02)00496-4] [PMID: 12656256]

[63]   Persoone G, Janssen C, De Coen W. Cyst-based toxicity tests X: comparison of the sensitivity of the acute Daphnia magna test and two crustacean microbiotests for chemicals and wastes. Chemosphere 1994; 29(12): 2701-10.
[http://dx.doi.org/10.1016/0045-6535(94)90068-X]

[64]   Cardoso DN, Santos MJ, Soares AM, Loureiro S. Molluscicide baits impair the life traits of Folsomia candida (Collembola): Possible hazard to the population level and soil function. Chemosphere 2015; 132: 1-7.
[http://dx.doi.org/10.1016/j.chemosphere.2015.02.035] [PMID: 25769136]

[65]   Pasteris A, Vecchi M, Reynoldson TB, Bonomi G. Toxicity of copper-spiked sediments to Tubifex

tubifex (Oligochaeta, Tubificidae): a comparison of the 28-day reproductive bioassay with a 6-month cohort experiment. Aquat Toxicol 2003; 65(3): 253-65.
[http://dx.doi.org/10.1016/S0166-445X(03)00136-X] [PMID: 13678845]

[66]   Thompson HM, Fryday SL, Harkin S, Milner S. Potential impacts of synergism in honeybees (Apismellifera) of exposure to neonicotinoids and sprayed fungicides in crops. Apidologie (Celle) 2014; 45(5): 545-53.
[http://dx.doi.org/10.1007/s13592-014-0273-6]

[67]   Gómez-Oliván LM, Neri-Cruz N, Galar-Martínez M, *et al.* Assessing the oxidative stress induced by paracetamol spiked in artificial sediment on Hyalellaazteca. Water Air Soil Pollut 2012; 223(8): 5097-104.
[http://dx.doi.org/10.1007/s11270-012-1261-y]

[68]   Kokkali V, van Delft W. Overview of commercially available bioassays for assessing chemical toxicity in aqueous samples. TrAC Trends Anal Chem 2014; 61: 133-55.
[http://dx.doi.org/10.1016/j.trac.2014.08.001]

[69]   Hsu HH, Lin LY, Tseng YC, Horng JL, Hwang PP. A new model for fish ion regulation: identification of ionocytes in freshwater- and seawater-acclimated medaka (Oryzias latipes). Cell Tissue Res 2014; 357(1): 225-43.
[http://dx.doi.org/10.1007/s00441-014-1883-z] [PMID: 24842048]

[70]   Czech B, Jośko I, Oleszczuk P. Ecotoxicological evaluation of selected pharmaceuticals to Vibrio fischeri and Daphnia magna before and after photooxidation process. Ecotoxicol Environ Saf 2014; 104: 247-53.
[http://dx.doi.org/10.1016/j.ecoenv.2014.03.024] [PMID: 24726936]

[71]   Vilar VJ, Rocha EM, Mota FS, Fonseca A, Saraiva I, Boaventura RA. Treatment of a sanitary landfill leachate using combined solar photo-Fenton and biological immobilized biomass reactor at a pilot scale. Water Res 2011; 45(8): 2647-58.
[http://dx.doi.org/10.1016/j.watres.2011.02.019] [PMID: 21411117]

[72]   Wang Y, Cang T, Zhao X, *et al.* Comparative acute toxicity of twenty-four insecticides to earthworm, Eisenia fetida. Ecotoxicol Environ Saf 2012; 79: 122-8.
[http://dx.doi.org/10.1016/j.ecoenv.2011.12.016] [PMID: 22244824]

[73]   Leusch FD, Khan SJ, Gagnon MM, *et al.* Assessment of wastewater and recycled water quality: a comparison of lines of evidence from *in vitro, in vivo* and chemical analyses. Water Res 2014; 50: 420-31.
[http://dx.doi.org/10.1016/j.watres.2013.10.056] [PMID: 24210511]

[74]   Tigini V, Giansanti P, Mangiavillano A, Pannocchia A, Varese GC. Evaluation of toxicity, genotoxicity and environmental risk of simulated textile and tannery wastewaters with a battery of biotests. Ecotoxicol Environ Saf 2011; 74(4): 866-73.
[http://dx.doi.org/10.1016/j.ecoenv.2010.12.001] [PMID: 21176963]

Environmental Contaminants and Natural Products, 2019, 65-93

Allium cepa Root Chromosomal Aberration Assay: A Tool to Assess Genotoxicity of Environmental Contaminants

Mandeep Kaur[1], Ashita Sharma[2], Rajneet Kour Soodan[3], Vanita Chahal[4], Vaneet Kumar[3], Jatinder Kaur Katnoria[3] and Avinash Kaur Nagpal[3,*]

[1] *Commerce College, Jammu, J&K, India*

[2] *Department of Civil Engineering., Chandigarh University, Chandigarh, Punjab, India*

[3] *Department of Botanical and Environmental Sciences, Guru Nanak Dev University, Amritsar-143005, Punjab, India*

[4] *Department of Sciences, Kamla Nehru College for Women, Phagwara, Punjab, India*

Abstract: Various anthropogenic activities have resulted in constant increase in contamination of ecosystem and degradation of human health due to bio-accumulation of different types of chemical pollutants. Higher plants being sensitive can be used as one of the rapid biotools for assessment and screening of chemical contaminants present in air, water and soil ecosystems. Different toxicity assessment assays employing a range of living organisms such as microbes, plants, animals and even human cell cultures have been widely used for environmental monitoring. Among all, *Allium cepa* root chromosomal aberration assay is recognised as one of the important biological test systems being used as a better bio-indicator for different types of pollutants (heavy metals, polycyclic aromatic hydrocarbons, and pesticides). Literature survey has revealed the assessment of genotoxic effects by induction of different types of mitotic abnormalities in *Allium cepa* root tips. From the literature surveying, it was clear that before exploration of any chemical compound/pesticide/heavy metal for its medicinal or other commercial use, genotoxic study is the first and mandatory step to be carried out. Present chapter is an attempt to compile studies on genotoxic effects of chemical compounds, pesticides, heavy metals, soil samples and wastewater effluents using *Allium cepa* root chromosomal aberration assay.

Keywords: Aberrations, Aceto-orecin, Agrochemicals, *Allium Cepa*, Bio-indicator, Clastogenic Aberrations, Colchicine, Farmer's Fluid, Fixation, Genotoxin, Heavy metals, PAHs, Pollutant, Pesticides, Physiological Aberrations, Root Chromosomal Aberration Assay, Root Dip, Spindle Proteins, Toxicity.

---

* **Corresponding author Avinash Kaur Nagpal:** Department of Botanical and Environmental Sciences, Guru Nanak Dev University, Amritsar-143005, Punjab, India; +911832451048; E-mail: avnagpal@yahoo.co.in

Ashita Sharma, Manish Kumar, Satwinderjeet Kaur & Avinash Kaur Nagpal (Eds.)

INTRODUCTION

Phenomenal industrial growth in the past few decades has led to exponential economic and social growth. At the same time, it has also magnified environmental contamination and has led to release of many toxic contaminants in the air we breathe, water we drink and the food we eat. The health hazards associated with the exposure to these contaminants are no less. The large-scale release of these contaminants into our environment has resulted in increasing disease incidences in terms of both magnitude and variety. Many degenerative diseases associated with the exposure of environmental contaminants have come into existence. Increase in environmental contaminants is also known to cause various genotoxic effects resulting in the distortion of gene pool. Studies have proved that environmental contaminants not only alter genetic makeup but are also known to cause cancer [1, 2].

Scientific community across the world is bothered about the increase in carcinogenic substances in the environment. Efforts are being made to frame the policies and regulations to safeguard human health from these deadly contaminants increasing in ambient environment. For this purpose, there is need to analyze the magnitude of exposure and the risk to human health associated with the exposure level. The risk assessment of humans is not considering the ethical issues possible directly, hence there is a need to identify some assays which can be used for the assessment of genotoxic risk associated with the exposure to the environmental contaminants. Various plants, animal and microbial assays are used for this purpose [1, 3, 4].

Among plant bioassays, *Allium cepa* root chromosomal aberration assay is most commonly used to determine the genotoxicity of chemicals. The assay was introduced in 1938, to examine mitotic aberrations of root tips cells of *Allium cepa* by the action of colchicine. The assay is popular and widely accepted due to inexpensive and simple methodology involved. Also, the number (2n=16) and size of chromosomes (8-16 μm) enhance the feasibility of assay [5]. The correlation of results of this assay with other assays is well established [4, 6, 7].

Methodology

Protocol for estimation of genotoxicity of various chemicals is well established and can be slightly modified according to the need. Young and fresh onion bulbs of uniform size are used. Treatment of onion bulbs can be of two types, *in situ* and root dip. In both modes of treatment, the primary roots are first removed. In *in situ* treatment, the denuded bulbs are placed directly on solution of chemical to be studied to root while in root dip treatment, the bulbs are placed on couplin jars filled with water for 24–36 h for rooting. After 24–36 h, bulbs with freshly emerged roots of size 1–2 cm are treated with different concentrations of test sample for 2-5 h (mostly 3 h) [8]. The root tips are then harvested from thoroughly washed bulbs and fixed in Farmer's fluid (ethanol and glacial acetic acid; 3:1). After at least 24 h of fixation period, slides are prepared by squashing root tips in aceto-orecin and observed under microscope.

Genotoxic effects triggered by physical and chemical agents lead to chromosomal aberrations, thus making chromosomal aberrations as estimate of exposure of various organisms to these agents. These aberrations result because of alteration in DNA synthesis or replication or nucleoproteins which result in either breakage of chromosome of abnormal segregation. Broadly, aberrations can be clastogenic aberrations (direct breaking effect on chromosomes) and physiological (effect on spindle proteins). Fig. (**1**) represents effects of a genotoxin on the root tip cells of *Allium cepa*.

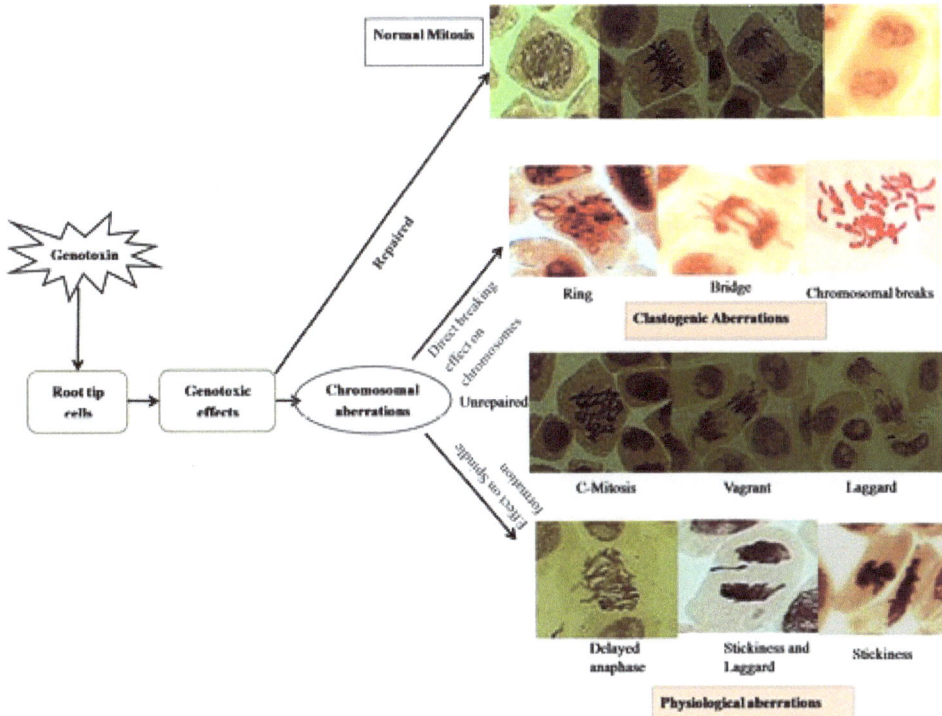

Fig. (1). Representation of effects of a genotoxin on the root tip cells of *Allium cepa* [8].

Applications

Allium cepa root chromosomal aberration assay is widely used to assess the genotoxicity of various genotoxins in environment. Table **1** summarizes literature on genotoxicity of chemical compounds and their mixtures employing *Allium* *cepa* root chromosomal aberration assay.

Table 1. Summary of literature on genotoxicity of chemical compounds and their mixtures employing *Allium cepa* root chromosomal aberration assay

S. No.	Chemical compound/s	Nature	Types of chromosomal aberrations	Reference
1.	Hydrogen peroxide and formalin	Oxidizer, bleaching agent and antiseptic	Spindle disturbances at prophases	[9]
2.	Bromoform and chloroform	Trihalomethanes	Laggards, stickiness, anaphasic bridges, disturbed anaphases and telophases	[10]
3.	Indium tin oxide	Transparent conducting oxide	Laggards, stickiness, anaphasic bridges, c-metaphase, binuclear cells and disturbed anaphases and telophases	[11]
4.	Sodium selenite, sodium hydrogen selenite, sodium selenate and sodium ammonium selenate	Components of plant foods, meat, seafood and supplements	Reduction in the mitotic index	[12]
5.	Rhodamine B	Xanthene dye	Chromosomal breaks at metaphases, abnormal nuclei, bridged nuclei, nuclear buds in interphase cells, single and double nuclear buds during prophase, multiple nuclear buds, nuclear buds with broken connection to nuclei	[13]
6.	Erythrosine, brilliant blue and red 40	Food dyes	Anaphasic and telophasic bridges, micronucleated cells; significant reduction in cell division	[14]
7.	Zinc oxide eugenol	Dental cement	Anaphasic bridges, chromosomal breakages, micronuclei	[15]
8.	Diphenyl-ether	Flavouring ingredient	C-mitosis, vagrant chromosomes, fragments, anaphasic and telophasic bridges, multipolar anaphases	[16]
9.	Mercuric chloride	Corrosive sublimate used as antiseptic salves	Sticky chromosomes, vagrants, c-anaphases, bridges, chromosomal fragments, multipolarity	[17]

(Table 1) cont.....

S. No.	Chemical compound/s	Nature	Types of chromosomal aberrations	Reference
10.	Ethyl methanesulphonate	Carcinogenic organic compound	Chromosomal fragments, multipolarity, sticky chromosomes, vagrants, c-anaphases, bridges,	[17]
11.	Benzo(a)pyrene	Polycyclic aromatic hydrocarbon found in coal tar	Stickiness, vagrants, laggards, acentric fragments, anaphasic bridges	[18]
12.	N-nitrosodiethylamine	Nitrosamine compound used as lubricant additive and stabilizer for industry materials	Laggards, stickiness, anaphasic bridges, micronuclei	[19]
13.	Peracetic acid	Organic peroxide used as cleanser and disinfectant	Laggards, chromatin bridges, fragments, micronucleated anaphases	[20]
14.	Ochratoxin A	Toxin produced by *Aspergillus* and *Penicillum* spp.	Breaks, bridges, stickiness, disturbed, meta and anaphases, binucleated cells	[21]
15.	Sodium azide and acetaminophen	Chemical compounds	Sticky metaphases, bridges	[22]
16.	Sodium chloride	Salt or halite used as electrolyte	Decrease in mitotic index, c-mitosis, stickiness, anaphasic bridges	[23]
17.	Mixture of chlorite and chlorate	Chemical mixtures	Laggards, chromatin bridges, chromosomal breaks	[24]
18.	Curcumin	Spice and food coloring agent	Reduction in mitotic index, chromosomal breaks and gaps, iso-chromatid breaks and exchanges	[25]
19.	Sodium benzoate, boric acid, citric acid, potassium citrate and sodium citrate	Food preservatives	Decreased mitotic index, c-mitosis, anaphasic bridges, micronuclei, laggards, stickiness, broken and unequal distributions	[26]
20.	N,N–Bis(3–Aminopropylo) Dodecyloamine	Disinfectants (derivative of fatty amines)	Stickiness, vagrants, fragments, bridges, c-anaphases, multipolarity	[27]
21.	Fumonisins (B1, B2 and B3; FB1, FB2 and FB3)	Toxins produced by moulds	Chromatid breaks, fragments and acentric chromatids, sister chromatid exchange, micronuclei	[28]

(Table 1) cont.....

S. No.	Chemical compound/s	Nature	Types of chromosomal aberrations	Reference
22.	Sodium metabisulfite	Food preservative	Decrease mitotic index and percentage of anaphase and telophase stages; c-mitosis, lagging chromosomes, sticky chromosomes, bridges	[29]
23.	N-methyl-N-nitrosourea and ethyl methanesulfonate	Alkalylating agents	Chromosomal bridges, fragments and vagrants	[30]
24.	Sodium azide	Mutagenic compound	Chromosomal bridges, fragments and vagrants	[30]
25.	Defoamex, idthin 400, magco thin and slick pipe	Oil and gas chemicals	Spindle disturbances, chromosomal breaks, irregular metaphases and anaphases with delay in spindle formation, multipolar spindles, clumping, stickiness, laggard chromosomes, c-mitosis, polyploid nuclei, sticky subchromatid connections and blurred chromosome borderlines, decreased mitotic index	[31]
26.	Methyl methanesulfonate	Chemical compound	Stickiness, vagrants. c-mitosis, bridges, fragments, decreased mitotic index	[32]
27.	Alkylbenzene sulphonate and citowett	Surfactants	Chromatin bridges, stickiness, distorted nuclei	[33]
28.	2-methoxy ethyl mercury chloride Phenyl mercury acetate	Organometallic compound Organomercury compound used as preservative in paints	C-mitosis, chromosomal breaks, chromatin bridges	[34]
29.	Tetracycline	Antibiotic used for treating infections	Stickiness, c-mitosis, chromatin bridges	[35]
30.	Cyanein Griseofulvin	Antibiotic isolated from *Penicillium cyaneum* Antifungal agent	C-mitosis, chromosomal breaks, chromatin bridges	[36]
31.	Trifluralin (a, a, a-trifluoro-2, 6-dinitro-n, n-dipropyl-p-toluidine)	Fluorinated benzene compound	C-mitosis, stickiness	[37]
32.	6-methylcoumarin	Polyaromatic compound used as food additive	Chromatin bridges, chromosomal breaks	[38]

(Table 1) cont.....

S. No.	Chemical compound/s	Nature	Types of chromosomal aberrations	Reference
33.	2,4 Dichlorophenoxy-acetic acid and 2,4,5-Trichlorophenoxy acetic acid	Chlorinated phenoxyacetic acid growth substances	Laggards, bridges, stickiness and irregular metaphases, tripolar anaphases; tumour formation in root cells	[39]
34.	Colchicine and acenaphthene	Anti-mitotic compounds	Chromosomal extrusion, clumping, fragmentation, c-tumors	[40]
35.	Colchicine	Anti-mitotic compound	No growth of root tips, intensified clumping, fragmentation and extrusion of chromosomes	[3]

This assay is also widely used to assess the genotoxicity of various agrochemicals. Use of pesticides has increased drastically on Indian soil after green revolution. Being persistent these chemicals degrade after a long time and thus cause ecological problems. Pesticides are also known to cause genetic damage. In order to attain food security many kinds of steps are being taken to produce sufficient food. Injudicious use of agrochemicals *viz.* fertilizers and pesticides not only cause damage to the ecosystem but the health effects associated with these chemicals is also well known. The term food security not only means sufficient food but safety of food is also an important aspect to be considered while achieving food security for all. Thus, it is important to assess the hazards associated with injudicious use of these chemicals. *Allium cepa* root chromosomal aberration assay is widely used to assess the toxicity associated with these chemicals. Table **2** summarizes literature on genotoxicity of pesticides (herbicides, fungicides, insecticides, weedicides) employing *Allium cepa* root chromosomal aberration assay.

Table 2. Summary of literature on genotoxicity of pesticides (herbicides, fungicides, insecticides, weedicides) employing *Allium cepa* root chromosomal aberration assay

S. No.	Pesticides	Chemical class	Nature	Type of chromosomal aberrations	Reference
1.	Aldrin	Organochlorine	Insecticide	Stickiness, chromosomal breaks, micronuclei, inhibition of mitotic index	[41]
2.	Alpha-thrin	Pyrethroid	Insecticide	Stickiness, laggards, bridges, multipolar anaphases, pulverized chromosomes	[42]

(Table 2) cont.....

S. No.	Pesticides	Chemical class	Nature	Type of chromosomal aberrations	Reference
3.	Aminotriazole	Heterocyclic organic compound	Herbicide	Chromosomal breaks and bridges	[43]
4.	Anilofos	Organophosphate	Herbicide	C-metaphase, disturbed nucleus, binuclear cells, laggard, stickiness, anaphase bridge and decrease in mitotic index	[44]
5.	Aphicide	Active compound	Herbicide	Stickiness, vagrant chromosomes, chromosomal fragments, chromatin bridges, C-anaphases, multipolarity	[17]
6.	Basagran	Contain dimethoate as the active ingredient	Emulsifiable herbicide and insecticide	Stickiness, chromosomal fragments, chromatin bridges, C-anaphases, vagrant chromosome, multipolarity	[17]
7.	Benomyl	Carbanilate compound	Herbicide	Stickiness, chromosomal breaks, chromatin bridges, micronuclei	[45]
8.	Benomyl (Benlate)	Carbamate	Fungicide	Stickiness, chromosomal breaks, bridges, decreased mitotic index	[46]
9.	Bexadust	Organochlorine	Insecticide	Stickiness, vagrant chromosomes, C-anaphase, chromosomal fragments, bridges, multipolarity	[17]
10.	Butachlor	Chloroacetanilide compound	Herbicide	Vagrant chromosomes, chromatin bridges, chromosomal fragments	[47]

(Table 2) cont.....

S. No.	Pesticides	Chemical class	Nature	Type of chromosomal aberrations	Reference
11.	Chlorpyrifos	Synthetic pyrethroid	Insecticide	Sticky chromosomes, vagrants, c-anaphases, chromosomal fragments, bridges, multipolarity	[48]
12.	Carbofuran	Carbamate compound	Insecticide	C-mitosis, laggards, chromatin bridges and multipolar anaphases	[49]
13.	Chlorpyrifos	Synthetic pyrethroid	Insecticide	Sticky chromosomes, lagging chromosomes, bridges, multipolar anaphase, pulverized chromosomes	[42]
14.	Carbaryl	Carbamate compound	Insecticide	Vagrants, fragments, bridges, stickiness, c-mitosis	[50]
15.	Cypermethrin	Pyrethroid	Insecticide	C-metaphases, stickiness, multipolar anaphases, laggards, binucleated cells	[51]
16.	Carbetamide	Carbanilate compound	Herbicide	Stickiness, chromosomal breaks, chromatin bridges, micronuclei	[45]
17.	Carbendazim	Benzimidazoles	Fungicide	Chromatin bridges and chromosomal breaks	[52]
18.	Chlorophenoxy acid	Synthetic compound similar to auxins	Herbicide	C-mitosis, delayed anaphases, stickiness, vagrants, multipolar chromosomes, reduction in root growth	[53]
19.	Dioxacarb	Methylcarbamate compound	Insecticide	Chromosomal and chromatid breaks, dicentric chromosomes, fragmentation, polyploidy and chromatid exchange	[54]

(Table 2) cont.....

S. No.	Pesticides	Chemical class	Nature	Type of chromosomal aberrations	Reference
20.	Dithane	Dithiocarbamate	Fungicide	Lagging chromosomes, chromosomal bridges, multipolarity at anaphases and telophases	[55]
21.	Dithane	Dithiocarbamate	Fungicide	C-mitosis, stickiness and chromatin bridges	[56]
22.	Dinoseb	Dinitrophenol compound	Herbicide	C-mitosis, stickiness, chromatin bridges, chromosomal breaks	[57]
23.	Efekto virikop	Active ingredient as copper oxychloride	Insecticide	Lagging chromosomes, bridges, multipolar anaphase, pulverized and sticky chromosomes	[42]
24.	Fenvalerate	Pyrethroid	Insecticide	C-metaphases, stickiness, multipolar anaphases, laggards, binucleated cells	[51]
25.	Fungi-nil	Contain active ingredient as chlorothalonil	Fungicide	Vagrant chromosomes, sticky chromosomes. c-anaphases, chromosomal fragments, bridges, multipolarity at anaphases and telophases	[17]
26.	Garden ripcord	Organophosphate	Insecticide	Vagrants, c-anaphases, multipolarity, sticky chromosomes, chromosomal fragments, bridges	[48]
27.	Garden ripcord	Organophosphate	Insecticide	Lagging chromosomes, chromosomal bridges, multipolarity at anaphases and telophases.	[55]

(Table 2) cont.....

S. No.	Pesticides	Chemical class	Nature	Type of chromosomal aberrations	Reference
28.	Homorban	Dicamba and acetic acid as active ingredients	Fungicide	Vagrant and sticky chromosomes, c-anaphases, fragments, bridges, multipolarity at anaphases and telophases	[17]
29.	Illoxan	Diclofop-methyl compound	Herbicide	C-mitosis, stickiness, laggards, fragments, bridges, multipolarity	[58]
30.	Imazethapyr	Imidazolinone compound	Herbicide	Stickiness, laggard chromosomes, chromatin bridges, disturbed anaphases and telophases	[59]
31.	Isopuroturon	Phenylurea compound	Weedicide	C-mitosis, stickiness, chromosomal breaks and chromatin bridges	[60]
32.	Isoproturon	Phenylurea compound	Weedicide	Stickiness, laggards, chromatin breaks, bridges, multipolar anaphases, bi- and micro-nucleated cells	[61]
33.	Karbadust	Active ingredient as flucoumafen	Rodenticide	Vagrant chromosomes, c-anaphases, sticky chromosomes. chromosomal fragments and bridges, multipolarity at anaphases and telophases	[17]
34.	Malathion	Organothiophosphate	Insecticide	Lagging chromosomes, chromosomal bridges, multipolarity at anaphases and telophases	[55]
35.	Malathion	Organothio-phosphate	Insecticide	Chromosomal bridges	[62]

(Table 2) cont.....

S. No.	Pesticides	Chemical class	Nature	Type of chromosomal aberrations	Reference
36.	Maleic hydrazide	Isomer of uracil	Herbicide and growth regulator	Stickiness at metaphase/anaphases, chromatin bridges, micro-nucleated cells, decreased mitotic index	[63]
37.	Maleic hydrazide	Isomer of uracil	Herbicide	Chromosomal bridges, fragments and vagrants, decreased mitotic index	[30]
38.	Metasystox-R	Oxydemeton methyl compound	Insecticide	C-mitosis, chromosomal breaks and chromatin bridges	[64]
39.	Parathion	Organothiophosphate	Insecticide and ascaricide	Chromosomal breaks and bridges	[65]
40.	Pendimethalin	Dinitroaniline	Herbicide	C-mitosis, laggards, chromatin bridges, chromosomal fragments, multipolar micronuclei	[66]
41.	Pentachlorophenol	Organochlorine	Insecticide	Chromatin bridges, chromosomal fragments and vagrant chromosomes	[47]
42.	Phosphine	Organophosphorus	Fumigant	Laggards, chromatin bridges, fragmentation	[67]
43.	Phenyl mercury acetate	Organomercuric compound	Fungicide	C-mitosis	[68]
44.	Prometryne	Triazine compound	Herbicide	Micronuclei induction	[69]
45.	Quizalofop-P-ethyl	Aryloxyphenoxy-propionic compound	Herbicide	Stickiness, vagrant chromosomes, c-anaphases, chromosomal bridges and fragments,	[70]
46.	Raxil	Contains tebuconazole as an active ingredient	Fungicide	C-mitosis, laggard chromosomes, disturbed metaphases	[71]

(Table 2) cont.....

S. No.	Pesticides	Chemical class	Nature	Type of chromosomal aberrations	Reference
47.	Springbok	Active ingredient as glyphosate	Insecticide	Lagging chromosomes, bridges, multipolar anaphase, pulverized and stick chromosomes	[42]
48.	Thimet	Organophosphate	Insecticide	Stickiness, c-mitosis, vagrants, fragments, bridges	[50]
49.	Thimet	Organophosphate	Insecticide	Chromatin bridges, stickiness, chromosomal breaks, micronuclei	[72]
50.	Villa	Synthetic pyrethroid	Insecticide	C-anaphases, stickiness, vagrants, chromosomal fragments and bridges, multipolarity at anaphases and telophases	[17]
51.	Vinclozolin	Dicarboximide compound	Fungicide	C-mitosis, chromosomal breaks and chromatin bridges	[73]
52.	γ - HCH	Organochlorine compound	Insecticide	Chromosomal breaks	[74]
53.	4, 6-dinitro-*o*-cresol	Dinitrophenolic compound	Herbicide	C-metaphases, stickiness and chromosomal breaks	[75]
54.	2,4,5-trichlorophenoxyacetic acid	Phenoxyacetic chlorinated compound	Herbicide	Inhibition in mitotic index, c-mitosis, stickiness, laggards, multipolarity of chromosomes	[4]
55.	2,4-D and 2,4,5-T	Phenoxyacetic chlorinated compounds	Herbicide	Chromatin bridges, c-mitosis, micronuclei	[39]

Heavy metals also known as toxic metals are present in earth's atmosphere naturally but increase in anthropogenic activity has led to increase in heavy metal in our biosphere. These metals are known to exhibit toxic properties. Metals like lead and cadmium are identified as Class1 carcinogen by USEPA. This assay can also be used to estimate the genotoxicity associated with exposure to heavy metals. Table **3** sums up literature on genotoxicity of elements, heavy metals their salts using this assay.

Table 3. Summary of literature on genotoxicity of elements, heavy metals their salts employing *Allium cepa* root chromosomal aberration assay

S. No.	Heavy metals and salts	Nature	Type of chromosomal aberrations	Reference
1.	Aluminum	Element	DNA damage and cells death	[76]
2.	Arsenic	Heavy metal	Micronuclei induction, decreased mitotic index	[77]
3.	Bismuth (III) oxide	Heavy metal salt	Stickiness, laggards, c- metaphases, anaphase bridges, disturbed anaphases and telophases	[78]
4.	Boron	Element	Disturbed anaphases, c-metaphases, telophase-anaphase bridges	[79]
5.	Butyl mercury bromide	Heavy metal salt	C-mitosis and c-tumors	[80]
6.	Cadmium	Heavy metal	Stickiness, lagging chromosomes, bridges, breaks, micronuclei induction, multipolar arrangement of chromosome, inhibition of mitotic index	[81]
7.	Cadmium	Heavy metal	Cell vacuolization, decreased nucleoplasmic ratio, nucleus and cytoplasmic condensation, nucleus margination	[82]
8.	Chromium	Heavy metal	Formation of micronuclei, decreased mitotic index	[83]
9.	Copper	Heavy metal	Root length effected	[84]
10.	Ethyl mercury chloride	Heavy metal salt	C-mitosis and c-tumors	[80]
11.	Lead	Heavy metal	Stickiness and micronucleated cells	[85]
12.	Lead	Heavy metal	Root length effected	[84]
13.	Lead	Heavy metal	Mitotic cells decreased, DNA synthesis reduced and nuclei synthesizing DNA declined	[86]
14.	Lead	Heavy metal	Stickiness, laggard chromosomes, chromosomal bridges with chromosomal rupture, micronuclei cells	[84]
15.	Metal ions (Hg, Cu, Ni, Cd, Be, Al, Mn and Li)	Heavy metal salts	Stickiness, c-mitosis, chromatin bridges, fragmentation	[87]
16.	Methyl mercury chloride	Heavy metal salt	C-mitosis and c-tumors	[80]
17.	Mercury	Heavy metal	Delayed anaphases, c-mitosis, stickiness, bridges, fragments, multipolarity	[88]
18.	Selenium	Heavy metal	C-mitosis, delayed anaphases, stickiness, bridges, fragments, multipolarity	[88]
19.	Silver	Element	Stickiness, chromosomal breaks and disturbed metaphases	[89]

(Table 3) cont.....

S. No.	Heavy metals and salts	Nature	Type of chromosomal aberrations	Reference
20.	Titanium dioxide	Heavy metal salt	Stickiness, laggards, chromosomal breaks, multipolarity, micronuclei formation	[90]
21.	Vanadium	Element	C-mitosis, stickiness, anaphasic bridges, decreased mitotic index	[91]
22.	Zn(II) and Cd(II)	Heavy metal salts	Chromosomal bridges and breaks	[92]
23.	Zinc oxide	Heavy metal salt	Stickiness, laggards, anaphasic bridges, multipolar and binucleated cells, disturbed anaphases and metaphases	[93]
24.	Zinc	Heavy metal	Root length effected	[84]

Wide application of this assay is attributed to the fact that this assay can also be used to analyze genotoxicity of environmental samples. Soil is the sink of all pollutants and plants grown on polluted soil can be affected by the toxicity of soil. Also, presence of genotoxins in soil can degrade soil fauna. Literature summarizing the use of *Allium cepa* assay is given in Table **4**.

Table 4. Summary of literature on genotoxicity of soil samples and their leacheates employing *Allium cepa* root chromosomal aberration assay

S. No.	Soil samples	Nature	Type of chromosomal aberrations	Reference
1.	Soil from different fields of Amritsar, Punjab, India	Agricultural soil	C-mitosis, delayed anaphases, laggards, vagrant, stickiness, chromosomal breaks, chromatin bridges, abnormal anaphases/metaphases	[94]
2.	Soil from road and railway track side areas of Amritsar, Punjab, India	Roadside soil	C-mitosis, delayed anaphases, laggards, vagrant, stickiness, chromosomal breaks, chromatin bridges, ring chromosomes	[95]
3.	Soil samples from roadsides (Golden Temple and Putlighar chowk area) of Amritsar, Punjab, India	Roadside soil	C-mitosis, abnormal anaphases/metaphases delayed anaphases, stickiness, laggards, vagrant, chromosomal breaks, chromatin bridges	[96]
4.	Soil collected from Rio Grande do Sul region, Brazil	Heavy metals contaminated soil	Multipolar anaphases, binucleated cells, chromatin bridges, chromosomal losses	[97]

(Table 4) cont.....

S. No.	Soil samples	Nature	Type of chromosomal aberrations	Reference
5.	Tannery industry effluents contaminated soil collected from three sites of Jajmau, Kanpur (India)	Agricultural soil	C-mitosis, laggards, stickiness, anaphasis bridges, binucleated cells	[98]
6.	Soil collected from Nangli village of Amritsar, Punjab	Agricultural soil	C-mitosis, delayed anaphases, stickiness, chromosomal breaks, chromatin bridges	[99]
7.	Soil collected from outskirts of two industries *viz.* zinc coating industry and copper sulphate manufacturing industry, Amritsar	Industrial soil	Laggards, chromosomal breaks, chromatin bridges, vagrants	[100]
8.	Soil samples from fields of Bulgaria	Agricultural soil	Vagrant chromosomes, chromosomal fragments at anaphase/telophase and multipolar anaphases	[101]
9.	Spent /soil leachates from aluminium industry of Brazil	Waste soil	Bridges, stickiness, fragments, C-metaphase, multipolar anaphases	[102]
10.	Leachates from hazardous solid wastes of metal and dye industries	Heavy metal contaminated soil	Stickiness, laggards, chromatin bridges, micronucleated cells, inhibition of mitotic index	[103]
11.	Soil from four areas of Slovenia	Polluted soil	Bridges and c-mitosis	[104]
12.	Soil sample (leachetes) from Queretaro, Mexico	Soil irrigated with wastewater	Decreased mitotic index, chromosomal damage	[105]
13.	Soil sample from industrial and cokeworks waste sites in France	Waste soil	Reduction in mitotic index	[106]
14.	Contaminated soil from Allison's and Monroe's farms and soil sample from Vienna, Austria	Pesticide-contaminated soil	Bridges and fragments	[107]
15.	Soil from Chernobyl accident areas of Ukraine	Radioactive compounds contaminated soil	Sticky chromosomes, c-mitosis, vagrant chromosomes, bridges, fragments, multipolar anaphases, decreased mitotic index	[108]

Like soil, water bodies are also exposed to various kinds of pollutants. Genotoxic compounds present in water sample can affect humans and aquatic life. Table **5** summarizes literature on genotoxicity of water samples and effluent samples estimated using this assay.

Table 5. Summary of literature on genotoxicity of water samples and effluent samples employing *Allium cepa* root chromosomal aberration assay

S. No.	Water and effluent samples	Types of chromosomal aberrations	Reference
1.	Water samples collected from the mining areas of Sorex Barobo, Surigao del Sur and Rosario, Agusan del Sur, Philippines	Spindle disturbances at prophases	[9]
2.	Surface water from Quatorze River, Francisco Beltrao, Parana, Brazil	No chromosomal alterations	[109]
3.	Effluent of water constructed wetlands	Centromeric breaks, single and double chromatid breaks, chromatid gaps	[110]
4.	Hospital laundry waste water	Bridges, multipolar anaphase with chromosomal bridges, micronucleated cells, metaphase with spindle viscosity, anaphase with adhesion and multiple chromosomal anaphase and telophase with chromosomal delay, metaphase with chromosomal loss	[111]
5.	Water samples from Paraiba do Sul River at the Brazilian cities of Tremembe and Aparecida	Micronuclei formation	[112]
6.	Polluted surface and waste water from Sava river, Croatia	Stickiness, c-mitosis, lagging chromosomes, chromatin bridges	[113]
7.	Pharmaceutical effluents	C-mitosis, chromatin bridges, multipolar nuclei	[114]
8.	Effluent from processing unit of cassava mill in Usela Quarters, Benin city, Edo state in Nigeria	Vagrant chromosomes, anaphasic bridges, fragments, polar deviations	[115]
9.	Three industrial wastewaters (brewery, rubber and bottling industry) collected from Benin city metropolis, Nigeria	C-mitosis, stickiness, laggards, vagrant chromosomes, bridges, fragments, micronuclei, disturbed spindle formation	[116]
10.	Surface wastewater	Inhibition of root growth, decreased mitotic index	[113]
11.	Water samples of paint and textile units	Sticky chromosomes, vagrant chromosomes, bridges, fragments	[117]
12.	Effluents from six indigenous Pharmaceutical companies in Nigeria	Stickiness, c-mitosis, vagrant chromosomes, bridges, multipolar fragments	[118]

(Table 5) cont.....

S. No.	Water and effluent samples	Types of chromosomal aberrations	Reference
13.	Deficit irrigation water from deep well drilled in the area of Turkey	Effect on onion bulb height, diameter and weight	[119]
14.	Atibaia river water	C-mitosis, vagrant chromosomes, chromatin bridges, chromosomal fragmentation, multipolarity	[120]
15.	Raw effluent from a pharmaceutical plant in Lagos state, Nigeria	Sticky chromosomes, breaks, ring chromosomes, dicentric chromosomes	[121]
16.	Effluent from hospital in the municipality of Santa Maria, Rio Grande do Sul State (RS), Brazil	Vagrant chromosomes, stickiness, breaks, anaphasic bridges, ring chromosomes	[122]
17.	Water samples from the river Paraiba do Sul, at the cities of Tremembe and Aparecida, (Sao Paulo state), Brazil	Decreased mitotic index, inhibition and delay of root growth	[123]
18.	Waste water from eight gasoline stations in Brasilia, Brazil	Lagging chromosomes, bridges, fragments	[124]
19.	River water polluted with hexavalent chromium	Inhibition of root growth	[125]
20.	Spent potliners/ soil leachates from aluminium industry of Brazil	Stickiness, c-metaphases, bridges, fragments, multipolar anaphases	[102]
21.	Textile industry effluents	C-mitosis, chromatin bridges, decreased mitotic index	[126]
22.	Contaminated river water from downstream and upstream of river Alamuyo	Lagging chromosomes, chromatin bridges, chromosomal fragments	[127]
23.	Petroleum polluted water	Laggards, c-mitosis, bridges, fragments	[128]
24.	Water from oil spill	C-mitosis, stickiness, laggards, vagrant chromosomes, bridges, fragments	[129]
25.	Leachates from highly polluted lowland river sediments polluted with complex chemical mixture of pollutants	Vagrant chromosomes, chromatin bridges	[130]
26.	Heavy metals contaminated industrial effluent	Stickiness, lagging chromosomes, bridges, micro-nucleated cells, late-separating chromosomes, unoriented chromosomes at metaphases and anaphases, mitotic index decreased	[131]
27.	Water samples collected along the river Rasina in Serbia	C-mitosis, vagrant chromosomes, bridges, fragments and multipolarity	[132]
28.	Industrial wastewater	Effect on anti-oxidative enzyme activity	[133]
29.	Municipal wastewater effluents	Vagrant chromosomes, fragments, chromatin bridges	[134]

(Table 5) cont.....

S. No.	Water and effluent samples	Types of chromosomal aberrations	Reference
30.	Municipal sludge and vermin-composted sludge from disposal site situated on Lucknow–Kanpur highway in the state of Uttar Pradesh (India)	Chromosomal bridges, fragments and gaps, increased mitotic index	[135]
31.	Water sample collected from the region of Panagjurishte, southwest Bulgaria	Vagrant chromosomes, chromatin bridges, fragments	[136]
32.	Polluted water samples of Slovenia river	Chromatin bridges, chromosomal breaks, decreased mitotic index	[104]
33.	Water samples from industrial Sandub area in Mansoura district	Chromosomal bridges, fragments, chromosomal rings, asynchrony	[137]
34.	Water from natural reservoirs located near the radium production industry storage cells	Vagrant chromosomes, chromatin bridges, fragments	[138]
35.	Sewage water	Stickiness, ring chromosomes, binucleated cells, inhibition of root growth	[139]
36.	Non-carbonated mineral waters	C-mitosis, laggards, bridges, fragments	[140]
37.	Wastewater from the phosphoric gypsum depot	C-mitosis, laggards, bridges, fragments	[141]
38.	Sewage and industrial effluent	Chromosomal bridges, chromosomal breaks, micronucleus induction	[142]
39.	Shallow well water	Chromatin bridges, fragmentation, decreased mitotic index	[107]
40.	Municipal wastewater sludge	C-mitosis, vagrants, bridges, fragments	[143]
41.	Water samples of various pollutant levels like biological treatment plant output water-undiluted, industrial and municipal wastewater	Stickiness, laggards, chromosomal fragments and c-metaphases, chromosomal clumping, anaphasic disruption and multipolar spindles	[144]
42.	Non-diluted wastewater effluents sample from slaughter house and dye house	Stickiness, vagrants, c-mitosis, bridges, fragments, decreased mitotic index	[32]
43.	Drilling fluids from reserve pit of oil and gas drilling activity	C-mitosis, stickiness, multipolarity, micronuclei formation, reduced mitotic index	[145]

SUMMARY

Increase in anthropogenic activities lead to acceleration of many pollutants in environment. There is a need to assess the toxic potential of these pollutants, so as to frame policies and framework. Thus, it is required to follow a cost effective yet efficient tool for estimation of genotoxicity. *Allium cepa* root chromosomal aberration assay is an answer to all these issues. The wide applicability and ease of performance makes this assay popular amongst scientific community.

CONSENT FOR PUBLICATION

Not applicable.

CONFLICT OF INTEREST

The authors declare no conflict of interest, financial or otherwise.

ACKNOWLEDGEMENTS

Authors would like to acknowledge University Grants Commission (UGC) New Delhi, for financial assistance.

REFERENCES

[1]   Ames BN, Lee FD, Durston WE. An improved bacterial test system for the detection and classification of mutagens and carcinogens. Proc Natl Acad Sci USA 1973; 70(3): 782-6.
        [http://dx.doi.org/10.1073/pnas.70.3.782] [PMID: 4577135]

[2]   Xia Z, Duan X, Tao S, *et al.* Pollution level, inhalation exposure and lung cancer risk of ambient atmospheric polycyclic aromatic hydrocarbons (PAHs) in Taiyuan, China. Environ Pollut 2013; 173: 150-6.
        [http://dx.doi.org/10.1016/j.envpol.2012.10.009] [PMID: 23202645]

[3]   Levan A. The effect of colchicine on root mitosis in *Allium.* Hereditas 1938; 24: 471-86.
        [http://dx.doi.org/10.1111/j.1601-5223.1938.tb03221.x]

[4]   Grant WF. The genotoxic effects of 2,4,5-T. Mutat Res 1979; 65(2): 83-119.
        [http://dx.doi.org/10.1016/0165-1110(79)90001-0] [PMID: 470970]

[5]   Bolle P, Mastrangelo S, Tucci P, Evandri MG. Clastogenicity of atrazine assessed with the *Allium cepa* test. Environ Mol Mutagen 2004; 43(2): 137-41.
        [http://dx.doi.org/10.1002/em.20007] [PMID: 14991755]

[6]   Hazarika A, Sarkar SN. Effect of isoproturon pretreatment on the biochemical toxicodynamics of anilofos in male rats. Toxicology 2001; 165(2-3): 87-95.
        [http://dx.doi.org/10.1016/S0300-483X(01)00411-5] [PMID: 11522367]

[7]   Özkara A, Akyıl D, Eren Y, Erdoğmuş SF. Potential cytotoxic effect of Anilofos by using *Allium cepa* assay. Cytotechnology 2015; 67(5): 783-91.
        [http://dx.doi.org/10.1007/s10616-014-9716-1] [PMID: 24838422]

[8]   Soodan RK, Sharma A, Kaur M, Katnoria JK, Nagpal AK. *Allium cepa* Root Chromosomal Aberration Assay: An Application in Assessing Anti-genotoxic Potential of Ashwagandha. In: Kaul S, Wadhwa R, Eds. Science of Ashwagandha: Preventive and Therapeutic Potentials. Cham: Springer 2017.
        [http://dx.doi.org/10.1007/978-3-319-59192-6_3]

[9]   Cabuga CC Jr, Abelada JJZ, Apostado RRQ, *et al. Allium cepa* test: An evaluation of genotoxicity. Proc of the Int Acad of Ecol and Environ Sci 2017; 7(1): 12-9.

[10]   Khallef M, Liman R, Konuk M, *et al.* Genotoxicity of drinking water disinfection by-products (bromoform and chloroform) by using both *Allium* anaphase-telophase and comet tests. Cytotechnology 2015; 67(2): 207-13.
        [http://dx.doi.org/10.1007/s10616-013-9675-y] [PMID: 24363168]

[11]   Ciğerci IH, Liman R, Özgül E, Konuk M. Genotoxicity of indium tin oxide by *Allium* and Comet tests. Cytotechnology 2015; 67(1): 157-63.
        [http://dx.doi.org/10.1007/s10616-013-9673-0] [PMID: 24337653]

[12]   Michalska-Kacymirow M, Kurek E, Smolis A, Wierzbicka M, Bulska E. Biological and chemical investigation of *Allium cepa* L. response to selenium inorganic compounds. Anal Bioanal Chem 2014; 406(15): 3717-22.
       [http://dx.doi.org/10.1007/s00216-014-7742-7] [PMID: 24652154]

[13]   Tan D, Bai B, Jiang D, *et al.* Rhodamine B induces long nucleoplasmic bridges and other nuclear anomalies in *Allium cepa* root tip cells. Environ Sci Pollut Res Int 2014; 21(5): 3363-70.
       [http://dx.doi.org/10.1007/s11356-013-2282-9] [PMID: 24234815]

[14]   Aguiar de Oliveira MA, Alves DDL, Guedes de Morais Lima LH. Marcelo de Castro -e- Souza J, Peron AP. Cytotoxicity of erythrosine (E-127), brilliant blue (E-133) and red 40 (E-129) food dyes in a plant test system. Acta Scientiarum 2013; 35: 557-62.

[15]   Rezende E, Mendes-Costa MC, Fonseca JC, Ribeiro AO. Evaluation of the genotoxicity of zinc oxide-eugenol cement to *Allium cepa* L. Acta Scientiarum 2013; 35: 563-9.

[16]   Dragoeva A, Koleva V, Hasanova N, Slanev S. Cyrotoxic and genotoxic effect of diphenyl-ether herbicide GOAL (oxyfluorfen) using the *Allium cepa* test. Res J Mutagen 2012; 2: 1-9.
       [http://dx.doi.org/10.3923/rjmutag.2012.1.9]

[17]   Asita AO, Matebesi LP. Genotoxicity of hormoban and seven other pesticides to onion root tip meristematic cells. Afr J Biotechnol 2010; 9: 4225-32.

[18]   Cabaravdic M. Induction of chromosome aberrations in the *Allium cepa* test system caused by the exposure of cells to benzo(a) pyrene. Med Arh 2010; 64(4): 215-8.
       [PMID: 21246918]

[19]   de Rainho CR, Kaezer A, Aiub CA, Felzenszwalb I, Felzenszwa I. Ability of *Allium cepa* L. root tips and Tradescantia pallida var. purpurea in N-nitrosodiethylamine genotoxicity and mutagenicity evaluation. An Acad Bras Cienc 2010; 82(4): 925-32.
       [http://dx.doi.org/10.1590/S0001-37652010000400015] [PMID: 21152767]

[20]   Rathore HS, Rathore M, Panchal S, Makwana M, Sharma A, Shrivastava S. Can genotoxicity effect be model dependent in *Allium* test? An evidence. Environ Asia 2010; 3: 29-33.

[21]   Lerda D, Biagi Bistoni M, Pelliccioni P, Litterio N. *Allium cepa* as a biomonitor of ochratoxin A toxicity and genotoxicity. Plant Biol (Stuttg) 2010; 12(4): 685-8.
       [PMID: 20636912]

[22]   Singh M, Solanke P, Rathore H, Sharma A, Makwana M, Shrivastava S. Influence of decoction of seeds of *Cassia tora* Linn (Leguminosae) on the genotoxicity of sodium azide and acetaminophen in *Allium cepa* model. Internet J Toxicol 2009; 6: 1-8.

[23]   Teerarek M, Bhinija K, Thitavasanta S, Laosinwattana C. The impact of sodium chloride on root growth, cell division and interphase silver-stained nucleolar organizer regions (AgNORs) in root tip cells of *Allium cepa*. Sci Hortic (Amsterdam) 2009; 121: 228-32.
       [http://dx.doi.org/10.1016/j.scienta.2009.01.040]

[24]   Feretti D, Zerbini I, Ceretti E, *et al.* Evaluation of chlorite and chlorate genotoxicity using plant bioassays and *in vitro* DNA damage tests. Water Res 2008; 42(15): 4075-82.
       [http://dx.doi.org/10.1016/j.watres.2008.06.018] [PMID: 18718628]

[25]   Ragunathan I, Panneerselvam N. Antimutagenic potential of curcumin on chromosomal aberrations in *Allium cepa*. J Zhejiang Univ Sci B 2007; 8(7): 470-5.
       [http://dx.doi.org/10.1631/jzus.2007.B0470] [PMID: 17610326]

[26]   Türkoğlu S. Genotoxicity of five food preservatives tested on root tips of *Allium cepa* L. Mutat Res 2007; 626(1-2): 4-14.
       [http://dx.doi.org/10.1016/j.mrgentox.2006.07.006] [PMID: 17005441]

[27]   Gabara B, Kalwinek J, Koziróg A, Żakowska Z, Brycki B. Influence of n,n–bis(3–aminopropylo) dodecyloamine on the ultrastructure of nuclei in Aspergillus niger mycelium and on cell proliferation

and mitotic disturbances in *Allium cepa* L. root meristem. Acta Biol Cracov Ser; Bot 2006; 48: 45-52.

[28] Lerda D, Biaggi Bistoni M, Peralta N, Ychari S, Vazquez M, Bosio G. Fumonisins in foods from Cordoba (Argentina), presence and genotoxicity. Food Chem Toxicol 2005; 43(5): 691-8. [http://dx.doi.org/10.1016/j.fct.2004.12.019] [PMID: 15778008]

[29] Rencuzogullari E, Kayraldiz A. Ila HS, Cakmak T, Topaktas M. The cytogenetic effects of sodium metasulfide a food preservative in root tip cells of *Allium cepa*. Turk J Biol 2001; 25: 361-70.

[30] Rank J, Nielsen MH. *Allium cepa* anaphase-telophase root tip chromosome aberration assay on N-methyl-N-nitrosourea, maleic hydrazide, sodium azide, and ethyl methanesulfonate. Mutat Res 1997; 390(1-2): 121-7. [http://dx.doi.org/10.1016/S0165-1218(97)00008-6] [PMID: 9150760]

[31] Zoldos V, Cifrek ZV, Tomic M, Papes D. Oil and gas industrial chemicals cytotoxicity studied by *Allium* test. Water Air Soil Pollut 1997; 383: 232-6.

[32] Rank J, Nielsen MH. A modified *Allium* test as a tool in the screening of genotoxicity of complex mixtures. Hereditas 1993; 118: 49-53. [http://dx.doi.org/10.1111/j.1601-5223.1993.t01-3-00049.x]

[33] Bellani LM, Rinallo C, Bennici A. Cyto-morphological alternations in *Allium* roots induced by surfactants. Environ Exp Biol 1991; 31: 179-85.

[34] Nandi S. Studies on the cytogenetic effect of some mercuric fungicides. Cytologia (Tokyo) 1985; 50: 913-9. [http://dx.doi.org/10.1508/cytologia.50.921]

[35] Mann SK. Interaction of tetracycline (TCA) with chromosomes in *Allium cepa*. Environ Exp Bot 1978; 18: 201-5. [http://dx.doi.org/10.1016/0098-8472(78)90039-4]

[36] Frank V. Restoration of mitotic and differentiation processes in the root apices of *Allium cepa* treated with cyanein and griseofulvin. Biol Plant 1974; 16: 28-34. [http://dx.doi.org/10.1007/BF02920817]

[37] Bayer DE, Foy CL, Mallory TE, Cutter EG. Morphological and histological effects of trifluralin on root development. Am J Bot 1967; 54: 945-52. [http://dx.doi.org/10.1002/j.1537-2197.1967.tb10719.x]

[38] Ronchi VN, Arcara PG. The chromosome breaking effect of 6-methylcoumarin in *Allium cepa* in relation to the mitotic cycle. Mutat Res 1967; 4(6): 791-6. [http://dx.doi.org/10.1016/0027-5107(67)90088-7] [PMID: 5591290]

[39] Croker BH. Effects of 2, 4-dichlorophenoxyacetic acid and 2,4,5-trichlorophenoxyacetic acid on mitosis in *Allium cepa*. Bot Gaz 1953; 114: 274-84. [http://dx.doi.org/10.1086/335769]

[40] Levine M. The effect of colchicine and acenaphthene in combination with X-Rays on Plant Tissue-III. Bull Torrey Bot Club 1946; 73(2): 167-83. [http://dx.doi.org/10.2307/2481614]

[41] Scholes ME. The effects of aldrin, dieldrin, isodrin, endrin and DDT on mitosis of the onion *(Allium cepa L.)*. J Horti Sci 1955; 30: 181-7.

[42] Asita AO, Makhalemele R. Genotoxicity of chlorpyrifos, alpha-thrin, efekto virikop and springbok to onion root tip cells. Afr J Biotechnol 2008; 7(23): 4244-50.

[43] Mohandas T, Grant WF. Cytogenetic effects of 2, 4-D and amitrole in relation to nuclear volume and DNA content in some higher plants. Can J Genet Cytol 1972; 14: 773-83. [http://dx.doi.org/10.1139/g72-095]

[44] Ozkara A, Akyil D, Eren Y, Erdogmus SF. Potential cytotoxic effect of Anilofos by using *Allium cepa* assay. Cytotechnology 2014.

[http://dx.doi.org/10.1007/s10616-014-9716-1] [PMID: 24838422]

[45]   Badr A. Mitodepressive and chromotoxic activities of 2 herbicides in *Allium cepa*. Cytologia (Tokyo) 1983; 48: 451-8.
[http://dx.doi.org/10.1508/cytologia.48.451]

[46]   Dane F, Dalgiç O. The effects of fungicide benomyl (benlate) on growth and mitosis in onion (*Allium cepa* L.) root apical meristem. Acta Biol Hung 2005; 56(1-2): 119-28.
[http://dx.doi.org/10.1556/ABiol.56.2005.1-2.12] [PMID: 15813220]

[47]   Ateeq B, Abul Farah M, Niamat Ali M, Ahmad W. Clastogenicity of pentachlorophenol, 2,4-D and butachlor evaluated by Allium root tip test. Mutat Res 2002; 514(1-2): 105-13.
[http://dx.doi.org/10.1016/S1383-5718(01)00327-8] [PMID: 11815249]

[48]   Asita OA, Matobole RM. Comparative study of the sensitivities of onion and broad bean root tip meristematic cells to genotoxins. Afr J Biotechnol 2010; 9: 4465-70.

[49]   Saxena PN, Gupta SK, Murthy RC. Carbofunan induced cytogenetic effects in root meristem cells of *Allium cepa* and *Allium* sativum: A spectroscopic approach for chromosome damage. Pestic Biochem Physiol 2009; 96: 93-100.
[http://dx.doi.org/10.1016/j.pestbp.2009.09.006]

[50]   Nagpal A, Grover IS. Genotoxic evaluation of some systemic pesticides in *Allium cepa* following *in situ* and direct treatments I mitotic effects. Nucleus 1994; 37: 99-105.

[51]   Chauhan LKS, Saxena PN, Gupta SK. Cytogenetic effects of cypermethrin and fenvalerate on the root meristem cells of *Allium cepa*. J Environ Exp Bot 1999; 42: 181-9.
[http://dx.doi.org/10.1016/S0098-8472(99)00033-7]

[52]   Sahu RK, Behera BN, Sharma CBSR. Cytogenetic effects from agricultural chemicals-V. Clastogenic effects of some benzimidazole fungicides on root meristems. Environ Exp Bot 1983; 23: 79-83.
[http://dx.doi.org/10.1016/0098-8472(83)90023-0]

[53]   Fiskesjo G. Benzo(a)pyrene and N-methyl-N-nitroN-nitrosoguanidine in the *Allium* test. Hereditas 1981; 95: 155-62.
[http://dx.doi.org/10.1111/j.1601-5223.1981.tb01334.x]

[54]   Eren Y, Erdogmus SF, Akyil D, Ozkara A, Konuk M, Saglam E. Cytotoxic and genotoxic effects of dioxacarb by human peripheral blood lymphocytes CAs and *Allium* test. Cytotechnology 2014.
[http://dx.doi.org/10.1007/s10616-014-9741-0] [PMID: 24848210]

[55]   Asita AO, Makhalemele R. Genotoxic effects of dithane, malathion and garden ripcord on onion root tip cells. AJFAND online 2009; 9(4): 1191-209.
[http://dx.doi.org/10.4314/ajfand.v9i5.45096]

[56]   Mann SK. Cytological and genetic effects of dithane fungicides on *Allium cepa*. Environ Exp Bot 1977; 17: 7-12.
[http://dx.doi.org/10.1016/0098-8472(77)90014-4]

[57]   Sawamura S. Cytological studies on the effect of herbicides on plant cells *in vivo*. II. Non-hormonic herbicides. Cytologia (Tokyo) 1965; 30: 325-48.
[http://dx.doi.org/10.1508/cytologia.30.325]

[58]   Yuzbasioglu D, Unal F, Sancak C. Genotoxic effects of herbicide Illoxan (diclofop methyl) on *Allium cepa* L. Tubitak J Biol 2009; 33: 283-90.

[59]   Liman R, Ciğerci IH, Öztürk NS. Determination of genotoxic effects of Imazethapyr herbicide in *Allium cepa* root cells by mitotic activity, chromosome aberration, and comet assay. Pestic Biochem Physiol 2015; 118: 38-42.
[http://dx.doi.org/10.1016/j.pestbp.2014.11.007] [PMID: 25752428]

[60]   Badr A, Elkington TT. Antimitotic and chromotoxic activities of isoproturon in *Allium cepa* and *Hordeum vulgare*. Environ Exp Bot 1982; 22: 265-70.

[http://dx.doi.org/10.1016/0098-8472(82)90017-X]

[61]   Chauhan LKS, Sundararaman V. Effects of substituted ureas on plant cells. I. Cytological effects of isopruturon on the root meristem cells of *Allium cepa.* Cytologia (Tokyo) 1990; 55: 91-8.
         [http://dx.doi.org/10.1508/cytologia.55.91]

[62]   Bianchi J, Mantovani MS, Marin-Morales MA. Analysis of the genotoxic potential of low concentrations of Malathion on the *Allium cepa* cells and rat hepatoma tissue culture. J Environ Sci (China) 2015; 36: 102-11.
         [http://dx.doi.org/10.1016/j.jes.2015.03.034] [PMID: 26456612]

[63]   Marcano L, Carruyo I, Del Campo A, Montiel X. Cytotoxicity and mode of action of maleic hydrazide in root tips of *Allium cepa* L. Environ Res 2004; 94(2): 221-6.
         [http://dx.doi.org/10.1016/S0013-9351(03)00121-X] [PMID: 14757385]

[64]   Pandita TK. Mutagenic studies on the insecticides metasystox-R with *Allium cepa.* Cytologia (Tokyo) 1986; 51: 387-92.
         [http://dx.doi.org/10.1508/cytologia.51.387]

[65]   Ravindran RN. Cytological effects of parathion. Cytologia (Tokyo) 1971; 36: 504-8.
         [http://dx.doi.org/10.1508/cytologia.36.504]

[66]   Promkaew N, Soontornchainaksaeng P, Jampatong S, Rojanavipart P. Toxicity and genotoxicity of pendimethalin in maize and onion. Kasetstart J Nat Sci 2010; 44: 1010-5.

[67]   Younis SA, Al-Hakkak S, Al-Rawi FI, Hagop EG. Physiological and cytogenetic effects of phosphine gas in *Allium cepa.* J Seed Prod Res 1989; 25: 25-30.
         [http://dx.doi.org/10.1016/0022-474X(89)90005-2]

[68]   Bielecki E. The influence of phenyl mercury acetate on mitosis and chromosome structure of *Allium cepa.* Acta Biol Cracov Ser; Bot 1974; 17: 119-32.

[69]   Mousa M. Mitotic inhibition and chromosomal aberrations induced by some herbicides in root tips of *Allium cepa.* Egypt J Genet Cytol 1982; 11: 193-208.

[70]   Mustafa Y, Arikan ES. Genotoxicity testing of quizalofop-P-ethyl herbicide using the *Allium cepa* anaphase-telophase chromosome aberration assay. Caryologia 2008; 61: 45-52.
         [http://dx.doi.org/10.1080/00087114.2008.10589608]

[71]   Fisun K, Rasgele PG. Genotoxic Effects of raxil on root tips and anthers of *Allium cepa* L. Caryologia 2009; 62: 1-9.
         [http://dx.doi.org/10.1080/00087114.2004.10589659]

[72]   Pandita TK, Khoshoo TN. Mutagencity testing of thimet 10-G. Nucleus 1984; 27: 168-71.

[73]   Escelza P, Cortes F, Lopez-Campos JL. Action of vinclozolin on cell division and its effectiveness in the production of chromosomal aberrations but not sister chromatid exchanges. Cytobios 1983; 38: 149-58.

[74]   Quidet P, Hitier H. The production of polyploidy plants by treatment with hexachlorocyclohexane sulphide. Comptes Rendus de Academie des Sciences 1948; 226: 833-5.

[75]   Aydemir N, Celikler S, Summak S, Yilmaz D, Ozer O. Evaluation of clastogenecity of 4, 6-Dinitro-o-cresol (DNOC) in *Allium* root tip test. J Environ Bot 2008; 2: 59-63.

[76]   Achary VMM, Jena S, Panda KK, Panda BB. Aluminium induced oxidative stress and DNA damage in root cells of *Allium cepa* L. Ecotoxicol Environ Saf 2008; 70(2): 300-10.
         [http://dx.doi.org/10.1016/j.ecoenv.2007.10.022] [PMID: 18068230]

[77]   Yi H, Wu L, Jiang L. Genotoxicity of arsenic evaluated by *Allium*-root micronucleus assay. Sci Total Environ 2007; 383(1-3): 232-6.
         [http://dx.doi.org/10.1016/j.scitotenv.2007.05.015] [PMID: 17574654]

[78]   Liman R. Genotoxic effects of Bismuth (III) oxide nanoparticles by *Allium* and Comet assay.

Chemosphere 2013; 93(2): 269-73.
[http://dx.doi.org/10.1016/j.chemosphere.2013.04.076] [PMID: 23790828]

[79] Konuk M, Liman R, Cigerci IH. Determination of genotoxic effect of boron on *Allium cepa* root meristamatic cells. Pak J Bot 2007; 39: 73-9.

[80] Fiskesjö G. Some results from *allium* tests with organic mercury halogenides. Hereditas 1969; 62(3): 314-22.
[http://dx.doi.org/10.1111/j.1601-5223.1969.tb02241.x] [PMID: 5400476]

[81] Seth CS, Misra V, Chauhan LKS, Singh RR. Genotoxicity of cadmium on root meristem cells of *Allium cepa*: cytogenetic and Comet assay approach. Ecotoxicol Environ Saf 2008; 71(3): 711-6.
[http://dx.doi.org/10.1016/j.ecoenv.2008.02.003] [PMID: 18358534]

[82] Behboodi BS, Samadi L. Morphological study of cadmium induced changes on root apex of *Allium cepa*. Iran Int J Sci 2002; 3: 11-22.

[83] Patnaik AR, Mohan V, Achary M, Panda BB. Chromium (VI)-induced hormesis and genotoxicity are mediated through oxidative stress in root cells of *Allium cepa* L. Plant Growth Regul 2013; 71: 157-70.
[http://dx.doi.org/10.1007/s10725-013-9816-5]

[84] Arambašić MB, Bjelić S, Subakov G. Acute toxicity of heavy metals (copper, lead, zinc), phenol and sodium on *Allium cepa* L., *Lepidium sativum* L. and *Daphnia magna* St: Comparative investigations and the practical applications. Water Res 1995; 29: 497-503.
[http://dx.doi.org/10.1016/0043-1354(94)00178-A]

[85] Carruyo I, Fernández Y, Marcano L, Montiel X, Torrealba Z. Correlation of toxicity with lead content in root tip cells (*Allium cepa* L.). Biol Trace Elem Res 2008; 125(3): 276-85.
[http://dx.doi.org/10.1007/s12011-008-8175-y] [PMID: 18636231]

[86] Wierzbicka M. Comparison of lead tolerance in *Allium cepa* with other plant species. Environ Pollut 1999; 104: 41-52.
[http://dx.doi.org/10.1016/S0269-7491(98)00156-0]

[87] Fiskesjö G. The *Allium* test--an alternative in environmental studies: the relative toxicity of metal ions. Mutat Res 1988; 197(2): 243-60.
[http://dx.doi.org/10.1016/0027-5107(88)90096-6] [PMID: 3340086]

[88] Fiskesjo G. Mercury and selenium in a modified *Allium* test. Hereditas 1979; 91: 169-78.
[http://dx.doi.org/10.1111/j.1601-5223.1979.tb01659.x]

[89] Kumari M, Mukherjee A, Chandrasekaran N. Genotoxicity of silver nanoparticles in *Allium cepa*. Sci Total Environ 2009; 407(19): 5243-6.
[http://dx.doi.org/10.1016/j.scitotenv.2009.06.024] [PMID: 19616276]

[90] Pakrashi S, Jain N, Dalai S, *et al. In vivo* genotoxicity assessment of titanium dioxide nanoparticles by *Allium cepa* root tip assay at high exposure concentrations. PLoS One 2014; 9(2): e87789.
[http://dx.doi.org/10.1371/journal.pone.0087789] [PMID: 24504252]

[91] Marcano L, Carruyo I, Fernández Y, Montiel X, Torrealba Z. Determination of vanadium accumulation in onion root cells (*Allium cepa* L.) and its correlation with toxicity. Biocell 2006; 30(2): 259-67.
[PMID: 16972550]

[92] Borboa L, Torre CD. The genotoxicity of Zii(II) and Cd(I) *Allium cepa* root meristematic cells. New Phytol 1996; 134: 481-6.
[http://dx.doi.org/10.1111/j.1469-8137.1996.tb04365.x]

[93] Kumari M, Khan SS, Pakrashi S, Mukherjee A, Chandrasekaran N. Cytogenetic and genotoxic effects of zinc oxide nanoparticles on root cells of *Allium cepa*. J Hazard Mater 2011; 190(1-3): 613-21.
[http://dx.doi.org/10.1016/j.jhazmat.2011.03.095] [PMID: 21501923]

[94]   Kaur M, Soodan RK, Katnoria JK, Bhardwaj R, Pakade YB, Nagpal AK. Analysis of physico-chemical parameters, genotoxicity and oxidative stress inducing potential of soils of some agricultural fields under rice cultivation. Trop Plant Res 2014; 1(3): 49-61.

[95]   Kaur R, Pakade YB, Katnoria JK. A study on physicochemical analysis of road and railway track side soil samples of Amritsar (Punjab) and their genotoxic effects. Int J Environ Chem Ecol Geo and Geophy Eng 2014; 8: 510-3.

[96]   Kaur R, Pakade YB, Katnoria JK. Genotoxicity and tumor inducing potential of roadside soil samples exposed to heavy traffic emissions at Amritsar (Punjab), India. J Appl Nat Sci 2013; 5: 382-7.
[http://dx.doi.org/10.31018/jans.v5i2.337]

[97]   Pohren P, Thatiana C, Vargas VMF. Investigation of sensitivity of the *Allium cepa* test as an alert system to evaluate the genotoxic potential of soil contaminated by heavy metals. Water Air Soil Pollut 2013; 224: 1460-70.
[http://dx.doi.org/10.1007/s11270-013-1460-1]

[98]   Masood F, Malik A. Cytotoxic and genotoxic potential of tannery waste contaminated soils. Sci Total Environ 2013; 444: 153-60.
[http://dx.doi.org/10.1016/j.scitotenv.2012.11.049] [PMID: 23268142]

[99]   Chahal V, Nagpal A, Katnoria JK. Genotoxicity evaluation of soil sample from agricultural field under wheat cultivation. Bot Res Int 2012; 5: 1-3.

[100]   Katnoria JK, Arora S, Bhardwaj R, Nagpal A. Evaluation of genotoxic potential of industrial waste contaminated soil extracts of Amritsar, India. J Environ Biol 2011; 32(3): 363-7.
[PMID: 22167950]

[101]   Dragoeva A, Kalcheva V, Slanev ST. Genotoxicity of agricultural soils after one year of conversion period and under convential agriculture. J Appl Sci Environ Manag 2009; 13: 81-3.

[102]   Andrade LF, Campos JMS, Davide LC. Cytogenetic alterations induced by SPL (spent potliners) in meristematic cells of plant bioassays. Ecotoxicol Environ Saf 2008; 71(3): 706-10.
[http://dx.doi.org/10.1016/j.ecoenv.2008.02.018] [PMID: 18395259]

[103]   Chandra S, Chauhan LKS, Murthy RC, Saxena PN, Pande PN, Gupta SK. Comparative biomonitoring of leachates from hazardous solid waste of two industries using *Allium* test. Sci Total Environ 2005; 347(1-3): 46-52.
[http://dx.doi.org/10.1016/j.scitotenv.2005.01.002] [PMID: 16084966]

[104]   Glasencnik E, Ribaric-Lasnik C, Savinek K, Zalubersek M, Mueller M, Batic F. Impact of air pollution on genetic material of shallot (*Allium cepa* L. var. ascalonicum) exposed at differentially polluted sites in Slovenia. J Atmos Chem 2004; 49: 363-76.
[http://dx.doi.org/10.1007/s10874-004-1252-5]

[105]   Cabrera GL, Rodriguez DM. Genotoxicity of leachates from a landfill using three bioassays. Mutat Res 1999; 426(2): 207-10.
[http://dx.doi.org/10.1016/S0027-5107(99)00069-X] [PMID: 10350599]

[106]   Cotelle S, Masfaraud JF, Férard JF. Assessment of the genotoxicity of contaminated soil with the Allium/Vicia-micronucleus and the Tradescantia-micronucleus assays. Mutat Res 1999; 426(2): 167-71.
[http://dx.doi.org/10.1016/S0027-5107(99)00063-9] [PMID: 10350593]

[107]   Kong MS, Ma TH. Genotoxicity of contaminated soil and shallow well water detected by plant bioassays. Mutat Res 1999; 426(2): 221-8.
[http://dx.doi.org/10.1016/S0027-5107(99)00072-X] [PMID: 10350602]

[108]   Kovalchuk O, Kovalchuk I, Arkhipov A, Telyuk P, Hohn B, Kovalchuk L. The *Allium cepa* chromosome aberration test reliably measures genotoxicity of soils of inhabited areas in the Ukraine contaminated by the Chernobyl accident. Mutat Res 1998; 415(1-2): 47-57.
[http://dx.doi.org/10.1016/S1383-5718(98)00053-9] [PMID: 9711261]

[109] Düsman E, Luzza M, Savegnago L, *et al. Allium cepa* L. as a bioindicator to measure cytotoxicity of surface water of the Quatorze River, located in Francisco Beltrão, Paraná, Brazil. Environ Monit Assess 2014; 186(3): 1793-800.
[http://dx.doi.org/10.1007/s10661-013-3493-8] [PMID: 24162370]

[110] Firbas P, Amon T. *Allium* chromosome aberration test for evaluation effect of cleaning municipal water with constructed wetland (CW) in Sveti Tomaz, Slovenia. Biorem Biodeg 2013; 4: 1-5.

[111] Kern DI, Schwaickhardt RdeO, Mohr G, Lobo EA, Kist LT, Machado EL. Toxicity and genotoxicity of hospital laundry wastewaters treated with photocatalytic ozonation. Sci Total Environ 2013; 443: 566-72.
[http://dx.doi.org/10.1016/j.scitotenv.2012.11.023] [PMID: 23220390]

[112] Barbério A, Voltolini JC, Mello ML. Standardization of bulb and root sample sizes for the *Allium cepa* test. Ecotoxicology 2011; 20(4): 927-35.
[http://dx.doi.org/10.1007/s10646-011-0602-8] [PMID: 21298340]

[113] Radić S, Stipanicev D, Vujćić V, Rajcić MM, Sirac S, Pevalek-Kozlina B. The evaluation of surface and wastewater genotoxicity using the *Allium cepa* test. Sci Total Environ 2010; 408(5): 1228-33.
[http://dx.doi.org/10.1016/j.scitotenv.2009.11.055] [PMID: 20018345]

[114] Abu NE, Mba KC. Mutagenecity testing of phamarceutical effluents on *Allium cepa* root tip meristems. J Toxicol Environ Health Sci 2011; 3: 44-51.

[115] Olorunfemi DI, Okoloko GE, Bakare AA, Akinboro A. Cytotoxic and genotoxic effects of cassava effluents using the *Allium cepa* assay. Res J Mutagen 2011; 1: 1-9.
[http://dx.doi.org/10.3923/rjmutag.2011.1.9]

[116] Olorunfemi DI, Iogieseri UM, Akinboro A. Genotoxicity screening of industrial effluents using Onion bulbs (*Allium cepa* L.). J Appl Sci Environ Manag 2011; 15: 211-6.
[http://dx.doi.org/10.4314/jasem.v15i1.65700]

[117] Samuel OB, Osuala FI, Odeigah PGC. Cytogenotoxicity evaluation of two industrial effluents using *Allium cepa* assay. Afr J Environ Sci Technol 2010; 4: 21-7.

[118] Akintonwa A, Awodele O, Olofinnade AT, Anyakora C, Afolayan GO, Coker HAB. Assessment of the mutagenicity of some pharmaceutical effluents. Am J Pharmacol Toxicol 2009; 4: 144-50.
[http://dx.doi.org/10.3844/ajptsp.2009.144.150]

[119] Ayas S, Demirtas C. Deficit irrigation effects on cucumber (*Cucumis sativus* L. Maraton) yield in unheated greenhouse condition. J Food Agric Environ 2009; 7(3&4). 645-9.

[120] Hoshina MM, Marin-Morales MA. Micronucleus and chromosome aberrations induced in onion (*Allium cepa*) by a petroleum refinery effluent and by river water that receives this effluent. Ecotoxicol Environ Saf 2009; 72(8): 2090-5.
[http://dx.doi.org/10.1016/j.ecoenv.2009.07.002] [PMID: 19647317]

[121] Bakare AA, Okunola AA, Adetunji OA, Jenmi HB. Genotoxicity assessment of a pharmaceutical effluent using four bioassays. Genet Mol Biol 2009; 32(2): 373-81.
[http://dx.doi.org/10.1590/S1415-47572009000200026] [PMID: 21637694]

[122] Bagatini MD, Vasconcelos TG, Laughinghouse HD IV, Martins AF, Tedesco SB. Biomonitoring hospital effluents by the *Allium cepa* L. test. Bull Environ Contam Toxicol 2009; 82(5): 590-2.
[http://dx.doi.org/10.1007/s00128-009-9666-z] [PMID: 19224103]

[123] Barbério A, Barros L, Voltolini JC, Mello ML. Evaluation of the cytotoxic and genotoxic potential of water from the River Paraíba do Sul, in Brazil, with the *Allium cepa* L. test. Braz J Biol 2009; 69(3): 837-42.
[http://dx.doi.org/10.1590/S1519-69842009000400010] [PMID: 19802442]

[124] Cynthia R, Oliveria-Martins CR, Grisolia CK. Toxicity and genotoxicity of waste water from gasoline stations. Genet Mol Biol 2009; 32: 1415-4757.

[125] Espinoza-Quiñones FR, Szymanski N, Palácio SM, *et al.* Inhibition effect on the Allium cepa L. root growth when using hexavalent chromium-doped river waters. Bull Environ Contam Toxicol 2009; 82(6): 767-71.
[http://dx.doi.org/10.1007/s00128-009-9682-z] [PMID: 19280093]

[126] Caritá R, Marin-Morales MA. Induction of chromosome aberrations in the *Allium cepa* test system caused by the exposure of seeds to industrial effluents contaminated with azo dyes. Chemosphere 2008; 72(5): 722-5.
[http://dx.doi.org/10.1016/j.chemosphere.2008.03.056] [PMID: 18495201]

[127] Fawole OO, Yekeen TA, Ayandele AA, Akinboro A, Azeez MA, Adewoye SO. Polluted Alamuyo river: Impacts on surrounding wells, microbial attributes and toxic effects on *Allium cepa* root cells. Afr J Biotechnol 2008; 7: 450-8.

[128] Leme DM, Marin-Morales MA. Chromosome aberration and micronucleus frequencies in *Allium cepa* cells exposed to petroleum polluted water--a case study. Mutat Res 2008; 650(1): 80-6.
[http://dx.doi.org/10.1016/j.mrgentox.2007.10.006] [PMID: 18068420]

[129] Leme DM, de Angelis DdeF, Marin-Morales MA. Action mechanisms of petroleum hydrocarbons present in waters impacted by an oil spill on the genetic material of *Allium cepa* root cells. Aquat Toxicol 2008; 88(4): 214-9.
[http://dx.doi.org/10.1016/j.aquatox.2008.04.012] [PMID: 18556073]

[130] Magdaleno A, Mendelson A, de Iorio AF, Rendina A, Moretton J. Genotoxicity of leachates from highly polluted lowland river sediments destined for disposal in landfill. Waste Manag 2008; 28(11): 2134-9.
[http://dx.doi.org/10.1016/j.wasman.2007.09.027] [PMID: 18440215]

[131] Abdel Migid HM, Azab YA, Ibrahim WM. Use of plant genotoxicity bioassay for the evaluation of efficiency of algal biofilters in bioremediation of toxic industrial effluent. Ecotoxicol Environ Saf 2007; 66(1): 57-64.
[http://dx.doi.org/10.1016/j.ecoenv.2005.10.011] [PMID: 16376989]

[132] Vujosević M, Andelković S, Savić G, Blagojević J. Genotoxicity screening of the river Rasina in Serbia using the Allium anaphase-telophase test. Environ Monit Assess 2008; 147(1-3): 75-81.
[http://dx.doi.org/10.1007/s10661-007-0099-z] [PMID: 18080777]

[133] Fatima RA, Ahmad M. Certain antioxidant enzymes of *Allium cepa* as biomarkers for the detection of toxic heavy metals in wastewater. Sci Total Environ 2005; 346(1-3): 256-73.
[http://dx.doi.org/10.1016/j.scitotenv.2004.12.004] [PMID: 15993699]

[134] Grisolia CK, Bilich MR, Formigli LM. A comparative toxicologic and genotoxic study of the herbicide arsenal, its active ingredient imazapyr, and the surfactant nonylphenol ethoxylate. Ecotoxicol Environ Saf 2004; 59(1): 123-6.
[http://dx.doi.org/10.1016/j.ecoenv.2004.01.014] [PMID: 15261733]

[135] Srivastava R, Kumar D, Gupta SK. Bioremediation of municipal sludge by vermitechnology and toxicity assessment by *Allium cepa.* Bioresour Technol 2005; 96(17): 1867-71.
[http://dx.doi.org/10.1016/j.biortech.2005.01.029] [PMID: 15927461]

[136] Staykova TA, Ivanava EN, Velcheva IG. Cytogenetic effect of heavy metal and cyanide in contaminated waters from the region of southwest Bulgaria. J Cell Mol Biol 2005; 4: 41-6.

[137] El-Shahaby OA, Abdel-Migid HM, Soliman MI, Mashaly IA. Genotoxicity screening of industrial wastewater using the *Allium cepa* chromosome aberration assay. Pak J Biol Sci 2003; 6: 23-8.
[http://dx.doi.org/10.3923/pjbs.2003.23.28]

[138] Evseeva TI, Geras'kin SA, Shuktomova II, Shuktomova I. Genotoxicity and toxicity assay of water sampled from a radium production industry storage cell territory by means of *Allium*-test. J Environ Radioact 2003; 68(3): 235-48.
[http://dx.doi.org/10.1016/S0265-931X(03)00054-7] [PMID: 12782475]

[139] Amin AW. Cytotoxicity testing of sewage water treatment using *Allium cepa* chromosomal aberrations assay. Pak J Biol Sci 2002; 5: 184-8.
[http://dx.doi.org/10.3923/pjbs.2002.184.188]

[140] Evandri MG, Tucci P, Bolle P. Toxicological evaluation of commercial mineral water bottled in polyethylene terephthalate: a cytogenetic approach with *Allium cepa.* Food Addit Contam 2000; 17(12): 1037-45.
[http://dx.doi.org/10.1080/02652030010014411] [PMID: 11271838]

[141] Pavlica M, Besendorfer V, Rosa J, Papes D. The cytotoxic effect of wastewater from the phosphoric gypsum depot on common oak (*Quercus robur* L.) and shallot (*Allium cepa* var. ascalonicum). Chemosphere 2000; 41(10): 1519-27.
[http://dx.doi.org/10.1016/S0045-6535(00)00106-5] [PMID: 11057676]

[142] Grover IS, Kaur S. Genotoxicity of wastewater samples from sewage and industrial effluent detected by the *Allium* root anaphase aberration and micronucleus assays. Mutat Res 1999; 426(2): 183-8.
[http://dx.doi.org/10.1016/S0027-5107(99)00065-2] [PMID: 10350595]

[143] Rank J, Nielsen MH. Genotoxicity testing of wastewater sludge using the *Allium cepa* anaphase-telophase chromosome aberration assay. Mutat Res 1998; 418(2-3): 113-9.
[http://dx.doi.org/10.1016/S1383-5718(98)00118-1] [PMID: 9757013]

[144] Smaka-Kincl V, Stegnar P, Lovka M, Toman MJ. The evaluation of waste, surface and ground water quality using the *Allium* test procedure. Mutat Res 1996; 368(3-4): 171-9.
[http://dx.doi.org/10.1016/S0165-1218(96)90059-2] [PMID: 8692223]

[145] Vidakovic Z, Papes D. Toxicity of waste drilling fluids in modified *Allium cepa* test. Water Air Soil Pollut 1993; 69: 413-23.
[http://dx.doi.org/10.1007/BF00478174]

CHAPTER 5

Pesticides: Problems and Remedial Measures

Sonal Yadav and **Satyawati Sharma**[*]

Centre for Rural Development and Technology, Indian Institute of Technology Delhi, Hauz Khas, New Delhi, India

Abstract: The population is increasing at a tremendous rate and to feed them from the exploited land would be of great concern. Presently, the major concern is the use of hybrid seed, genetically modified crops, chemical pesticides, and fertilizer. The purpose of this chapter is mainly to focus on the soil contamination problem caused by pesticides and present a review of the existing methods for remediation of contaminated soils. First, a brief discussion of the pesticides along with its classification, and the impact caused by the pesticides on the environment and human health are examined. Then, the current practices and evolving techniques for soil remediation, which are mainly used for the eradication of pesticides are discussed, along with their advantages and disadvantages. Among the different existing methods, bioremediation is the most promising technique. Bioremediation of contaminated soil using pesticides by the use of microorganisms is eco-friendly, most effective and economical method of detoxification.

Keywords: Biopesticides, Bioairsparging, Biopiles, Biostimulation, Bioaugmentation, Composting, Carbamates, DDT, DDE, Fungicides, Herbicides, Landfarming, Microbial Remediation, Organochlorines, Organophosphates, Pyrethrins, Pyrethroids, Plant Incorporated Protectants, Phytoremediation, Soil Fertility.

INTRODUCTION

Since 1950, the arable land used to provide food to human population is getting reduced three times which is creating a huge pressure to provide food, at low price as the nutrients are stripped from the soil making land unfertile. Dependency on other sources that is fertilisers and pesticides remain to be the temporary solution for commercial agricultural systems that runs on large scale. As reported by Green Peace International laboratory in 2015, chemical pesticides are considerably used in agriculture across worldwide. Due to the extensive repeated use, many chemical pesticides become tremendously persistent in the environment. There

[*] **Corresponding author Satyawati Sharma:** Centre for Rural Development and Technology, Indian Institute of Technology Delhi, Hauz Khas, New Delhi, India; Tel: 01126591116; E-mail: ssharma@rdat.iitd.ac.in

Ashita Sharma, Manish Kumar, Satwinderjeet Kaur & Avinash Kaur Nagpal (Eds.)

are some chemical substances that takes extremely long time to degrade due to which they are banned decades ago, including DDT but still it's residues are consistently found in the environment. Study related to the influence of pesticides on environment has increased immensely over the past 30 years due to its persistence and possible hazards to wildlife [1]. Therefore, it is clear that the effects caused by chemical pesticide are wide and varied. Scientific understanding of mechanisms of action of pesticides and their effects on human health has also prolonged rapidly by pesticide exposure causing neurological and immunological illnesses along with some cancers. Though, pesticidal exposure affects human body and it is examined that there is no crowd in the human groups which is entirely not exposed to agrochemicals and maximum disorders significantly give complexity to public health assessments [2]. Pesticides contribute further to this toxic burden. The prime-most objective in world is to increase the production of food, as world population is expected to grow to nearly 10 billion by 2050. Based on evidence, every year the human population is increasing by ~97 million [3]. The food and agricultural organization (FAO) of the United Nations has circulated a serious urge that globally the food production demands to increase by 70%, so that the requirement of rapidly increasing population can be met [9]. In order to overcome the pressure created by growing demand of food on the current agricultural system led to the overuse of herbicides, fungicides, nematicides, insecticides, chemical fertilizers and soil amendments. Before the chemical pesticide was introduced, the majority of weeds, insects, pests and diseases were handled by physical and mechanical control strategies which were considered as sustainable practice. Pesticides are not only used for protecting the agricultural land but also eradicates the pests that transmits the dangerous contagious diseases. Every year approximately outflow of $38 billion is done on pesticides [4].

New formulations of pesticides are designed by manufacturers and researchers to meet the comprehensive demand. For an ideal pesticide, they should be toxic to the target organisms, eco-friendly and biodegradable [5]. Various chemical pesticides are found to be non-specific in nature therefore they may kill organisms that are useful to the ecosystem or that do not harm. Nearly 0.1% of agrochemicals grasp the target whereas the left behind one adulterates the surroundings [6]. The repetitive use of chemical pesticide that are persistent in nature and non-biodegradable pollute marine, air and soil environment. Pesticides have also bioaccumulated in the higher trophic level through the food chain. Due to the continuous exposure of pesticides, large number of cases in human population related to acute and chronic illnesses has been reported [7]. Uninterrupted use of pesticides to the soil ecosystem and marine system caused health hazards and environmental pollution which is the major public concern today.

Pattern of Pesticide Usage

As per the agrochemicals knowledge report, 2016, Indian market is mainly conquered by insecticides, comprising almost 60% of total chemical pesticide with major applications in cotton and rice crops as shown in Fig. (1) [8]. Fungicides and herbicides comprises 18% and 16% respectively of total pesticide respectively. The trade of herbicides is seasonal due to the growth pattern observed in weeds, growing in warm weather and die in cold season. Herbicides are majorly used for wheat and rice crops whereas the fungicides have vast application in vegetables, fruits, and rice. Bio-pesticides are the biological constituent organisms, that can be applied to regulate pests. At present, biopesticides covers only 3% of Indian pesticide market.

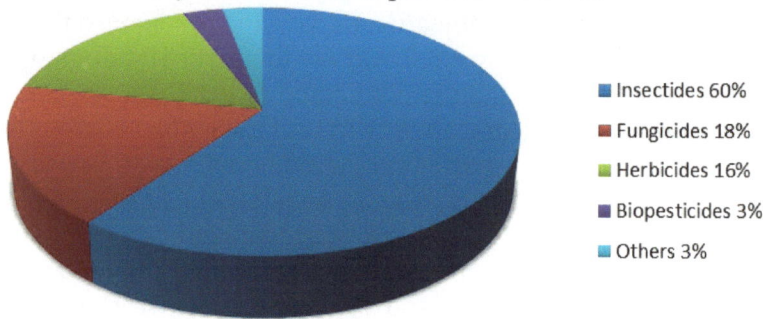

Fig. (1). A Pattern of pesticide usage in Indian crop protection market.
Source: Industry reports, Analysis by Tata Strategic

The top 3 states that constitute 45% of pesticide consumption in India are Andhra Pradesh including Seemandhra and Telangana sharing 24%, Maharashtra, and Punjab. Whereas rest top seven states in India together comprise above 70% of total pesticide usage in India [8] as shown in Fig. (2).

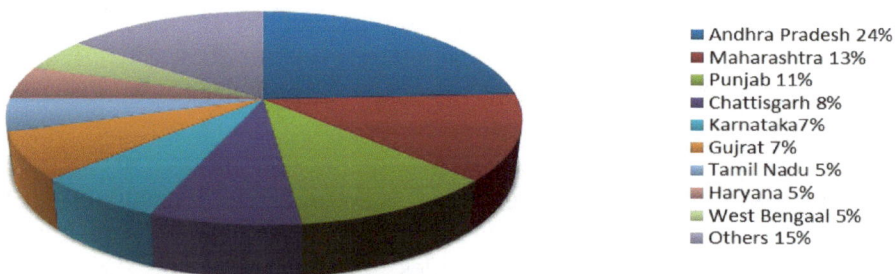

Fig. (2). A pattern of state wise consumption of pesticide.
Source: Industry reports, Analysis by Tata Strategic

Pesticide

These are the chemical substances used to kill pests. In the perspective of soil, pests can be bacteria, worms, fungi, insects, and nematodes *etc.* that triggers destruction to agricultural crops. Thus, on the basis of the target, pesticides are categorized into fungicides, insecticides, herbicides, bactericides, and nematicides. They are employed in controlling or inhibiting plant diseases and insect pests. By the use of pesticide, crop yield can be enhanced significantly but its extreme use leads to the microbial imbalance polluting the environment along with the health hazards. An ideal pesticide should be target specific with the capability to destroy them rapidly along with the ability to degrade toxic substances into harmless substances as rapidly as possible.

A "pesticide" is defined similarly as given by the United Nation's (UN) Food and Agriculture Organization (FAO) [9] was implemented:

"… any substance or mixture of substances intended for preventing, destroying, or controlling any pest, including vectors of human or animal disease, unwanted species of plants or animals, causing harm during or otherwise interfering with the production, processing, storage, transport, or marketing of food, agricultural commodities, wood and wood products or animal feedstuffs, or substances that may be administered to animals for the control of insects, arachnids, or other pests in or on their bodies. The term includes substances intended for use as a plant growth regulator, defoliant, desiccant, or agent for thinning fruit or preventing the premature fall of fruit and substances applied to crops either before or after harvest to protect the commodity from deterioration during storage and transport."

Classification of Pesticides

One traditional classification of pesticides places them in one of two groups: chemical and biological pesticide as shown in Fig. (**3**).

Organochlorines (Chlorinated Hydrocarbons):

These are organic compounds with several atoms of chlorine per molecule. It is the first group of chemical pesticides that were synthesized for *e.g.* DDT. They are resistant to natural breakdown processes due to which these pesticides are very stable making them persistent in the environment [10]. Some of the common organochlorines are carbon tetrachloride; chlordane; DDT (Dichloro diphenyl trichloroethylene); DDE (Dichloro diphenyl dichloroethylene); dieldrin; heptachlor; β-HCH; γ-HCH. Most other organochlorines used for arthropod control, including chlordane, dieldrin, and lindane. Termites attack can be controlled by use of Aldrin. All these substances are lipophilic and get bioaccumulated in the fatty tissue of animals.

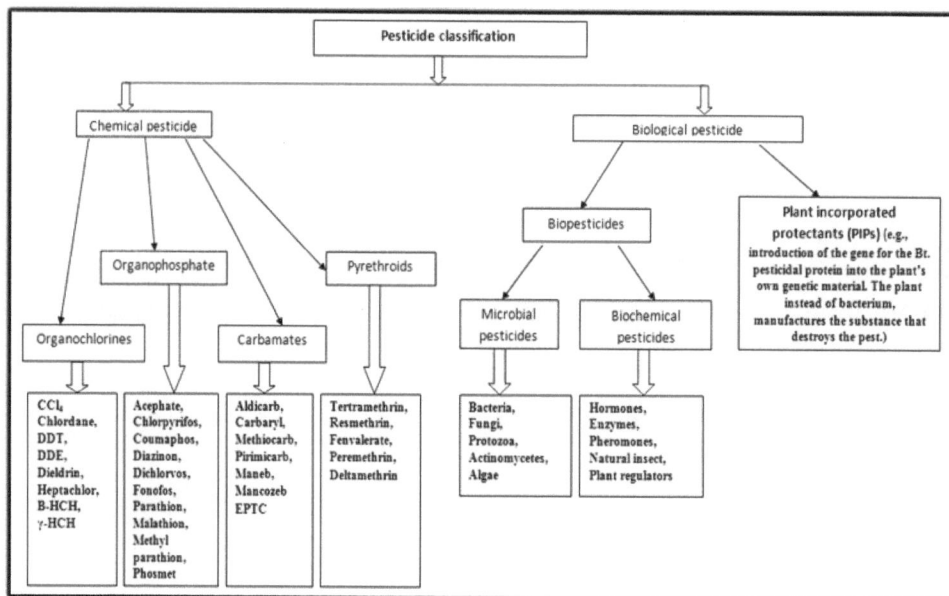

Fig. (3). Pesticide classification.

Organophosphates (OP):

Organophosphate compounds are derived from phosphoric acid and the commonly used group of pesticides across the worldwide. Its application is varied like in agricultural farms, homes, gardens and veterinary practices. There are several well-known organophosphates which have been discontinued for practice, including parathion, due to their toxicity and persistent nature. All contribute to a common phenomenon of cholinesterase inhibition (nerve poisons which destroy the target pest, commonly insects) and can cause similar symptoms. Most commonly used organophosphates are chlorpyrifos; parathion; methyl parathion; coumaphos; acephate; diazinon; dichlorvos; fonofos; malathion; phosmet.

Carbamates

They are the derivates from carbamic acid. They are basically used for vector control like carbaryl (Sevin®), used for dusting rodent burrows to control fleas, propoxur (Baygon®), used against insect pests. Carbamates show relatively high risk for human health. Some carbamates come under the category of herbicides. In general, carbamates are neurotoxic and inhibits the acetylcholinesterase. They also affect babies and children by showing harmful effects on human development [11]. Commonly used carbamates are aldicarb; pirimicarb; methiocarb; carbaryl;

mancozeb or maneb (both dithiocarbamates); Methiocarb; pirimicarb; S-Ethyl-N, N-dipropylthiocarbamate.

Pyrethrins and Pyrethroids

Pyrethrins are botanical pesticides and derivatives of chrysanthemum flowers. They are most commonly found in Africa and Australia. They alter nerve role causing paralysis in insect pests, ultimately leads to death. Pyrethroids are (synthetic) chemical insecticides modified from the chemical structures of the pyrethrins acting similarly to pyrethrins. Pyrethroids are altered to increase their strength in sunlight. At present, both insecticides are used commonly for the public health, mainly to control adult mosquitoes. Their use now far exceeds that of established synthetic pesticides like organochlorines and organophosphates.

Biopesticides (Biorationals)

These are comparatively less toxic to human population and are environmentally friendly. EPA (Environment Protection Authority) states biorationals as "certain types of pesticides derived from such natural materials as animals, plants, bacteria, and certain minerals". Biorationals can be divided into two categories:

1. Biochemical (pheromones, enzymes, hormones, plant regulators and natural insect)
2. Microbial (protozoa, bacteria, viruses, fungi, and nematodes).

Biochemical biorationals comprises plant growth regulators, insect growth regulators (IGRs), chemosterilants and chitin inhibitors. It interrupts biological growth phenomenon of arthropods and is little toxic for vertebrates, that includes people too. Diflubenzuron (Dimilin®) is an example of insect growth regulators.

Microbial pesticides eradicate arthropods by the phenomenon of releasing toxins by microorganisms, or due to the infection caused by the organisms. The bacterial toxin produced by Bti, and *Bacillus sphaericus* (Bs) are the common pesticides that come under this group. These bacteria act against mosquito larvae and are potent in killing black fly larvae as well. A large number of microbial pesticides are likely to be in practice in comparison of biochemical pesticides.

Plant-incorporated protectants (PIPs)

They are plants that produce the pesticides in its own tissue for which a gene is being incorporated. As the plant is genetically modified, the pesticide produced is regulated as per Environmental Protection Agency (EPA).

Effects of Pesticides

Effect on Soil

A huge number of transformed products from a wide-ranging of pesticides have been recognized [12]. Movement with the persistent nature of these pesticides and their transformed products can be measured by the parameters like the octanol/water partition coefficient (K_{ow}), solubility in water, soil-sorption constant and half-life in soil (DT_{50}) [13].

Chemical pesticides with their transformed products can be categorized into two groups:

a. Bioaccumulable, hydrophobic and persistent pesticides which are firmly tied to the soil. Pesticides illustrating such kind of performance are DDT, endrin, lindane, endosulfan, heptachlor and their transformed products. Long ago, many pesticides among them are banned now in agricultural farms but pesticidal residues still exist.

b. Polar pesticides show the mobility from the soil by the process of leaching and runoff, which cause a problem in delivering the drinking water to the public. They are mainly represented by herbicides along with some carbamates, organophosphorus insecticide, fungicides and their transformed products.

Undoubtedly, the most explored pesticide transformed product in the soil is seen from herbicides. Chemical pesticides and their transformed products are well-kept *via* soils to different amount relying on the interactions of soil and pesticide properties. Soil pH also plays a significant role as it decreases for ionizable pesticides, the adsorption increases (*e.g.* picloram, 2,4,5-T, atrazine, and 2,4-D) [14]. Organic matter is the most significant soil characteristic, therefore, higher the amount of the organic matter, the greater will be the adsorption of pesticides along with their TPs.

Effect on Soil Fertility

Substantial usage of pesticides in the soil can cause favorable soil microbes population to downfall. According to the American microbiologist and soil scientist Dr. Elaine Ingham, "If we lose both bacteria and fungi, then the soil degrades. Overuse of chemical fertilizers and pesticides greatly affects the soil organisms. Indiscriminate use of chemicals might work for a few years, but after a while, there aren't enough beneficial as they do not let soil organisms to hold the nutrients" [14]. Many herbicides interfere in the process for *e.g.*

- Triclopyr which prevents the conversion of ammonia into nitrite with the help of soil bacteria [15].
- Glyphosate lowers the action of free-living nitrogen-fixing microbes in soil and the growth [72].
- 2,4-Dichlorophenoxyacetic acid reduces nitrogen fixation by the microbes living on the roots of bean plants [16 - 18], lowers the growth and action of nitrogen-fixing BGA (blue-green algae) [19, 21] and also obstructs the conversion of ammonia into nitrates or nitrites by microorganisms [2, 22].

Mycorrhizal fungi grow with the roots of various plants and help in the uptake of nutrient. But due to the presence of herbicides in the soil, these fungi can be damaged. It has been found that trifluralin and oryzalin inhibit the growth of some groups of mycorrhizal fungi [23]. Even triclopyr was reported as lethal to many species of mycorrhizal fungi that converts ammonia into nitrite [24].

Effect on Humans

The human health is greatly affected by pesticides because of their toxic nature and persistence in the surrounding. Pesticides have the capability to penetrate in the food chain. Therefore, these can greatly affect the body of human either by direct contact *via* chemicals or through food particularly vegetables, fruits, polluted air or polluted water. Chronic or acute illness can be the outcome from the exposure of chemical pesticide which is reviewed below:

Acute Illness

It usually emerges for short period after exposure to the pesticide or by the direct contact. Many symptoms like body aches, cramps, headaches, dizziness, skin rashes, nausea, impaired vision, poor concentration, panic attacks, however, in serious circumstances coma and death could happen because of the pesticide poisoning [4]. Due to pesticidal poisoning, approximately 3 million incidents are testified globally each year, and out of which, 2 million cases were seen as suicide attempts and rest were due to the occupational or accidental poisoning cases [70]. Several policies are recommended to lessen the occurrences that arise due to poisoning caused by pesticide such as regulating the accessibility of pesticides, replacing the toxic pesticide with non-toxic one [71].

Chronic Illness

Due to continuous exposure to pesticides from a sustained period of time, chronic illness is reported in humans. Symptoms are not immediately outward and noticeable in the future phase. Workers at farms get expose to pesticide due to their occupation found to be at higher risk to get affected, however regular public also suffer due to contaminated food and water [4].

Various incidences of human chronic illness affecting the nervous system, reproductive system, renal system, cardiovascular system, and respiratory systems are reported due to the exposure of pesticide for long period of time [7]. The list of chronic diseases that are related to prolonged pesticide exposure by various studies is mentioned in Table **1**.

Table 1. List of chronic illness linked to the exposure to pesticides.

DISEASES
1. Cancer: Prostate Cancer, brain cancer, lymphocytic leukemia *etc*
2. Neurodegenerative disorders like Alzheimer disease, Parkinson disease *etc.*
3. Cardio-vascular disorders
4. Diabetes (Type ii)
5. Reproductive disorders
6. Hormonal imbalances like infertility *etc.*
7. Respiratory diseases like asthma, chronic obstructive pulmonary disease (COPD)
8. Birth defects

Effects on Plants

The insecticides dichloro diphenyl trichloroethylene (DDT), methyl pentachlorophenol and parathion are reported to interfere with legume-rhizobium chemical signalling which affects the nitrogen fixation thus reducing crop yield [25]. Pesticides directly affect the root hair development of the plant, yellowing of plant shoot and reduce plant growth [26]. The USDA (United States Department of Agriculture) and USFWS (United States Fish and Wildlife Service) evaluated that farmers lost around $200 million per year from reduced crop pollination [70].

Effect on Animals

Pesticides immigrate into the ecosystems by different ways relying upon their solubility. Firstly, water-soluble pesticides that get easily dissolved in water enters into the groundwater, rivers, streams, and lakes hence affecting damage to non-targeted species. Secondly, fat-soluble pesticides which enter into the animal bodies by the phenomenon known as "bioamplification". They easily get absorbed into the fatty tissues and therefore cause persistence of pesticide in animals through food chains for the prolonged period of time [27].

Effect on Birds

Pesticides that are available in granular form are camouflaged as food grains by birds due to birds and mammal populations are directly affected. Organophosphate insecticides are very toxic to birds and raptors in the agricultural fields. Sub-lethal amount of pesticides can also damage nervous system resulting in the behavioural changes [28].

Effect on Aquatic Life

About $1/3^{rd}$ of 6,000 amphibian species globally are disappearing due to the toxicity of pesticides. Amphibians are suffered by the impurity found in surface waters with fertilizers and pesticides as reported by IUCN (International Union for Conservation of Nature), Asian amphibian crisis, 2009 [29]. Atrazine is harmful to particular fish species and can damage the aquatic plants indirectly affecting aquatic ecosystems. Urea herbicides such as diuron and isoproturon usually pollute lakes, rivers, and groundwater. Chlorpyrifos and endosulfan also have the ability to cause severe damage to amphibians even at concentrations existing in the environment when used under normal conditions [29]. Most transformed products of diuron are highly toxic to microorganisms than diuron itself [30] as reported by EFSA (European Food Safety Authority), 2008. Fungicides (based on copper) are examined more toxic to aquatic ecosystems. A study conducted using 261 pesticides to an aquatic community in field gutter, ~ 95% of the predicted risk was only caused by seven pesticides [31]. Surface water is regularly contaminated with pesticides through its normal level usage are known to aquatic invertebrates and harm fish.

Pesticide Degrading Strategies

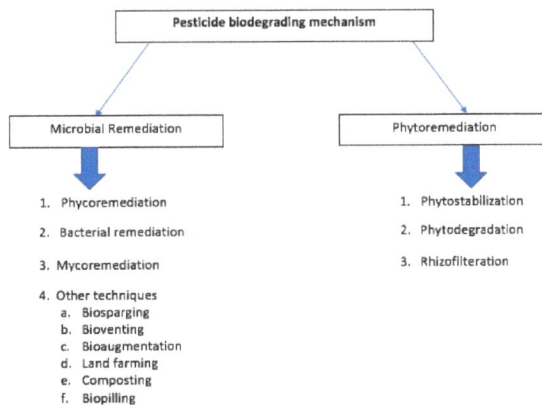

```
                    ┌─────────────────────────────────┐
                    │  Pesticide biodegrading mechanism │
                    └─────────────────────────────────┘
               ┌──────────────────────┐        ┌──────────────────────┐
               │  Microbial Remediation │        │    Phytoremediation    │
               └──────────────────────┘        └──────────────────────┘
```

Microbial Remediation
1. Phycoremediation
2. Bacterial remediation
3. Mycoremediation
4. Other techniques
   a. Biosparging
   b. Bioventing
   c. Bioaugmentation
   d. Land farming
   e. Composting
   f. Biopilling

Phytoremediation
1. Phytostabilization
2. Phytodegradation
3. Rhizofilteration

Fig. (4). Bioremediation strategies for pesticide degradation.

Microbial Remediation

Role of Algae in Biodegradation

The contamination by pesticides in aquatic systems is one of the serious problems since they have the ability to create adverse effects on ecosystem [32]. Recent studies showed that the of aquatic plants including *Leman minor, Cabomba aquatica and Elodea canadensis* have the potential to eradicate and assimilate pesticides like flazasulfuron (herbicide), dimethomorph (fungicide) and copper sulphate (fungicide) [33]. The microalgae *Scenedesmus quadricauda* was found to be more efficient in the removal of fungicides (*i.e.* pyrimethanil and dimethomorph) and herbicide like isoproturon. The green alga *Chlamydomonas reinhardtii* showed a great capability to gather and remediate prometryne (herbicide) [14]. List of organic pollutants accumulated by algae is listed in Table **2**.

Table 2. List of organic pollutants accumulated by algae.

Sr. No.	Species	Organic Pollutant
a.	*Monoraphidium braunii*	Bisphenol
b.	*Agmenellum quadruplicatum*	Bisphenol
c.	*Selenastrum capricornutum*	Benzene, phenanthrene, naphthalene, pyrene, toluene
d.	*Scenedesmus obliquus* GH2	Crude oil degradation
e.	Consortium of *Chlorella sorokiniana and Pseudomonas migulae*	Phenanthrene
f.	*Scenedesmus quadricauda*	dimethomrph and pyrimethanil- Fungicides, isoproturon- herbicide
g.	Phytoplanton	Chlorinated hydrocarbons
h.	*Chlamydomonas reinhardtii*	Herbicide (prometryne, fluxopyr)

Role of Fungi in Biodegradation

Fungi have the potential to degrade chemical pesticides by doing some minor structural alterations to the pesticides and translating them into non-toxic compounds. They are further released into the soil, where they are subjected to bacterial degradation [34]. Several fungi such as *Pleurotus ostreatus, Auricularia auricula, Agrocybe semiorbicularis, Flammulina velupites, Hypholoma fasciculare, Dichomitus squalens, Coriolus versicolor, Stereum hirsutum, and Avatha* sp. showed their capability to remediate several pesticide groups such as organophosphate pesticides, phenylurea, dicarboximide, triazine and phenylamide [35]. Fungi releases enzymes extracellularly that leads to degradation. Several groups of pesticides (dieldrin, atrazine, terbuthylazine, diuron, aldrin, heptachlor, lindane, metalaxyl, DDT, chlordane, lindane, gamma-hexachlorocyclohexane and mirex) have been remediated to the distinctive amount by white-rot fungi [36].

Role of Bacteria in Biodegradation

For degradation and detoxification of several toxic chemical pesticides bacteria has played a good role to bioremediate the polluted sites. Successful removal of Pesticides like chlorpyrifos, endosulfan, malathion, parathion, quinalphos, ethoprop, and atrazine has been removed successfully with the aid of bacteria [37]. Several bacteria species that has the ability to degrade, belong to the classes *Arthrobacter, Burkholderia, Flavobacterium, Azotobacter, and Pseudomonas.* Along with the degradative enzymes present in bacteria that helps in degradation, several environmental parameters like pH, nutrients, temperature, water potential, and the quantity of pesticide or metabolite) in soil may act as limiting factor for biological species used for remediating the contaminated soil with pesticide [38]. *Klebsiella pneumoniae and Pseudomonas* sp. have hydrolytic enzymes that have the capability to break down s-triazine herbicides (*e.g.* atrazine). Likewise, various enzymes such as hydroxylases, oxygenases, isomerases and hydrolases present in *Pseudomonas* and *Alcaligenes* sp. that showed the ability to remediate herbicide 2,4-Dichlorophenoxyacetic acid, neonicotinoids and organophosphorus compounds have been reported [39]. Microbial population, not only degrade the pesticides completely but also enhance the degradation of pesticide metabolites in the soil by developing the optimized the environmental condition. Some of the pesticides are easily degraded but some due to the presence of anionic species in the chemical compound are recalcitrant [71]. Besides are degraded by the Pseudomonas species.

Other Microbial Remediation Techniques

Bioairsparging

The concentration of volatile compounds that get absorbed into the soil or dissolved in the underground water or into the saturated zone can be reduced with the help of the bioairsparging technique [40, 41]. It consists the introduction of nutrients and oxygen into the saturated zone, from time to time to enhance the action of microorganisms. It can also be used for the removal of contamination from unsaturated zone.

Biostimulation

In situ (treatment of soil in place) biostimulation technique usually comprise the process of bioventing where nutrients and oxygen are injected *via* injection wells into the soil. Throughout the contaminated soil, it is essential that distribution of oxygen and nutrients should be uniform. The permeability of soil to air and soil to water is a function of soil texture, therefore, the surface of soil directly affects the process of bioventing. Due to the low permeability of fine-textured soils like clay, blocks the biovented nutrients and oxygen from dissipating throughout the soil. Moisture content is also hard to regulate in fine-textured soils because of their smaller pores and high surface area that allow it to hold water. Fine textured soils are slow to drain from water-saturated soil conditions, which results in restricting the oxygen from reaching soil microbes throughout the polluted area [42, 43]. Bioventing is compatible with well-drained and coarse-textured soils.

Bioaugmentation

It is regarded as one of the most relevant bioremediation approaches [44]. This technique involves the augmentation of the target contaminants by introducing the specific microorganisms (indigenous or non-indigenous) to the polluted matrix. The use of microbial inocula isolated from aquifer material to stimulate atrazine degradation in aquifers has been shown in laboratory experiments [45 - 47]. Remediation of dinoseb (herbicide in the dinitrophenol family) in soil was proven with this approach [48]. The laboratory scale experiments, application of bioaugmentation approaches to a field site points out the concerns with respect to its efficiency. It may be due to the non-availability of optimum conditions for the microorganisms or competition occurs between inoculated and other microorganisms present on the field site.

Landfarming

It is a practice in which excavation of soil is done followed by its mechanical separation with the help of a process called sieving. The impure soil which is contaminated is then placed in stacks not < 0.4 meters thick. A synthetic, clay or concrete membrane is built to envelop the polluted soil layer. Oxygen is supplemented followed by mixing which is done by the process of harrowing, plowing or milling. Moisture and nutrients may also be supplemented to aid the bioremediation process. The soil pH is regulated (~7.0) using agricultural lime or crushed limestone [49]. Landfarming is most successful in eradicating the pentachlorophenol (PCP) and polycyclic aromatic hydrocarbons (PAH). Fig. (**4**) demonstrates the landfarming technique [42].

Composting

Composting is an under-controlled, biotic technique through which organic biodegradable pollutants are converted into safe and stabilized by-products, by the action of microorganisms (following either aerobic or anaerobic environment).

The composting of soils polluted with chemical pesticides, thermophilic conditions (54–65°C) are kept. The soil that needs to be remediated is excavated and allowed to mix with organic dispersants along with the remedial agents, such as vegetal wastes, animal wastes, sawdust and their residues to enhance the permeability of the material that is under treatment. By keeping the concentration of oxygen, humidity, temperature, pH and carbon: nitrogen (C: N) ratio constant, maximum degradation efficiency can be achieved. Composting is most efficient in removing PAH, TNT and RDX [49].

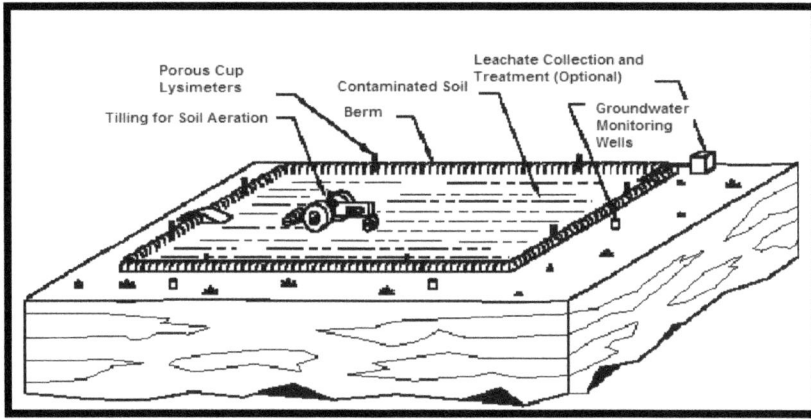

Fig. (5). Landfarming Technique (EPA, 2006).

Biopiles

Biopiles (also known as the heap technique) is an *in situ* process. Initially, this process is performed at pilot scale to determine the bioremediation efficiency followed by the mechanical separation of the soil, that not only homogenize but also removes the disruptive material for *e.g.* plastics, metals, and stones. Homogenization of soil is done in order to average out pollution concentration within the complete sample of soil. Homogenization permits the biopiling as the more potent technique [50]. After the soil is piled, biological degradation of the contaminants occurs with the aid of nutrients, microbes, oxygen, and substrate. Mineral fertilizers in the form of nutrients and microorganisms such as bacteria, bacteria, or enzymes are also being supplemented [50]. Static piles are generally found in the shape of trapezoids or pyramids. Depending upon the kind of aeration used (either active or passive), their height may vary from 0.8 to 2m.

Biopiling is very efficient in treating pollutants such as phenols, PAHs, BTEX (Mixture of benzene, toluene, ethylbenzene and xylene), and explosives such as 2,4,6-trinitrotoluene and royal demolition explosive [49, 50]. This technique is very cost-effective because of its low set up cost. A large number of treatment plants are constructed that reduced the conveyance costs, but since the government regulation has become stringent making it costlier with respect to transportation and disposal [50]. For optimal biodegradation, the conditions must be supervised and adjusted regularly. The problem of supervising and containing volatilization of contaminants can be seen in land farming and biopiles technique.

Phytoremediation

Phytoremediation is cost-effective, advanced and socio-economically advantageous technology that remediates, metabolize, incorporate or detoxify metals, pesticides and hydrocarbons with the aid of plants at polluted sites [51 - 53, 55]. Phytoremediation involves several processes, such as phytodegradation (uptake of the organic contaminant by the plant enzymes), phytotransformation (conversion of toxic organic pollutants into a non-toxic, more stable form with less mobility), rhizoremediation and phytovolatilization (volatilization of organic pollutants *via* plant leaves). The plant roots release organic compounds in at the soil-root interface that encourages the microbial population, improving the degradation of pollutants in the rhizosphere.

Pesticides Uptake by Plant

The uptake of chemical compounds by plants is managed by various plant and soil features [55 - 57]. Various parameters like organic carbon, moisture content and clay content, the residence time of pesticide in the soil affect the accessibility for transport and plant uptake [58]. Along with the soil and pesticide properties, chemical pesticides also experience chemical or biological alterations in water and soil environments and tend to form metabolites that are frequently weak electrolytes and may be easily taken up by plants [59]. Soil polluted with diversified chemical pollutants have heavy metals that affect the process of uptake in plants.

From soil to plant pesticides uptake takes place. Plants can also absorb airborne pesticides. Plant species that are potential to bioaccumulate pesticides are summarized in Table **3**.

Table 3. Plant species potential to bioaccumulate pesticides.

Pesticide	Plant species	Comments	References
Atrazine	Hybrid poplars *(Populus deltoides x nigra)*	Uptake of atrazine by poplars from sandy was high in comparison of normal soil and was found to be remediated in plant tissues.	[54]

(Table 3) cont.....

Pesticide	Plant species	Comments	References
Atrazine, Terbutryn, Trifluralin, Cycloxidim	Parrot feather *(Myriophyllum aquaticum)*	Uptake of cycloxidim and atrazine was more than trifluralin and terbutryn.	[60]
Arbofuran, Terbuthylazin	Barley (*Hordeum vulgare* L.), wheat (*Triticum aestivum* L.)	wheat grown on loamy soil barley on sandy soil removed significant amount of pesticides.	[61]
Chlordane	*Cucurbita pepo* L. (zucchini) *Cucumis sativus* L.	Chlordane was reported to be bioaccumulated highest in the root tissue and with some amount in detected in plant tissues (fruit, leaf, root and stem).	[62]
Chlorpyrifos, Atrazine	*Juncus effusus L.*	Uptake of both pesticides was observed but chlorpyrifos was absorbed faster in comparison to atrazine.	[63]
Tridemorph, Dodemorph	*Barley (Hordeum vulgare L.)*	Tridemorph was found to be gathered in roots and translocated to shoots at pH 8 while dodemorph was translocated to shoots.	[64]
Metolachlor, Atrazine	*Ceratophyllum demersum* L. *Elodea canadensis*	It was seen that both plants removed and metabolized more than ninty percent of metolachlor and a substantial quantity of atrazine from polluted waters while initial sixteen days.	[65]
Malathion, Demeton-S-methyl, Crufomate	*Spirodela oligorrhiza* L., *Elodea canadensis, Myriophyllumaquaticum*	*M. aquaticum* was able to remove 58 to 83% of the pesticides. *S. oligorrhiza* L. and *E. Canadensis* showed early lag phase and then began removing complete pesticides quickly except crufomate.	[66]
Chlorinated pesticides (OCPs)	*Daucus carota, Solanum tuberosum*	Potatoes and carrots peel were able to eradicate 52 to 100% of chlorinated residues from the soil.	[67]
DDT, DDD, DDE	Pumpkin, ryegrass, zucchini, alfalfa, tall fescue	Both pumpkin and zucchini showed the ability to remove a great number of pesticides.	[68]

SUMMARY

Contamination found in soil is a complicated issue because of the various potential contaminants caused by the use of pesticides. Practically, it is impossible to undoubtedly select a method which proves the most appropriate method to bioremediate soil contaminated with pesticides, as many parameters, like contaminant concentration, characteristics *etc.* are to be considered. For determining an efficient remediation tool for a particular contaminant, soil characteristics are also an important factor to be taken into consideration. Microorganisms have potential to remove pesticides present in the soil. Use of bacteria and fungi has been considered one of the suitable methods. With the degradation of wastes or pollutants, this process is efficient in removing the waste or pollutant. In contrast to other remediation techniques *i.e.* dechlorination, incineration, soil flushing, thermal disposition, solvent extraction *etc.*, the biological remediation is the promising approach for removal of chemical pollutants from a polluted site.

Hence, bioremediation is a much-assuring method to overcome the contamination caused by pesticide from soils. This technology has evidenced again and again its capability to degrade pesticides and other various chemical compounds that are harmful to the environment. So, now is the phase to employ this environmental-friendly technology for healthier and protected future.

CONSENT FOR PUBLICATION

Not applicable.

CONFLICT OF INTEREST

The authors declare no conflict of interest, financial or otherwise.

ACKNOWLEDGEMENTS

Authors duly acknowledge IIT Delhi, India for providing infrastructural facilities and financial support as research scholarship by Ministry of Human Resource Development, India.

REFERENCES

[1]   Köhler HR, Triebskorn R. Wildlife ecotoxicology of pesticides: can we track effects to the population level and beyond? Science 2013; 341(6147): 759-65.
       [http://dx.doi.org/10.1126/science.1237591] [PMID: 23950533]

[2]   Meyer-Baron M, Knapp G, Schäper M, van Thriel C. Meta-analysis on occupational exposure to pesticides--neurobehavioral impact and dose-response relationships. Environ Res 2015; 136: 234-45.
       [http://dx.doi.org/10.1016/j.envres.2014.09.030] [PMID: 25460642]

[3]   Saravi SS, Shokrzadeh M. Role of pesticides in human life in the modern age: a review Pesticides in the Modern World-Risks and Benefits. InTech 2011.
       [http://dx.doi.org/10.5772/18827]

[4]   Pan-Germany. Pesticide and health hazards Facts and figures , [accessed on 11june2016]; www. pan-germany. org/ download/Vergift_EN-201112-web.pdf

[5]   Rosell G, Quero C, Coll J, Guerrero A. Biorational insecticides in pest management. J Pest Sci 2008; 33(2): 103-21.
[http://dx.doi.org/10.1584/jpestics.R08-01]

[6]   Carriger JF, Rand GM, Gardinali PR, Perry WB, Tompkins MS, Fernandez AM. Pesticides of potential ecological concern in sediment from south Florida canals: an ecological risk prioritization for aquatic arthropods. Soil Sediment Contam 2006; 15(1): 21-45.
[http://dx.doi.org/10.1080/15320380500363095]

[7]   Mostafalou S, Abdollahi M. Concerns of the environmental persistence of pesticides and human chronic diseases. Clin Exp Pharmacol 2012; 2(3): 1000-108.

[8]   http://wwwbusiness-standardcom/content/b2b-chemicals/indian-agrochemicals market-to- reach-6-8-bn-by-fy17-tata-strategic-management-group 113081200449_1htm accessed on 28july 2016

[9]   United Nations Food and Agriculture Organization (UN/FAO) International Code of Conduct on the Distribution and Use of Pesticides 2003.accessed 16.06.16 http://www.fao.org/docrep/005/y4544e/y4544e00.htm

[10]   Willett KL, Ulrich EM, Hites RA. Differential toxicity and environmental fates of hexachlorocyclohexane isomers. Environ Sci Technol 1998; 32(15): 2197-207.
[http://dx.doi.org/10.1021/es9708530]

[11]   Morais S, Dias E, de Lourdes Pereira M. Carbamates: human exposure and health effects.The impact of pesticides. Cheyenne, WY: Academic Press 2012; pp. 21-38.

[12]   Barceló D. Trace Determination of Pesticides and their Degradation Products in Water. Elsevier 1997.

[13]   Andreu V, Picó Y. Determination of pesticides and their degradation products in soil: critical review and comparison of methods. TrAC Trends Anal Chem 2004; 23(10-11): 772-89.
[http://dx.doi.org/10.1016/j.trac.2004.07.008]

[14]   Jin ZP, Luo K, Zhang S, Zheng Q, Yang H. Bioaccumulation and catabolism of prometryne in green algae. Chemosphere 2012; 87(3): 278-84.
[http://dx.doi.org/10.1016/j.chemosphere.2011.12.071] [PMID: 22273183]

[15]   Savonen C. Soil microorganisms object of new OSU service Good Fruit Grower 1997. http://www. goodfruit. com/archive/1995/6other. html

[16]   Pell M, Stenberg B, Torstensson L. Potential denitrification and nitrification tests for evaluation of pesticide effects in soil. Ambio 1998; 24-8.

[17]   Arias RN, Fabra de Peretti A. Effects of 2,4-dichlorophenoxyacetic acid on Rhizobium sp. growth and characterization of its transport. Toxicol Lett 1993; 68(3): 267-73.
[http://dx.doi.org/10.1016/0378-4274(93)90017-R] [PMID: 8516779]

[18]   Fabra A, Duffard R, Evangelista de Duffard A. Toxicity of 2,4-dichlorophenoxyacetic acid to Rhizobium sp in pure culture. Bull Environ Contam Toxicol 1997; 59(4): 645-52.
[http://dx.doi.org/10.1007/s001289900528] [PMID: 9307432]

[19]   Singh JB, Singh S. Effect of 2, 4-dichlorophenoxyacetic acid and maleic hydrazide on growth of bluegreen algae (cyanobacteria) Anabaena doliolum and Anacystisnidulans. Sci Cult 1989; 55: 459-60.

[20]   Tözüm☐Çalgan SD, Sivaci-Güner S. Effects of 2, 4-d and methylparathion on growth and nitrogen fixation in cyanobacterium, gloeocapsa. Int J Environ Stud 1993; 43(4): 307-11.
[http://dx.doi.org/10.1080/00207239308710839]

[21]   Frankenberger WT, Tabatabai MA. Factors affecting L-asparaginase activity in soils. Biol Fertil Soils 1991; 11(1): 1-5.
[http://dx.doi.org/10.1007/BF00335825]

[22]   Martens DA, Bremner JM. Influence of herbicides on transformations of urea nitrogen in soil. J

Environ Sci Health B 1993; 28(4): 377-95.
[http://dx.doi.org/10.1080/03601239309372831]

[23]   Chakravarty P, Sidhu SS. Effect of glyphosate, hexazinone and triclopyr on *in vitro* growth of five species of ectomycorrhizal fungi. Eur J Forest Pathol 1987; 17(4-5): 204-10.
[http://dx.doi.org/10.1111/j.1439-0329.1987.tb01017.x]

[24]   Kelley WD, South DB. *In vitro* effects of selected herbicides on growth and mycorrhizal fungi. InWeed Sci Soc America Meeting. 38Auburn University, Auburn, Alabama. 1978; pp. : 74-6.

[25]   Johnston AE. Soil organic matter, effects on soils and crops. Soil Use Manage 1986; 2(3): 97-105.
[http://dx.doi.org/10.1111/j.1475-2743.1986.tb00690.x]

[26]   Wells M. Vanishing bees threaten US crops. BBC News 2007; p. 11.

[27]   Warsi F. How do pesticides affect ecosystems?. Pesticides Accessed July 16, 2015 http://farhanwarsi.tripod.com/id9.html

[28]   Isenring R. Pesticides reduce biodiversity. Pest Act Net Rep 2010; 1(88): 4-7.

[29]   Sparling DW, Fellers GM. Toxicity of two insecticides to California, USA, anurans and its relevance to declining amphibian populations. Environ Toxicol Chem 2009; 28(8): 1696-703.
[http://dx.doi.org/10.1897/08-336.1] [PMID: 19290680]

[30]   Bonnet JL, Bonnemoy F, Dusser M, Bohatier J. Assessment of the potential toxicity of herbicides and their degradation products to nontarget cells using two microorganisms, the bacteria Vibrio fischeri and the ciliate Tetrahymena pyriformis. Environ Toxicol 2007; 22(1): 78-91.
[http://dx.doi.org/10.1002/tox.20237] [PMID: 17295264]

[31]   de Zwart D. Ecological effects of pesticide use in The Netherlands: modeled and observed effects in the field ditch. Integr Environ Assess Manag 2005; 1(2): 123-34.
[http://dx.doi.org/10.1897/IEAM_2004-015.1] [PMID: 16639894]

[32]   Moore MT, Cooper CM, Smith S, *et al.* Diazinon mitigation in constructed wetlands: influence of vegetation. Water Air Soil Pollut 2007; 184(1-4): 313-21.
[http://dx.doi.org/10.1007/s11270-007-9418-9]

[33]   Olette R, Couderchet M, Biagianti S, Eullaffroy P. Toxicity and removal of pesticides by selected aquatic plants. Chemosphere 2008; 70(8): 1414-21.
[http://dx.doi.org/10.1016/j.chemosphere.2007.09.016] [PMID: 17980900]

[34]   Gianfreda L, Rao MA. Potential of extra cellular enzymes in remediation of polluted soils: a review. Enzyme Microb Technol 2004; 35(4): 339-54.
[http://dx.doi.org/10.1016/j.enzmictec.2004.05.006]

[35]   Bending GD, Friloux M, Walker A. Degradation of contrasting pesticides by white rot fungi and its relationship with ligninolytic potential. FEMS Microbiol Lett 2002; 212(1): 59-63.
[http://dx.doi.org/10.1111/j.1574-6968.2002.tb11245.x] [PMID: 12076788]

[36]   Quintero JC, Lu-Chau TA, Moreira MT, Feijoo G, Lema JM. Bioremediation of HCH present in soil by the white-rot fungus Bjerkanderaadusta in a slurry batch bioreactor. Int Biodeter Biodegrad 2007; 60(4): 319-26.
[http://dx.doi.org/10.1016/j.ibiod.2007.05.005]

[37]   Singh BK, Walker A, Morgan JA, Wright DJ. Biodegradation of chlorpyrifos by enterobacter strain B-14 and its use in bioremediation of contaminated soils. Appl Environ Microbiol 2004; 70(8): 4855-63.
[http://dx.doi.org/10.1128/AEM.70.8.4855-4863.2004] [PMID: 15294824]

[38]   Singh DK. Biodegradation and bioremediation of pesticide in soil: concept, method and recent developments. Indian J Microbiol 2008; 48(1): 35-40.
[http://dx.doi.org/10.1007/s12088-008-0004-7] [PMID: 23100698]

[39]   Mulbry W, Kearney PC. Degradation of pesticides by micro-organisms and the potential for genetic manipulation. Crop Prot 1991; 10(5): 334-46.

[http://dx.doi.org/10.1016/S0261-2194(06)80021-9]

[40]   Norris RD. Handbook of bioremediation. CRC press 1993.

[41]   Ronneau C, Bitchaeva O. Biotechnology for waste management and site restoration: Technological, educational, business, political aspects. Springer Science & Business Media 2012 Dec 6

[42]   United States Environmental Protection Agency. Bioventing. 2006 Nov 24; http://www.epa.gov/oust/cat/biovent.htm

[43]   Environmental Protection Agency. 2006. A Citizen's Guide to Bioremediation April 1996 2006 Nov 24; accessed on 21.7. 2016 http://www.epa.gov/tio/download/citizens/a_citizens_guide_to_bioremediation.pdf

[44]   El Fantroussi S, Agathos SN. Is bioaugmentation a feasible strategy for pollutant removal and site remediation? Curr Opin Microbiol 2005; 8(3): 268-75.
[http://dx.doi.org/10.1016/j.mib.2005.04.011] [PMID: 15939349]

[45]   Shapir N, Mandelbaum RT, Jacobsen CS. Rapid atrazine mineralization under denitrifing conditions by Pseudomonas sp. strain ADP in aquifer sediments. Environ Sci Technol 1998; 32(23): 3789-92.
[http://dx.doi.org/10.1021/es980625l]

[46]   Franzmann PD, Zappia LR, Tilbury AL, Patterson BM, Davis GB, Mandelbaum RT. Bioaugmentation of atrazine and fenamiphos impacted groundwater: laboratory evaluation. Bioremediat J 2000; 4(3): 237-48.
[http://dx.doi.org/10.1080/10588330008951112]

[47]   Kristensen GB, Johannesen H, Aamand J. Mineralization of aged atrazine and mecoprop in soil and aquifer chalk. Chemosphere 2001; 45(6-7): 927-34.
[http://dx.doi.org/10.1016/S0045-6535(01)00020-0] [PMID: 11695615]

[48]   Kaake RH, Roberts DJ, Stevens TO, Crawford RL, Crawford DL. Bioremediation of soils contaminated with the herbicide 2-sec-butyl-4,6-dinitrophenol (dinoseb). Appl Environ Microbiol 1992; 58(5): 1683-9.
[PMID: 1622239]

[49]   Williams J. Bioremediation of contaminated soils: A comparison of *in situ* and *ex situ* techniques. 2006.

[50]   Schulz Berendt V. Bioremediation with heap technique. 2nd ed. Biotechnology Set 2000; pp. 319-28.

[51]   Lewandowski I, Schmidt U, Londo M, Faaij A. The economic value of the phytoremediation function–assessed by the example of cadmium remediation by willow (Salix ssp). Agric Syst 2006; 89(1): 68-89.
[http://dx.doi.org/10.1016/j.agsy.2005.08.004]

[52]   Licht LA, Isebrands JG. Linking phytoremediated pollutant removal to biomass economic opportunities. Biomass Bioenergy 2005; 28(2): 203-18.
[http://dx.doi.org/10.1016/j.biombioe.2004.08.015]

[53]   Schnoor JL, Licht LA, McCutcheon SC, Wolfe NL, Carreira LH. Phytoremediation of organic and nutrient contaminants. Environ Sci Technol 1995; 29(7): 318A-23A.
[http://dx.doi.org/10.1021/es00007a747] [PMID: 22667744]

[54]   Burken JG, Schnoor JL. Uptake and metabolism of atrazine by poplar trees. Environ Sci Technol 1997; 31(5): 1399-406.
[http://dx.doi.org/10.1021/es960629v]

[55]   Susarla S, Medina VF, McCutcheon SC. Phytoremediation: An ecological solution to organic chemical contamination. Ecol Eng 2002; 18(5): 647-58.
[http://dx.doi.org/10.1016/S0925-8574(02)00026-5]

[56]   Dzantor EK, Beauchamp RG. Phytoremediation, Part I: Fundamental basis for the use of plants in remediation of organic and metal contamination. Environ Pract 2002; 4(2): 77-87.

[http://dx.doi.org/10.1017/S1466046602021087]

[57]   Gao YZ, Zhu LZ. Phytoremediation and its models for organic contaminated soils. J Environ Sci (China) 2003; 15(3): 302-10.
       [PMID: 12938977]

[58]   Koskinen WC, Calderón MJ, Rice PJ, Cornejo J. Sorption-desorption of flucarbazone and propoxycarbazone and their benzenesulfonamide and triazolinone metabolites in two soils. Pest Manag Sci 2006; 62(7): 598-602.
       [http://dx.doi.org/10.1002/ps.1196] [PMID: 16691543]

[59]   Trapp S. Modelling uptake into roots and subsequent translocation of neutral and ionisable organic compounds. Pest Manag Sci 2000; 56(9): 767-78.
       [http://dx.doi.org/10.1002/1526-4998(200009)56:9<767: AID-PS198>3.0.CO;2-Q]

[60]   Turgut C. Uptake and modeling of pesticides by roots and shoots of parrotfeather (Myriophyllum aquaticum). Environ Sci Pollut Res Int 2005; 12(6): 342-6.
       [http://dx.doi.org/10.1065/espr2005.05.256] [PMID: 16305140]

[61]   Matthies M, Behrendt H, Trapp S, McFarlane JC. Dynamics of leaching, uptake, and translocation: The Simulation Model Network Atmosphere–Plant–Soil (SNAPS).Plant Contamination: Modeling and Simulation of Organic Chemical Processes. Michigan: Lewis Publishers 1994.

[62]   Krol WJ, Arsenault T, Mattina MJ. Assessment of dermal exposure to pesticides under "pick your own" harvesting conditions. Bull Environ Contam Toxicol 2005; 75(2): 211-8.
       [http://dx.doi.org/10.1007/s00128-005-0740-x] [PMID: 16222489]

[63]   Anudechakul C, Vangnai AS, Ariyakanon N. Removal of chlorpyrifos by water hyacinth (Eichhorniacrassipes) and the role of a plant-associated bacterium. Int J Phytoremediation 2015; 17(7): 678-85.
       [http://dx.doi.org/10.1080/15226514.2014.964838] [PMID: 25976881]

[64]   Requena L, Bornemann S. Barley (Hordeum vulgare) oxalate oxidase is a manganese-containing enzyme. Biochem J 1999; 343(Pt 1): 185-90.
       [http://dx.doi.org/10.1042/bj3430185] [PMID: 10493928]

[65]   Rice PJ, Anderson TA, Coats JR. Phytoremediation of herbicide-contaminated surface water with aquatic plants.Phytoremediation of Soil and Water Contaminants. Washington, DC: American Chemical Society 1997; pp. 133-51.
       [http://dx.doi.org/10.1021/bk-1997-0664.ch010]

[66]   Gao J, Garrison AW, Hoehamer C, Mazur CS, Wolfe NL. Uptake and phytotransformation of organophosphorus pesticides by axenically cultivated aquatic plants. J Agric Food Chem 2000; 48(12): 6114-20.
       [http://dx.doi.org/10.1021/jf9904968] [PMID: 11312784]

[67]   Zohair A, Salim AB, Soyibo AA, Beck AJ. Residues of polycyclic aromatic hydrocarbons (PAHs), polychlorinated biphenyls (PCBs) and organochlorine pesticides in organically-farmed vegetables. Chemosphere 2006; 63(4): 541-53.
       [http://dx.doi.org/10.1016/j.chemosphere.2005.09.012] [PMID: 16297429]

[68]   Lunney AI, Zeeb BA, Reimer KJ. Uptake of weathered DDT in vascular plants: potential for phytoremediation. Environ Sci Technol 2004; 38(22): 6147-54.
       [http://dx.doi.org/10.1021/es030705b] [PMID: 15573619]

[69]   United Nations Food and Agriculture Organization (UN/FAO) International Code of Conduct on the Distribution and Use of Pesticides 2003 [accessed 16.06.16]; http://www.fao.org/docrep/005/y4544e/y4544e00.htm

[70]   Singh BA, Mandal KO. Environmental impact of pesticides belonging to newer chemistry Integrated pest management. Jodhpur, India: Scientific Publishers 2013; pp. 152-90.

[71]   Konradsen F, van der Hoek W, Cole DC, et al. Reducing acute poisoning in developing countries-

-options for restricting the availability of pesticides. Toxicology 2003; 192(2-3): 249-61.
[http://dx.doi.org/10.1016/S0300-483X(03)00339-1] [PMID: 14580791]

[72]   Santos A, Flores M. Effects of glyphosate on nitrogen fixation of free☐living heterotrophic bacteria.
Lett Appl Microbiol 1995; 20(6): 349-52.
[http://dx.doi.org/10.1111/j.1472-765X.1995.tb01318.x]

CHAPTER 6

Biphenyls: Health Impacts and Toxicity Evaluation

Sakshi Sharma[1], Avinash Kaur Nagpal[1,*] and Inderpreet Kaur[2,*]

[1] *Department of Botanical and Environmental Sciences, Guru Nanak Dev University, Amritsar-143005, Punjab, India*

[2] *Department of Chemistry, Centre for Advanced Studies, Guru Nanak Dev University, Amritsar - 143005, Punjab, India*

Abstract: Biphenyl is an aromatic hydrocarbon, occuring naturally in crude oil, coal tar and natural gas or produced synthetically as a by-product during benzene production. Synthetically manufactured polychlorinated (PCBs) and polybrominated (PBBs) biphenyls have serious health impacts on human beings. Therefore, biphenyls are banned in many countries including US, UK, India, Japan, *etc.*, but are still present in the environment. This chapter is a comprehensive discussion on biphenyls based on available literature. It includes a brief introduction about physical and chemical nature of biphenyls, their applications, health impacts and toxicity evaluation. Various entry routes of these chemicals in the environment *e.g.,* improper disposal/ incineration of industrial wastes containing PCBs and PBBs, leakage from old electronic instruments, accidental spillage of these chemicals during transportation/ handling, *etc*, have also been discussed. PCBs and PBBs do not degrade readily and stay in different environmental media for longer durations and enter food chain and bioaccumulate in human bodies *via* dermal, inhalation and oral exposure routes. High exposure to biphenyls leads to various health problems such as neurological, reproductive, hepatic, gastrointestinal and renal disorders, endocrine disruption, cancer, *etc*. Various techniques used for the identification or estimation of PCBs in the environment and biological samples such as gas chromatography coupled with electron capture detection (GC-ECD) or mass spectrometry (GC-MS), high resolution gas chromatography (HRGC), *etc.* have been discussed briefly. Moreover, various assays employed for the assessment of toxic effects of PCBs and PBBs, using prokaryotic/ eukaryotic models and different remediation techniques for these chemicals have been discussed.

Keywords: Aromatic hydrocarbons, Biphenyls, Cancer, CARDIA, *Daphnia magna*, ECNI, Electronic waste, NHANES, Organic pollutants, Phytoextraction, Phytotransformation, Polychlorinated biphenyls, Polybrominated biphenyls,

* **Corresponding authors Avinash Kaur Nagpal:** Department of Botanical and Environmental Sciences, Guru Nanak Dev University, Amritsar-143005, Punjab, India; Tel: +91-94174-26060(M), 2258802 to 09 Ext. 3423(O); E-mail: avnagpal@yahoo.co.in
Inderpreet Kaur: Department of Chemistry, Centre for Advanced Studies, Guru Nanak Dev University, Amritsar-143005, Punjab, India; Tel: +91-8427662766(M), +91-183-2258802 to 09, Ext. 3285(O); E-mail: inderpreet11@yahoo.co.in

PBBs, Rhizoremediation, *Salmonella typhimurium*, Soil pollution, *Vicia faba*, Water pollution, Yusho accident.

INTRODUCTION

Rapid economic and industrial development has led to pervasive pollution globally. Different organic pollutants present in the environment are highly persistent and stay in environment for very long durations (years or decades) due to their long half-lives [1]. Most of the persistent organic pollutants (POPs) are toxic in nature and are widespread among different compartments of the environment including soil, water, sediments, air, and different levels of food chain. Due to this, several POPs have been shunned from manufacture and usage in many countries [2]. Biphenyl and its derivatives, especially polychlorinated biphenyls (PCBs) and polybrominated bihenyls (PBBs) are important group of organic pollutants which are banned almost worldwide. Biphenyl is an aromatic hydrocarbon, which exists naturally in crude oil, natural gas and coal tar or can be produced synthetically by different methods such as, addition of copper powder catalyst while heating iodobenzene; or by stirring halobenzene and a reducing agent in the presence of a base and catalyst [3]. PCBs and PBBs are synthetic chlorine or bromine substituted biphenyl molecules and do not exist in nature. Biphenyl, PBBs and PCBs have been used for various purposes in past, as fungistat, pesticides, flame retardants, and in electronic devices, thermoplastics, *etc* [4 - 8]. Improper handling and disposal are major reasons for the release of these pollutants into the environment. These contaminants have affected almost every aspect of environment and have widely bioaccumulated and biomagnified in higher trophic levels of food chain and lead to the occurrence of various health problems in wildlife and human beings over the decades. Various sources of biphenyls, which lead to human exposure, are indoor air, drinking water, food items, coal tar, natural gas, heat transfer fluids, dyestuff carrier for textiles or copying paper, solvents used in the pharmaceutical industries, agrochemicals, wood preservatives, *etc*. [9]. This chapter is a brief account of physical and chemical aspects of biphenyl and its derivatives along with their sources; routes through which they enter and their spread in different compartments of environment; various applications and health hazards associated with biphenyl, PCBs and PBBs. A brief account of various techniques to detect and quantify the level of these contaminants, methods to estimate their toxicity and remediation techniques has also been presented.

Physical and Chemical Properties of Biphenyl, PCBs and PBBs

Biphenyl consists of two phenyl rings attached by a central C-C bond (Fig. **1**). with chemical formula: $C_{12}H_{10}$ [9].

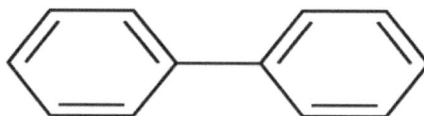

Fig. (1). Chemical structure of biphenyl.

Pure biphenyl is an aromatic hydrocarbon, solid at room temperature, normally exists as flakes and has strong and pleasant odor [9, 10]. Biphenyl is very stable white organic compound but it appears mildly yellow or pale when impure [10]. Major impurities in biphenyl are terphenyl, sulphur and benzene [9]. It has high boiling point, chemical stability, solvency and flash point but low polarity and reactivity. Biphenyl is semi-volatile and decomposes at higher temperatures; insoluble in water but soluble to varying degrees in organic solvents such as, ethanol, benzene and diethyl ether [9 - 11]. Biphenyl exists in planar confirmation when present in solid state, whereas as liquid or vapours, it exists as non-planar due to free rotation around the central C-C bond [10]. This becomes evident by dipole moment and X-ray measurements in different states. However, rotation gets restricted by substitution of ortho hydrogen atoms in biphenyl with different functional groups [10]. Some physical and chemical properties of biphenyl are enlisted in Table **1**.

Table 1. Physical and chemical properties of Biphenyl.

Molecular mass	154.21 g/mol.
Density	1.04 g/cm^3 at 20 °C
Melting point	69.20 °C
Boiling point	255.20 °C
Flash point	113 °C
Ignition temperature	570 °C
Vapour pressure	1.19 Pa at 25°C
Heat of vaporization	53.90 kJ/mol
Heat of fusion	18.60 kJ/mol
Henry's law constant	28 Pa.m^3/ mol
Octanol-water partition coefficient	4.01
Organic carbon partition coefficient	3.71

Sources [9, 10, 12]:

PCBs and PBBs are synthetic compounds having two phenyl rings joined by a central C-C bond and substituted with chlorine or bromine atoms, respectively (Fig. **2**). [13, 14].

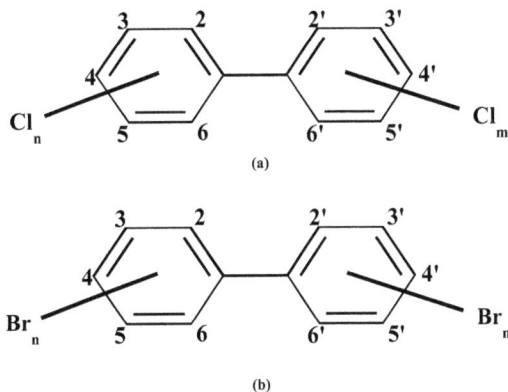

Fig. (2). Structures of (a) polychlorinated biphenyls (PCBs) and (b) polybrominated biphenyls (PBBs). Numbers denoted to carbon atoms are the possible positions of chlorine or bromine atoms on the phenyl rings in PCBs and PBBs, respectively.

In PCBs chlorine atoms have stronger association with phenyl rings as compared to association of bromine atoms in corresponding PBBs [15]. On the basis of substitution of phenyl rings with chlorine or bromine atoms, 209 compounds or congeners are possible for both PCBs (molar mass ranging from 188.66 g/mol to 498.66 g/mol) and PBBs (molar mass ranging from 232.90 g/mol to 943.00 g/mol) [13, 14, 16]. Heavy substitution, by chlorine and bromine atoms, on ortho positions in PCBs and PBBs, respectively, forms non-planar configuration for stability, as movement of phenyl rings around the central C-C bond gets hindered, however, planar configuration is found commonly in case of congeners which are not substituted at ortho positions [13, 14]. PCBs are yellow in colour and are oily to viscous in nature at room temperature, but at low temperatures they turn into solid resins instead of forming crystals [14, 17]. Similarly, PBBs are colourless or slightly yellowish solids [15, 16]. Both PCBs and PBBs have low vapour pressure and very low solubility in water, moreover, their vapour pressure and solubility in water are inversely proportional to degree of substitution by halogens in congeners [13 - 15, 17]. Both PCBs and PBBs are chemically and thermally stable, lipophilic, soluble in organic solvents [13 - 15, 17]. Similarly, both PCBs and PBBs are volatile in nature but PBBs have low volatility as compared to the corresponding PCBs [15, 17]. Mostly PCBs and PBBs are resistant to acids, alkalis, heat and oxidation but these properties can vary for different congeners depending on the degree of substitution [13, 15 - 17].

Applications of Biphenyls

PCBs and PBBs were manufactured commercially in many countries, such as, UK, former U.S.S.R., US and Czechoslovakia and Germany [16, 18]. Commercial PCBs were always manufactured in the form of technical mixture of various congeners under different trade names *e.g.,* Aroclor, Kanechlor, Phenochlor, Pyralene and Delor, *etc.* [14, 16]. Similarly, PBBs were also produced commercially and sold in the form of mixtures of different congeners under trade names such as, Firemaster FF-1, Firemaster BP-6, Bromkal 80, Bromkal 80–9D, Flammex B-10 and HFO 101, *etc.* [16]. Excessive usage, accidents, injudicious disposal practices and leakage of PCBs and PBBs from disposal sites and industries resulted in introduction of these toxic compounds into the environment [16, 18]. These biphenyl derivatives are stable toxic compounds, therefore, once enter the environment, they degrade very slowly and lead to bioaccumulation and biomagnification in the food chain [16, 18, 19]. Therefore, production and use of PCBs and PBBs has been banned in many countries. In US, maximum domestic applications of PCBs were stopped by 1977 [20, 21]. Presently, United States Environmental Protection Agency (US EPA) maintains a database containing information about the amount and location of PCBs in transformers used across US [20]. PCBs were firstly banned in 1976 through an EU initiative and thereafter, since 1986, sale of all electric equipments containing PCBs were banned [14]. Moreover, in 1995 usage of such PCBs containing electronics were completely forbidden [14]. By 1980s, several nations shunned the manufacture and usage of PCBs [18]. Stockholm Convention (2004) completely prohibited the manufacture of PCBs and made regulations regarding handling and disposal of PCBs, which has been accepted by many countries, worldwide [14]. PBBs also have been banned in USA [22]. In Michigan, the manufacturing of PBB "hexabromobiphenyl" was stopped in 1974 and few years later, production of other PBBs such as, octabromobiphenyl and decabromobiphenyl was also stopped. Similarly, production of PBBs have been stopped in other countries also, *e.g.,* UK, Germany and France [16]. Previous applications of biphenyls and its synthetic halogenated derivatives have been discussed below:

1. Biphenyl was used
   a. as fungicide by impregnating the wrappers used for citrus fruits [4].
   b. as component of coal tar due to its high boiling point and also present in creosote used for weather proofing of wood [23].
2. PCBs were used
   a. as heat transfer, hydraulic and dielectric fluid in electrical equipments [5, 7, 24].
   b. as industrial chemicals for insulation [24].
   c. in industry and electronic devices such as, capacitors or transformers, for lubrication or cooling [7, 24, 25].
   d. as additive in varnishes, paints, waxes and synthetic resins to improve their quality [24].
   e. for microencapsulation of dyes used for carbonless copying paper [20, 26].
   f. in inks, adhesives, pesticide extenders, polyolefin catalyst carriers, optical

brighteners, sealants, caulking agents, plasticizers, conveyor belts, cutting oils, metal coatings, surface polishes, pipeline condensates and cables [16, 19, 20, 23].

    g. as organic diluents, dust-reducing agents, laminating agents, repellents, organic diluents and adhesives [14, 18, 27].

3. PBBs were used

    a. as flame retardants due to their high flash point [6, 8].

    b. as additives in various commercial products such as, thermoplastics, lacquers, textiles, upholstery in vehicles, parts of microwave ovens, air conditioners, electric bulbs, computers, textiles, and televisions to make them fire resistant [19, 22].

Sources of Biphenyls and Their Distribution in Environment and Living Organisms; and Remediation Techniques

Biphenyl and its synthetic derivatives such as, PCBs and PBBs are widely distributed in different compartments of environment *i.e.*, soil, sediments, water bodies, air, plants and animals [28 - 30]. PCBs and PBBs have entered the environment over last few decades, through various routes, primarily during large-scale production and addition of congeners to the polymers [13, 31, 32]. Sometimes, these organic pollutants enter environment due to some accident. Major incident of PBB contamination occurred in 1973 in Michigan, where PBBs were accidentally added to animal feed which then entered the food chain through dairy and meat products from animals consuming that contaminated animal feed and caused toxicity in both animals and human beings [31]. In 1968, a PCB outbreak took place in Japan, which is known as Yusho accident and affected about 1000 residents with various health problems such as, blindness, jaundice, abdominal pain, skin discolouration, *etc.* [33]. Fumes emitted during incineration of different materials containing biphenyls and volatilisation from water and soil add PCBs and PBBs in the air [13, 31, 34, 35]. Atmospheric dispersion is a major source of contamination in other environmental reservoirs, especially in metropolitan cities due to larger industrial setups as compared to rural areas [36]. Contamination of soil with biphenyl and its congeners may be localised but is more intense as compared to other exposure routes [36]. Biphenyl, PBBs and PCBs entered the soil during their manufacturing processes, disposal in the landfills, recycling of scrap materials, auto repairing, from dye stuff carriers, wastes of creosote treated railway sleepers, plasticizers, lubricants and hydraulic fluids, photodecomposition of PBBs, leaching through municipal sludge and sewage disposal [9, 13, 34, 37]. Higher levels of PBBs and PCBs were found in soils from US, Brazil and Europe [13, 30]. These PBBs then accumulate in plants or get deposited on plant surface and get transferred to the humans and animals through food chain [13, 36]. PBBs and PCBs enter hydrosphere *via* various routes, such as, land runoff, release from electrical equipments, direct dumping of various products containing congeners, accidental release in water bodies, atmospheric deposition, *etc.* [31, 36]. Due to high mobility, these congeners get accumulated in aquatic organisms *e.g.,* in eggs of white tailed eagle from Norway, seal blubber in North Sea [13, 36]. Sediments also act as an important sink for congeners in hydrosphere and source of PCBs to aquatic flora and fauna [39]. PCBs and PBBs act as stable and persistent compounds and hence, when enter in any compartment of the environment, they spread in other environmental media also both regionally and globally [36]. Therefore, contamination by biphenyl and its synthetic derivatives has been reported in many areas where no source of contamination are present, *e.g.,* in snow from Antartica [36]. Probably, long-range transfer of these contaminants in the environment takes place *via* processes such as, evaporation and deposition [36].

Residues of PCBs and PBBs in soil, water, sediments and plants pose risk to human beings and wildlife [38]. Lipophilic nature of different congeners leads to their higher bioaccumulation, biomagnification and toxicity in various levels of food chain [36, 39]. Higher the degree of substitution, more stable and persistent is the congener in biological systems due to slower metabolization [36, 39, 40]. PCBs have been detected in tissues of birds, marine mammals and fish from Atlantic, Pacific and Baltic oceans [41, 42]. Exposure to biphenyl and its derivatives in human beings occurs through dermal, inhalation and ingestion pathways and may lead to neurological, mutagenic and carcinogenic health disorders [31, 37, 39, 43]. Different congeners were found in various food products such as, milk and other dairy products, meat products, animal and vegetable oils, seafood and commercial food products for children [13, 44 - 46]. About 90% exposure and health risk posed due to various congeners occur through food consumption, whereas, water consumption poses less health risk as water has low content of PCBs and PBBs due to their less solubility in water [14]. Moreover, biphenyl content is also low in groundwater as, it attaches itself to soil particles, therefore leaches in groundwater at a very slow rate [35]. Inhalation of dust particles through indoor and ambient air leads to higher absorption of PBBS and PCBs in body through respiratory and gastrointestinal tracts as compared to dermal exposure [14]. Children are exposed to these contaminants primarily during gestational period *via* transfer through blood across placenta or *via* breast milk after birth [31, 47, 48]. Due to less body weight, children, especially infants are more exposed to PBB and PCB congeners through milk and other dairy products as compared to other age groups [13]. PBBs and PCBs have been detected in hair, blood, serum, urine and adipose tissue of human beings [31, 36, 49]. Occupational exposure to PBBs and PCBs is very common in construction workers, farmers and labour working in chemical or electrical equipment production units [15, 40, 50, 51]. Due to this reason many countries have prohibited the production and usage of many congeners to stop their addition and distribution in the environment but PCBs and PBBs still enter the environment through previous reservoirs such as, old electrical equipments (still in use) and poorly managed waste disposal sites and cause toxic effects in living beings [31, 36, 39, 40].

To avoid the accumulation of biphenyls in environment, various countries have banned their usage and different alternatives are being used instead. For example, biphenyls used in dielectric fluids in transformers and capacitors, hydraulic fluids and heat exchange fluids have been replaced by biodegradable materials such as, silicone oil, mineral oil, ester-based materials, vegetable based oils, *etc* [52]. Moreover, it is very important to remove biphenyls and its variants from environment especially from soil. According to Chekol *et al.* and Aken *et al.* various cost-effective techniques such as, phytoextraction (uptake of contaminants by plants from soil), phytotransformation (enzymatic transformation of the contaminants by plants) and rhizoremediation (biodegradation of contaminants in soil by enhancement of microbial activity in root zone of plant) can also be used to get rid of these chemicals [53, 54]. Abraham *et al.* suggested that some microbial communities have ability to metabolize the complex PCB mixtures in soils, sediments or other natural systems and can be used for remediation of PCBs [55]. Singer *et al.* achieved partial bioremediation (30-36%) of PCBs in soil by repeatedly using PCB degrading bacteria *Arthrobacter sp.* Strain B1B and *Ralstonia eutrophus* H850 grown on carvone and a surfactant sorbitan trioleate for a period of 18 weeks [56]. Chen *et al.* demonstrated the anaerobic biodegradation of biphenyl mixture using bacteria isolated from the sediments in Hudson River at Fort Miller by enrichment technique [57]. Bedard *et al.* isolated *Alcaligenes eutrophus* H850, capable of degradation of a spectrum of PCBs from dredge spoils containing PCBs by enrichment technique of biphenyls, which degraded 81% of Aroclor 1242 and 35% of Aroclor 1254, in 2 days [58]. Enrichment technique is a method of culturing bacteria in such a culture media and environmental conditions, which favour the growth and enrichment of a certain microbial strain over others.

Health Implications

Biphenyl, PCBs and PCBs are persistent compounds with long biological half-lives. Therefore, they can get accumulated and biomagnified in higher trophic levels of a food chain. These compounds lead to various health problems in animals and human beings by interfering or disrupting various metabolic functions of the body. Biphenyl and its derivatives have been reported to have various harmful effects on human health such as inflammation in respiratory tract, headache, interruption in functioning of nervous system, immune system, hepatic system, reproductive system, liver damage, hepatitis, *etc.* [13, 36, 59]. Carpenter suggested that PCBs can promote tumours, and exposure to these contaminants during gestational and neonatal period leads to low IQ and body weight in children, causes altered reproductive and thyroid activities, liver disorders, and diabetes in both male and female individuals [60]. These synthetic organic compounds lead to endocrine disruptions, gastrointestinal disorders, renal disorders and cancer of liver, rectum, gall bladder and biliary tract along with various symptoms such as, weight loss, alopecia, skin eruptions, hypertrophy of glands, hyperplasia in epithelial lining of bile duct, urinary tract and gall bladder, atrophy in thymus and spleen, lymphoid involusion, hepatomegaly, altered thyroid and plasma levels, menstrual irregularities, *etc.* [13, 36, 50, 51, 61].

Fimm *et al.* conducted a study to identify the possible harmful effects of exposure to PCBs on neuropsychological parameters of employees and residents living around a capacitor and transformer recycling industry in Germany [40]. It was found that PCB exposure affects human sensorimotor processes. It was suggested that PCBs are capable to induce myelination in white matter of the brain and have adverse effects on functioning of cerebellum indicated by low dopamine production and imbalance in thyroid hormone [40].

Thyroid hormone homeostasis in people living near an electronic waste recycling site in China was found to be disturbed due to exposure to the PCBs released from the site [62]. Boas *et al.* suggested that as hydroxylated metabolites of PCBs are quite similar to T4, structurally, hence, they interfere with the functioning of thyroid gland in both humans and animals [63]. Bahn *et al.* medically evaluated workers of a brominated product manufacturing plant and found prevalence of hyperthyroidism, high amount of thyroid anti-microsomal antibody, increased serum levels of thyrotropin and high anti-thyroglobin level [64].

Bekesi *et al.* reported that immune system of Michigan residents who were exposed to PBBs *via* food intake was highly affected with symptoms such as, hypersensitivity to streptococci, hypergammaglobunemia, decrease in T and B lymphocyte contents, *etc.* [65]. Lipson *et al.* found that dairy farmers from Michigan consuming PBB contaminated dairy or meat products showed enhanced immunoglobuline IgG secretory levels indicating damaging effects of PBBs on immune system [66]. Furthermore, PBBs adversely affected the cell function. Oral exposure of guinea pigs to PCBs suppressed the cell-mediated and humoral immune response as number of cells producing the antitoxin for tetanus in lymph nodes was reduced [67]. Moreover, lymphopenia, cachexia, liver damage and destruction of lymphoid system were also observed. Moreover, skin reactions that take place due to tuberculination were also found to be reduced.

Lee *et al.* conducted a study on residents from Coronary Artery Risk Development in Young Adults (CARDIA) cohort to assess role of various Persistent Organic Pollutants as type 2 diabetes predictors. BB-153 and PCBs were found to pose higher risk of diabetes [68]. Similarly, Lim *et al.* analysed National Health and Nutrition Examination Survey (NHANES), cohort for risk of occurrence of diabetes due to PCB exposure [69].

Small *et al.* reported 3-fold increase in the occurrence of hernia in male offsprings of mothers with high serum PBB levels due to higher exposure to PBBs [70]. Sweeney and Symanski estimated the effect of age when women were exposed to PBBs on the length of pregnancy and the birth weight of offsprings in Michigan cohort [71]. It was found that PBB exposure had association with birth weight but not related with gestation period. Davis *et al.* estimated the menstrual cycle length of women from Michigan cohort with respect to self-reported length of menstrual cycles (20-35 days) [72]. An association was observed between duration of menstrual cycle and PBB exposure for women who had weight loss in the last one year indicating the impact of PBB exposure on ovarian function and also suggested the importance of mobilization of PBB reserved in lipids. Blanck *et al.* reported that women exposed to higher PBBs in the womb of mother or *via* breast milk had earlier onset of menarche as compared to women who had lower exposure to PBBs [73]. Pflieger-Bruss *et al.* reported that PCBs act as endocrine disruptors and have detrimental effects on male fertility [74].

Zhao *et al.* suggested that high burdens of PCBs and PBBs may be responsible for high occurrence of cancer in residents around the disassembly sites for electrical equipments in China [75]. Hoque *et al.* studied correlation between cancer risk and PBBs levels in serum of accidental victims of Michigan incident of PBB outbreak in 1973 and a positive association was found between PBBs concentration in serum and occurrence of lymphoma and cancer in digestive system [76]. Similarly, in a case controlled study by Henderson *et al.*, an association was observed between serum PBBs level and occurrence of breast cancer in women from Michigan [77]. Silver *et al.* conducted a study in women working in capacitor production plants in US for the duration of at least 1 year and found that the risk of occurrence of breast cancer in non-white employees had some association with PCB exposure [78].

Techniques for Estimation of Biphenyls in the Environmental and Biological Samples

Gas chromatography is a widely used technique for detection or analysis of PBBs and PCBs in different types of samples *e.g.,* soil, sediments, water, food, *etc.* Mostly organic polar solvents such as, hexane, acetone, iso-octane, dichloromethane, *etc.* are used for the extraction of PBB or PCB congeners from the samples and congener mixtures and later sample cleanup is done by various methods such as by gel permeation chromatography, by using Florisil columns, by passing through copper powder or silica impregnated with silver nitrate or acid impregnated silica [15, 79]. Recovery of analyte after extraction and clean up can be affected by the type of solvent used for extraction, therefore, most of the procedures used for extraction lead to 80 to 90% of recovery of analyte [15]. However, using Soxhlet apparatus for extraction with dichloromethane/n-pentane and thereafter cleaning up and fractionating the extract using alumina and silica gel columns can lead to over 95% recovery of congeners from different samples [15].

Soxhlet extraction is advantageous for extraction of PCBs and PBBs from solid matrices as equipment is easily available, cheap and easy to use [79]. However, other extraction methods such as, microwave assisted extraction (MAE), pressurized liquid extraction (PLE), supercritical fluid extraction (SFE) and ultrasonic assisted extraction (UAE) are also used as they not only have efficiency at par with Soxhlet extraction but are also automatically controlled with optimised pressure and temperature and have high sample throughput [79].

In complex mixtures of PCBs and PBBs, due to their differential volatility, it is very difficult to separate these compounds in gas chromatograph, moreover, there are high chances of overlapping of their spectra. In this case, mass spectrometer (MS) coupled with electron capture negative ionisation (ECNI) or electron impact (EI) methods are preferred over gas chromatography and for better resolution, it is suggested to use narrow bore capillary columns [80]. Electron capture negative ionisation (ECNI) can be combined with gas chromatography coupled with mass spectrometery (GC-MS) for detection of PCBs and PBBs in various environmental and biological samples as this detection method exhibits about 10 times higher sensitivity for congeners over other equipments such as, electron capture detector (ECD) [81].

Abdallah *et al.* analysed PCBs in solid matrices such as, indoor dust, sediment and soil samples using selective pressurized liquid extraction method followed by detection using gas chromatograph equipped with mass spectrometer (GC-MS) in electron impact ionization/ selected ion monitoring (EI/SIM) mode [79]. Hawthorne *et al.* determined PCB concentration in soil and sediment samples by subcritical water extraction along with solid-phase microextraction (SPME) followed by gas chromatograph–electron-capture detection (GC-ECD) method [82]. For determination of PBBs in soil and plant extracts, Jacobs *et al.* used gas chromatograph coupled with an electron capture detector (GC-ECD) with efficiency of PBB detection upto minimum concentration of 0.1 ppb for soil samples and 0.3 ppb for plant samples [83]. Comprehensive two-dimensional gas chromatography-isotope dilution time-of-flight mass spectrometry (GC×GC-IDTOFMS) was performed by Focant *et al.* for quantification of coplanar polychlorinated biphenyl in vegetation, fish, sediments and ash samples [84]. Digested fish and vegetation samples were extracted with hexane, whereas, sediment and ash samples were extracted with toluene using Soxhlet apparatus and cleaned up using anhydrous sodium sulphate and sulphuric acid-modified silica column.

Abbassy *et al.* extracted PCBs in water from Nile river, Egypt using extraction cartridge made of C-18 silica, then cleaned up and fractionated for analysis by gas chromatograph coupled with tritium electron capture detector (GC-3H-ECD) [85]. PCBs in surface waters collected from National Capital Region, India were extracted, purified and quantified with dichloromethane, column made of activated silica gel and gas chromatograph-Electron Capture Detector (GC-ECD), respectively [86].

Wang *et al.* determined the PCBs, and PBBs levels in atmosphere near a municipal solid waste incinerator in southern Taiwan [87]. Extraction was performed by Wang *et al.* using Soxhlet apparatus where toluene was used as solvent, cleaned up using multiple steps including treatment with conc. sulfuric acid, extraction with hexane in an acid silica gel column and alumina column, *etc* [87]. Then final extracts were concentrated by using nitrogen gas and analysed using a high-resolution gas chromatograph coupled with high-resolution mass spectrometer (HRGC/HRMS).

PCBs and PBBs in blood serum samples are extracted by denaturing the sample with methanol or hexane diethyl ether mixture followed by sample cleanup using adsorption chromatography or gel permeation chromatography and quantification is done by using gas chromatograph equipped with electron capture technique (GC-ECD) [15, 88, 89]. Similarly, PBBs and PCBs can be extracted from biological samples having fat followed by rigorous clean up process using one of the different methods discussed earlier and for estimation of different congeners high performance liquid chromatography (HPLC) or gas chromatograph coupled with electron capture detection (GC-ECD) can be used [15, 90]. Ferrario *et al.* determined PCB concentration in beef fat using HRGC/HRMS [91]. For sample preparation, Ferrario *et al.* extracted PCBs with methylene chloride and removed lipids using acid-impregnated silica gel and passed the extract through an acid/base silica gel column and a graphitized carbon column for clean up [91]. Takasuga *et al.* estimated PCBs in transformer oil (extracted and cleaned up by dimethyl sulfoxide saturated n-hexane and multilayer silica gel column, respectively) by high resolution gas chromatography-high resolution mass spectrometry (HRGC-HRMS) [92].

Enzyme-linked immunosorbent assay (ELISA) is used in commercial kits for analysis of PCBs [38]. However, detection limit for this method is quite low but these kits can be used easily for screening the samples with simpler sample preparation. There are some other techniques such as, flame ionisation detection (FID) and microwave-induced plasma atomic emission detection (MIP-AED) for detection of PCBs and PBBs but these methods have low sensitivity and specifity as compared to methods discussed earlier [93, 94].

Various Assays and Techniques Used to Assess Toxic Effects of Biphenyls

When PCBs and PBBs present in the environment enter biological systems, they interfere with physiology and metabolism of different organisms and lead to toxic effects causing changes or restrictions in normal functioning of the body. Toxicity of these chemicals have been studied widely in different prokaryotic and eukaryotic models and some examples are being discussed. Bacterial reverse mutation assay was performed by Tennant *et al.* to assess genotoxicity of Firemaster FF- I (PBB mixture) using *Salmonella typhimurium* strains TA97, TA98, TA100, TA1535 and TA1537 and no positive results were observed [95]. Polybrominated flame retardant PBB-153 and polychlorinated flame retardant PCB-153 were analysed for their toxicity by performing *an* acute toxicity test for duration of 24 hours and reproductive tests on a small planktonic crustacean *Daphnia magna* [96]. Test chemicals were found to be not toxic even at highest dose of 210 µg/L in case of 24 hr acute toxicity test. However, in case of reproductive tests it was found that at dose of 100 µg/L all organisms died by the day 4. At lower doses of 12.5 and 25 µg/L, PCB-153 had severe effects on reproduction in *Daphnia magna* as compared to PBB-153 [96].

Vicia faba roots tip micronucleus test was used by Song *et al.* to determine the genotoxicity of PCBs and it was found that frequency of micronuclei was higher by 2.2–48.4 times in comparison to control [97]. MTT assay was performed by Marabini *et al.* to assess the cytotoxicity of PCBs in a fish cell line (RTG-2) and cell viability was represented as percentage of viable cells left as compared to the negative control [98]. Marabini *et al.* used the same cell line for micronucleus assay, alkaline single-cell gel electrophoresis and modified comet assay to assess genotoxicity of PCBs and it was found that micronucleus frequency and DNA damage were highly increased in cells after treatment with PCBs. Babich and Borenfreund performed neutral red assay to assess *in vitro* cytotoxicity of PCB mixtures using bluegill sunfish BF-2 cells [99].

Gupta and Moore treated Fischer 344/N rats by giving them oral doses of Firemaster FF-1 (PBB mixture) for 4.5 weeks and kept them under observation for 90 days [100]. Pathological changes were observed in hepatic system, thymus and spleen of rats under treatment. 100% female rats and 38% male rats died by the days 53 and 73, respectively [100]. Rhesus monkeys were used by Allen *et al.* for assessing toxicity of PBBs and it was observed that PBB intake lead to weight loss, decrease in serum progesterone, alopecia, subcutaneous edema, erythropenia, leukopenia, hyperplastic gastroenteritis, hypoproteinemia and increased post conceptional bleeding [101].

Many other studies also include different techniques to study toxicity of PBBs and PCBs such as, their potential to induce gene mutations in mammalian cell lines such as Chinese hamster V79 and rat liver cells, chromosomal aberrations and sister chromatid exchanges in CHO cells, induction of micronuclei and unscheduled DNA synthesis in cells of B6C3F1 mice [13]. Moreover, different biomarkers including measurements of PCBs and PBBs in blood serum, adipose tissue, hair, breast milk and cord blood are also used for determination of level of exposure and toxicity of these chemicals [16].

SUMMARY

Biphenyl and its derivatives (PCBs and PBBs) are group of chemically and thermally stable compounds used as dielectric fluid in electronic instruments, lubricating agents, coolants, flame retardants, wood preservatives, fungicides, laminating agents, additives in inks, cutting oils, electric bulbs, lacquers, thermoplastics, *etc.* These organic compounds get introduced into the environment due to various anthropogenic processes such as, their injudicious application, accidental spillage, improper disposal, *etc.* Biphenyl and its derivatives are lipophilic and persistent in nature and degrade at a very slow rate in the environment, which leads to their biomagnification in food chain, therefore, manufacturing and usage of these compounds have been stopped in different countries. Biphenyl, PCBs and PBBs have been found in soil, sediments, water, flora and fauna. Residues of these toxic contaminants have also been found in tissues of living human beings exposed to these chemicals orally, dermally or through inhalation and lead to various health problems related to nervous, reproductive, respiratory and cardiovascular disorders, skin problems and cancer. Gas chromatography (GC), gas chromatograph–electron-capture detection (GC-ECD), gas chromatograph-mass spectrometer in electron impact ionization/ selected ion monitoring mode (GC-MS-EI/SIM), mass spectrometry with electron capture negative ionisation (GC-ECNI), high resolution gas chromatography with high resolution mass spectrometry (HRGC-HRMS) are few prominent techniques used to detect and estimate biphenyl, PCBs and PBBs contents in different compartments of environment, food, vegetation and animal and human tissues. Various bioassays used to assess the toxicity/ genotoxicity of these chemicals are bacterial reverse mutation assay, reproductive tests in *Daphnia magna, Vicia faba* roots tip micronucleus test, chromosomal aberrations, neutral red assay, MTT assay, alkaline single-cell gel electrophoresis and modified comet assay using different cell lines. Different bacteria and plants can be used to remediate biphenyl and its derivatives from environmental samples, especially, soil by various processes such as, phytoextraction, phytotransformation and rhizoreme-diation.

CONSENT FOR PUBLICATION

Not applicable.

CONFLICT OF INTEREST

The authors declare no conflict of interest, financial or otherwise.

ACKNOWLEDGEMENTS

Authors would like to thank Guru Nanak Dev University, Amritsar, Punjab, for providing library and internet facilities for survey of literature during preparation of this chapter.

REFERENCES

[1]   Jones KC, de Voogt P. Persistent organic pollutants (POPs): state of the science. Environ Pollut 1999; 100(1-3): 209-21.
        [http://dx.doi.org/10.1016/S0269-7491(99)00098-6] [PMID: 15093119]

[2]   Ang EL, Zhao H, Obbard JP. Recent advances in the bioremediation of persistent organic pollutants *via* biomolecular engineering. Enzyme Microb Technol 2005; 37(5): 487-96.
        [http://dx.doi.org/10.1016/j.enzmictec.2004.07.024]

[3]   Jain ZJ, Gide PS, Kankate RS. Biphenyls and their derivatives as synthetically and pharmacologically important aromatic structural moieties. Arab J Chem 2017; 10: S2051-66.
        [http://dx.doi.org/10.1016/j.arabjc.2013.07.035]

[4]   Ramsey GB, Smith MA, Heiberg BC. Fungistatic action of diphenyl on citrus fruit pathogens. Bot Gaz 1944; 106(1): 74-83.
        [http://dx.doi.org/10.1086/335270]

[5]   Westbom R, Thörneby L, Zorita S, Mathiasson L, Björklund E. Development of a solid-phase extraction method for the determination of polychlorinated biphenyls in water. J Chromatogr A 2004; 1033(1): 1-8.
        [http://dx.doi.org/10.1016/j.chroma.2004.01.022] [PMID: 15072285]

[6]   Vasiliu O, Cameron L, Gardiner J, Deguire P, Karmaus W. Polybrominated biphenyls, polychlorinated biphenyls, body weight, and incidence of adult-onset diabetes mellitus. Epidemiology 2006; 17(4): 352-9.
        [http://dx.doi.org/10.1097/01.ede.0000220553.84350.c5] [PMID: 16755267]

[7]   Wu JP, Luo XJ, Zhang Y, *et al.* Bioaccumulation of polybrominated diphenyl ethers (PBDEs) and

polychlorinated biphenyls (PCBs) in wild aquatic species from an electronic waste (e-waste) recycling site in South China. Environ Int 2008; 34(8): 1109-13.
[http://dx.doi.org/10.1016/j.envint.2008.04.001] [PMID: 18504055]

[8]   Yun SH, Addink R, McCabe JM, *et al.* Polybrominated diphenyl ethers and polybrominated biphenyls in sediment and floodplain soils of the Saginaw River watershed, Michigan, USA. Arch Environ Contam Toxicol 2008; 55(1): 1-10.
[http://dx.doi.org/10.1007/s00244-007-9084-3] [PMID: 18049786]

[9]   Maya-Drysdale L, Poulsen PB, Strandesen M. 2015. Survey of biphenyl (CAS no. 92-52-4). Copenhagen K Denmark: The Danish Environmental Protection Agency. 2015 [cited 07.11.2017]. Available from: https://www2.mst.dk/Udgiv/publications/2015/06/978-87-93352-35-3.pdf

[10]   Griesbaum K, Behr A, Biedenkapp D, *et al.* Hydrocarbons. Weinheim: Wiley-VCH Verlag GmbH and Co. KGaA: Ullmann's Encyclopedia of Industrial Chemistry. 2012.

[11]   Dow. Product Safety Assessment Biphenyl äDOW. The Dow Chemical Company. 2015 [cited 07.11.2017]. Available from: http://msdssearch.dow.com/PublishedLiteratureDOWCOM/dh_093f /0901b8038093fcdb.pdf?filepath=productsafety/pdfs/noreg/233-00584.pdf&fromPage=GetDoc

[12]   Environment Canada. Screening Assessment 1,1'-Biphenyl Chemical Abstracts Service Registry Number 92-52-4. Government of Canada Publications. 2014 [cited 07.11.2017]. Available from: http://publications.gc.ca/collections/collection_2014/ec/En14-197-2014-eng.pdf

[13]   Scientific opinion on polybrominated biphenyls(PBBs) in food. European Food Safety Authority. EFSA J 2010; 8(10): 1-151.

[14]   Larsen JC, Nielsen E, Boberg J, Petersen MA. Evaluation of health hazards by exposure to Polychlorinated biphenyls (PCB) and proposal of a health-based quality criterion for soil. Copenhagen K Denmark: The Danish Environmental Protection Agency. 2014 [cited 07.11.2017]. Available from: https://www2.mst.dk/Udgiv/publications/2014/03/978-87-93026-17-9.pdf

[15]   de Boer J, de Boer K, Boon JP. Polybrominated biphenyls and diphenylethers.Berlin, Springer Berlin Heidelberg: The Handbook of Environmental Chemistry Vol. 3 Part K New Types of Persistent Halogenated Compounds. 2000; pp. 61-96.
[http://dx.doi.org/10.1007/3-540-48915-0_4]

[16]   Polychlorinated Biphenyls And Polybrominated Biphenyls. Iarc Monographs On The Evaluation Of Carcinogenic Risks To Humans. Volume 107. Lyon: International Agency For Research On Cancer 2016 [cited 07.11.2017]. Available from: http://monographs. iarc. fr/ ENG/ Monographs /vol107 /mono107. pdf

[17]   Kodavanti PRS, Loganathan BG. Polychlorinated biphenyls, polybrominated biphenyls, and brominated flame retardants.Biomarkers in Toxicology. Oxford: Academic Press 2014; pp. 433-50.
[http://dx.doi.org/10.1016/B978-0-12-404630-6.00025-7]

[18]   Safe SH. Polychlorinated biphenyls (PCBs): environmental impact, biochemical and toxic responses, and implications for risk assessment. Crit Rev Toxicol 1994; 24(2): 87-149.
[http://dx.doi.org/10.3109/10408449409049308] [PMID: 8037844]

[19]   Mohapatra P, Dubey A. Polychlorinated biphenyls (PCBs) a persistent organic pollutant. Factsheet number 43 / March 2014 Toxic links , 2014 [cited 30.09.2017]; Available from: http://toxicslink. org/ docs/ PCB- Factsheet43.pdf

[20]   Toxicological profile for polychlorinated biphenyls. US Department of Health and Human Services, Public Health Service and Agency for Toxic Substances and Disease Registry , 2000 [cited 29.09.2017]; Available from: https://www.atsdr.cdc.gov/toxprofiles/tp17.pdf

[21]   Steenland K, Hein MJ, Cassinelli RT II, *et al.* Polychlorinated biphenyls and neurodegenerative disease mortality in an occupational cohort. Epidemiology 2006; 17(1): 8-13.
[http://dx.doi.org/10.1097/01.ede.0000190707.51536.2b] [PMID: 16357589]

[22]   Toxicological profile for polybrominated biphenyls. U.S. Department of Health and Human Services,

Public Health Service and Agency for Toxic Substances and Disease Registry. 2004a [cited 29.09.2017]. Available from: https://www.atsdr.cdc.gov/toxprofiles/tp68.pdf

[23] UK Marine Special Areas of Conservation Project Toxic substance profile: Biphenyl. n.d. [15.10.2107]. Available from: http://www.ukmarinesac.org.uk/activities/water-quality/wq8_30.htm

[24] Vos JG, de Roij T. Immunosuppressive activity of a polychlorinated diphenyl preparation on the humoral immune response in guinea pigs. Toxicol Appl Pharmacol 1972; 21(4): 549-55.
[http://dx.doi.org/10.1016/0041-008X(72)90011-7] [PMID: 4114810]

[25] Varanasi P, Fullana A, Sidhu S. Remediation of PCB contaminated soils using iron nano-particles. Chemosphere 2007; 66(6): 1031-8.
[http://dx.doi.org/10.1016/j.chemosphere.2006.07.036] [PMID: 16962632]

[26] Tröster AI, Ruff RM, Watson DP. Dementia as a neuropsychological consequence of chronic occupational exposure to polychlorinated biphenyls (PCBs). Arch Clin Neuropsychol 1991; 6(4): 301-18.
[http://dx.doi.org/10.1093/arclin/6.4.301] [PMID: 14589522]

[27] Safe S. Polychlorinated biphenyls (PCBs), dibenzo-p-dioxins (PCDDs), dibenzofurans (PCDFs), and related compounds: environmental and mechanistic considerations which support the development of toxic equivalency factors (TEFs). Crit Rev Toxicol 1990; 21(1): 51-88.
[http://dx.doi.org/10.3109/10408449009089873] [PMID: 2124811]

[28] Buckley EH. Accumulation of airborne polychlorinated biphenyls in foliage. Science 1982; 216(4545): 520-2.
[http://dx.doi.org/10.1126/science.216.4545.520] [PMID: 17735742]

[29] He MC, Sun Y, Li XR, Yang ZF. Distribution patterns of nitrobenzenes and polychlorinated biphenyls in water, suspended particulate matter and sediment from mid- and down-stream of the Yellow River (China). Chemosphere 2006; 65(3): 365-74.
[http://dx.doi.org/10.1016/j.chemosphere.2006.02.033] [PMID: 16580044]

[30] Rissato SR, Galhiane MS, Ximenes VF, *et al.* Organochlorine pesticides and polychlorinated biphenyls in soil and water samples in the Northeastern part of São Paulo State, Brazil. Chemosphere 2006; 65(11): 1949-58.
[http://dx.doi.org/10.1016/j.chemosphere.2006.07.011] [PMID: 16919310]

[31] Public Health Statement Polybrominated biphenyls. U.S. Department of Health and Human Services, Public Health Service and Agency for Toxic Substances and Disease Registry. 2004b [cited 29.09.2017]. Available from: https://www.atsdr.cdc.gov/toxprofiles/tp68-c1-b.pdf

[32] Sharma A. Environmentally sound management of polychlorinated biphenyls in India. Curr Sci 2014; 106(3): 358-9.

[33] Corbett TH, Beaudoin AR, Cornell RG, *et al.* Toxicity of polybrominated biphenyls Firemaster BP-6 in rodents. Environ Res 1975; 10(3): 390-6.
[http://dx.doi.org/10.1016/0013-9351(75)90034-1] [PMID: 1082414]

[34] de Cárcer DA, Martín M, Karlson U, Rivilla R. Changes in bacterial populations and in biphenyl dioxygenase gene diversity in a polychlorinated biphenyl-polluted soil after introduction of willow trees for rhizoremediation. Appl Environ Microbiol 2007; 73(19): 6224-32.
[http://dx.doi.org/10.1128/AEM.01254-07] [PMID: 17693557]

[35] Biphenyl (1,1-biphenyl). Department of the environment and energy, Australian government. 2017 [cited 30.09.2017]. Available from: http://www.npi.gov.au/resource/biphenyl-11-biphenyl

[36] Safe S. Polychlorinated biphenyls (PCBs): Mammalian and environmental toxicology. Berlin, Heidelberg: Springer-Verlag 1987.
[http://dx.doi.org/10.1007/978-3-642-70550-2]

[37] Arfaeinia H, Asadgol Z, Ahmadi E, Seifi M, Moradi M, Dobaradaran S. Characteristics, distribution and sources of polychlorinated biphenyls (PCBs) in coastal sediments from the heavily industrialized

area of Asalouyeh, Iran. Water Sci Technol 2017; 76(11-12): 3340-50.
[http://dx.doi.org/10.2166/wst.2017.500] [PMID: 29236013]

[38]   Muir D, Sverko E. Analytical methods for PCBs and organochlorine pesticides in environmental monitoring and surveillance: a critical appraisal. Anal Bioanal Chem 2006; 386(4): 769-89.
[http://dx.doi.org/10.1007/s00216-006-0765-y] [PMID: 17047943]

[39]   Ware GW, Ed. Reviews of environmental contamination and toxicology. New York: Springer-Verlag 2004.

[40]   Fimm B, Sturm W, Esser A, *et al.* Neuropsychological effects of occupational exposure to polychlorinated biphenyls. Neurotoxicology 2017; 63: 106-19.
[http://dx.doi.org/10.1016/j.neuro.2017.09.011] [PMID: 28947237]

[41]   Burreau S, Zebühr Y, Broman D, Ishaq R. Biomagnification of PBDEs and PCBs in food webs from the Baltic Sea and the northern Atlantic Ocean. Sci Total Environ 2006; 366(2-3): 659-72.
[http://dx.doi.org/10.1016/j.scitotenv.2006.02.005] [PMID: 16580050]

[42]   Loganathan BG, Masunaga S. PCBs, Dioxins, and Furans: Human Exposure and Health Effects.Handbook of Toxicology of Chemical Warfare Agents. Amsterdam: Elsevier 2009; pp. 245-53.
[http://dx.doi.org/10.1016/B978-0-12-374484-5.00018-3]

[43]   Kumar B, Verma VK, Singh SK, Kumar S, Sharma CS, Akolkar AB. Polychlorinated biphenyls in residential soils and their health risk and hazard in an industrial city in India. J Public Health Res 2014; 3(2): 252.
[http://dx.doi.org/10.4081/jphr.2014.252] [PMID: 25343135]

[44]   Di Carlo FJ, Seifter J, DeCarlo VJ. Assessment of the hazards of polybrominated biphenyls. Environ Health Perspect 1978; 23: 351-65.
[PMID: 209999]

[45]   Headrick ML, Hollinger K, Lovell RA, Matheson JC. PBBs, PCBs, and dioxins in food animals, their public health implications. Vet Clin North Am Food Anim Pract 1999; 15(1): 109-131, ix-x.
[http://dx.doi.org/10.1016/S0749-0720(15)30210-3] [PMID: 10088215]

[46]   Chovancová J, Kočan A, Jursa S. PCDDs, PCDFs and dioxin-like PCBs in food of animal origin (Slovakia). Chemosphere 2005; 61(9): 1305-11.
[http://dx.doi.org/10.1016/j.chemosphere.2005.03.057] [PMID: 16291405]

[47]   Jacobson JL, Fein GG, Jacobson SW, Schwartz PM, Dowler JK. The transfer of polychlorinated biphenyls (PCBs) and polybrominated biphenyls (PBBs) across the human placenta and into maternal milk. Am J Public Health 1984; 74(4): 378-9.
[http://dx.doi.org/10.2105/AJPH.74.4.378] [PMID: 6322600]

[48]   Koopman-Esseboom C, Weisglas-Kuperus N, de Ridder MA, Van der Paauw CG, Tuinstra LG, Sauer PJ. Effects of polychlorinated biphenyl/dioxin exposure and feeding type on infants' mental and psychomotor development. Pediatrics 1996; 97(5): 700-6.
[PMID: 8628610]

[49]   Zhang WH, Ying-Xin WU, Simonnot MO. Soil contamination due to e-waste disposal and recycling activities: a review with special focus on China. Pedosphere 2012; 22(4): 434-55.
[http://dx.doi.org/10.1016/S1002-0160(12)60030-7]

[50]   Brown DP, Jones M. Mortality and industrial hygiene study of workers exposed to polychlorinated biphenyls. Arch Environ Health 1981; 36(3): 120-9.
[http://dx.doi.org/10.1080/00039896.1981.10667615] [PMID: 6787990]

[51]   Brown DP. Mortality of workers exposed to polychlorinated biphenyls--an update. Arch Environ Health 1987; 42(6): 333-9.
[http://dx.doi.org/10.1080/00039896.1987.9934355] [PMID: 3125795]

[52]   Assessments of Technological Developments for the Production and Use, Including Exemptions, for

PCB Listed in Annex I in the POP protocol under the Convention on Long range transport of air pollution. The United Nations Economic Commission for Europe , 2005 [cited 03.06.2018]; Available from:
https://www.unece.org/fileadmin/DAM/env/lrtap/TaskForce/popsxg/2005/b_Norway_Alternatives%20to%20%20PCB_2.pdf

[53]   Chekol T, Vough LR, Chaney RL. Phytoremediation of polychlorinated biphenyl-contaminated soils: the rhizosphere effect. Environ Int 2004; 30(6): 799-804.
[http://dx.doi.org/10.1016/j.envint.2004.01.008] [PMID: 15120198]

[54]   Aken BV, Correa PA, Schnoor JL. Phytoremediation of polychlorinated biphenyls: new trends and promises. Environ Sci Technol 2010; 44(8): 2767-76.
[http://dx.doi.org/10.1021/es902514d] [PMID: 20384372]

[55]   Abraham WR, Nogales B, Golyshin PN, Pieper DH, Timmis KN. Polychlorinated biphenyl-degrading microbial communities in soils and sediments. Curr Opin Microbiol 2002; 5(3): 246-53.
[http://dx.doi.org/10.1016/S1369-5274(02)00323-5] [PMID: 12057677]

[56]   Singer AC, Gilbert ES, Luepromchai E, Crowley DE. Bioremediation of polychlorinated biphenyl-contaminated soil using carvone and surfactant-grown bacteria. Appl Microbiol Biotechnol 2000; 54(6): 838-43.
[http://dx.doi.org/10.1007/s002530000472] [PMID: 11152078]

[57]   Chen M, Hong CS, Bush B, Rhee GY. Anaerobic biodegradation of polychlorinated biphenyls by bacteria from Hudson River sediments. Ecotoxicol Environ Saf 1988; 16(2): 95-105.
[http://dx.doi.org/10.1016/0147-6513(88)90022-X] [PMID: 3148459]

[58]   Bedard DL, Wagner RE, Brennan MJ, Haberl ML, Brown JF Jr. Extensive degradation of Aroclors and environmentally transformed polychlorinated biphenyls by Alcaligenes eutrophus H850. Appl Environ Microbiol 1987; 53(5): 1094-102.
[PMID: 3111365]

[59]   Concise international chemical assessment document 6 BIPHENYL. Geneva: World Health Organization 1999.

[60]   Carpenter DO. Polychlorinated biphenyls (PCBs): routes of exposure and effects on human health. Rev Environ Health 2006; 21(1): 1-23.
[http://dx.doi.org/10.1515/REVEH.2006.21.1.1] [PMID: 16700427]

[61]   Li Z, Hogan KA, Cai C, Rieth S. Human health effects of biphenyl: key findings and scientific issues. Environ Health Perspect 2016; 124(6): 703-12.
[http://dx.doi.org/10.1289/ehp.1509730] [PMID: 26529796]

[62]   Zhang J, Jiang Y, Zhou J, *et al.* Elevated body burdens of PBDEs, dioxins, and PCBs on thyroid hormone homeostasis at an electronic waste recycling site in China. Environ Sci Technol 2010; 44(10): 3956-62.
[http://dx.doi.org/10.1021/es902883a] [PMID: 20408536]

[63]   Boas M, Feldt-Rasmussen U, Skakkebaek NE, Main KM. Environmental chemicals and thyroid function. Eur J Endocrinol 2006; 154(5): 599-611.
[http://dx.doi.org/10.1530/eje.1.02128] [PMID: 16645005]

[64]   Bahn AK, Mills JL, Snyder PJ, *et al.* Hypothyroidism in workers exposed to polybrominated biphenyls. N Engl J Med 1980; 302(1): 31-3.
[http://dx.doi.org/10.1056/NEJM198001033020105] [PMID: 6243165]

[65]   Bekesi JG, Holland JF, Anderson HA, *et al.* Lymphocyte function of Michigan dairy farmers exposed to polybrominated biphenyls. Science 1978; 199(4334): 1207-9.
[http://dx.doi.org/10.1126/science.204005] [PMID: 204005]

[66]   Lipson SM. Effect of polybrominated biphenyls on the growth and maturation of human peripheral blood lymphocytes. Clin Immunol Immunopathol 1987; 43(1): 65-72.

[http://dx.doi.org/10.1016/0090-1229(87)90157-7] [PMID: 3030592]

[67]   Vos JG, van Driel-Grootenhuis L. PCB-induced suppression of the humoral and cell-mediated immunity in guinea pigs. Sci Total Environ 1972; 1(3): 289-302.
[http://dx.doi.org/10.1016/0048-9697(72)90024-1] [PMID: 4633048]

[68]   Lee DH, Steffes MW, Sjödin A, Jones RS, Needham LL, Jacobs DR Jr. Low dose of some persistent organic pollutants predicts type 2 diabetes: a nested case-control study. Environ Health Perspect 2010; 118(9): 1235-42.
[http://dx.doi.org/10.1289/ehp.0901480] [PMID: 20444671]

[69]   Lim JS, Lee DH, Jacobs DR Jr. Association of brominated flame retardants with diabetes and metabolic syndrome in the U.S. population, 2003-2004. Diabetes Care 2008; 31(9): 1802-7.
[http://dx.doi.org/10.2337/dc08-0850] [PMID: 18559655]

[70]   Small CM, DeCaro JJ, Terrell ML, *et al.* Maternal exposure to a brominated flame retardant and genitourinary conditions in male offspring. Environ Health Perspect 2009; 117(7): 1175-9.
[http://dx.doi.org/10.1289/ehp.0800058] [PMID: 19654930]

[71]   Sweeney AM, Symanski E. The influence of age at exposure to PBBs on birth outcomes. Environ Res 2007; 105(3): 370-9.
[http://dx.doi.org/10.1016/j.envres.2007.03.006] [PMID: 17485077]

[72]   Davis SI, Blanck HM, Hertzberg VS, *et al.* Menstrual function among women exposed to polybrominated biphenyls: a follow-up prevalence study. Environ Health 2005; 4(1): 15.
[http://dx.doi.org/10.1186/1476-069X-4-15] [PMID: 16091135]

[73]   Blanck HM, Marcus M, Tolbert PE, *et al.* Age at menarche and tanner stage in girls exposed in utero and postnatally to polybrominated biphenyl. Epidemiology 2000; 11(6): 641-7.
[http://dx.doi.org/10.1097/00001648-200011000-00005] [PMID: 11055623]

[74]   Pflieger-Bruss S, Schuppe HC, Schill WB. The male reproductive system and its susceptibility to endocrine disrupting chemicals. Andrologia 2004; 36(6): 337-45.
[http://dx.doi.org/10.1111/j.1439-0272.2004.00641.x] [PMID: 15541049]

[75]   Zhao G, Wang Z, Zhou H, Zhao Q. Burdens of PBBs, PBDEs, and PCBs in tissues of the cancer patients in the e-waste disassembly sites in Zhejiang, China. Sci Total Environ 2009; 407(17): 4831-7.
[http://dx.doi.org/10.1016/j.scitotenv.2009.05.031] [PMID: 19539352]

[76]   Hoque A, Sigurdson AJ, Burau KD, Humphrey HE, Hess KR, Sweeney AM. Cancer among a Michigan cohort exposed to polybrominated biphenyls in 1973. Epidemiology 1998; 9(4): 373-8.
[http://dx.doi.org/10.1097/00001648-199807000-00005] [PMID: 9647899]

[77]   Henderson AK, Rosen D, Miller GL, *et al.* Breast cancer among women exposed to polybrominated biphenyls. Epidemiology 1995; 6(5): 544-6.
[http://dx.doi.org/10.1097/00001648-199509000-00014] [PMID: 8562633]

[78]   Silver SR, Whelan EA, Deddens JA, *et al.* Occupational exposure to polychlorinated biphenyls and risk of breast cancer. Environ Health Perspect 2009; 117(2): 276-82.
[http://dx.doi.org/10.1289/ehp.11774] [PMID: 19270799]

[79]   Abdallah MAE, Drage D, Harrad S. A one-step extraction/clean-up method for determination of PCBs, PBDEs and HBCDs in environmental solid matrices. Environ Sci Process Impacts 2013; 15(12): 2279-87.
[http://dx.doi.org/10.1039/c3em00395g] [PMID: 24145825]

[80]   Pijnenburg AM, Everts JW, De Boer J, Boon JP. Polybrominated biphenyl and diphenylether flame retardants: analysis, toxicity, and environmental occurrence.Reviews of environmental contamination and toxicology. New York: Springer 1995; pp. 1-26.
[http://dx.doi.org/10.1007/978-1-4612-2530-0_1]

[81]   de Boer J. Analysis and biomonitoring of complex mixtures of persistent halogenated micro-contaminants 1995.

[82]   Hawthorne SB, Grabanski CB, Hageman KJ, Miller DJ. Simple method for estimating polychlorinated biphenyl concentrations on soils and sediments using subcritical water extraction coupled with solid-phase microextraction. J Chromatogr A 1998; 814(1): 151-60.
[http://dx.doi.org/10.1016/S0021-9673(98)00418-X]

[83]   Jacobs LW, Chou SF, Tiedje JM. Field concentrations and persistence of polybrominated biphenyls in soils and solubility of PBB in natural waters. Environ Health Perspect 1978; 23: 1-8.
[http://dx.doi.org/10.1289/ehp.78231] [PMID: 209960]

[84]   Focant JF, Reiner EJ, Macpherson K, *et al.* Measurement of PCDDs, PCDFs, and non-ortho-PCBs by comprehensive two-dimensional gas chromatography-isotope dilution time-of-flight mass spectrometry (GC x GC-IDTOFMS). Talanta 2004; 63(5): 1231-40.
[http://dx.doi.org/10.1016/j.talanta.2004.05.043] [PMID: 18969552]

[85]   Abbassy MS, Ibrahim HZ, el-Amayem MM. Occurrence of pesticides and polychlorinated biphenyls in water of the Nile river at the estuaries of Rosetta and Damiatta branches, north of Delta, Egypt. J Environ Sci Health B 1999; 34(2): 255-67.
[http://dx.doi.org/10.1080/03601239909373196] [PMID: 10192956]

[86]   Kumar B, Singh SK, Verma VK, Gaur R, Kumar S, Sharma CS. A preliminary assessment of polychlorinated biphenyls in surface waters. Adv Appl Sci Res 2014; 5(1): 111-7.

[87]   Wang MS, Chen SJ, Huang KL, *et al.* Determination of levels of persistent organic pollutants (PCDD/Fs, PBDD/Fs, PBDEs, PCBs, and PBBs) in atmosphere near a municipal solid waste incinerator. Chemosphere 2010; 80(10): 1220-6.
[http://dx.doi.org/10.1016/j.chemosphere.2010.06.007] [PMID: 20598339]

[88]   Needham LL, Burse VW, Price HA. Temperature-programmed gas chromatographic determination of polychlorinated and polybrominated biphenyls in serum. J Assoc Off Anal Chem 1981; 64(5): 1131-7.
[PMID: 6270054]

[89]   Luotamo M, Järvisalo J, Aitio A. Analysis of polychlorinated biphenyls (PCBs) in human serum. Environ Health Perspect 1985; 60: 327-32.
[http://dx.doi.org/10.1289/ehp.8560327] [PMID: 3928361]

[90]   Toledo-Neira C, Enríquez P, Richter P. Integrated pressurized solvent extraction-cleanup for the rapid determination of polychlorinated biphenyls in meat samples. J Braz Chem Soc 2013; 24(5): 743-8.

[91]   Ferrario J, Byrne C, Dupuy AE Jr. Background contamination by coplanar polychlorinated biphenyls (PCBs) in trace level high resolution gas chromatography/high resolution mass spectrometry (HRGC/HRMS) analytical procedures. Chemosphere 1997; 34(11): 2451-65.
[http://dx.doi.org/10.1016/S0045-6535(97)00083-0] [PMID: 9192469]

[92]   Takasuga T, Senthilkumar K, Matsumura T, Shiozaki K, Sakai S. Isotope dilution analysis of polychlorinated biphenyls (PCBs) in transformer oil and global commercial PCB formulations by high resolution gas chromatography-high resolution mass spectrometry. Chemosphere 2006; 62(3): 469-84.
[http://dx.doi.org/10.1016/j.chemosphere.2005.04.034] [PMID: 15946725]

[93]   Krüger C. Polybrominated biphenyls and polybrominated diphenyl ethers–detection and quantitation in selected foods. Germany: University of Munster 1988.

[94]   Polybrominated Biphenyls IPCS, Environmental Health Criteria. Geneva: World Health Organization 1994.

[95]   Tennant RW, Stasiewicz S, Spalding JW. Comparison of multiple parameters of rodent carcinogenicity and *in vitro* genetic toxicity. Environ Mutagen 1986; 8(2): 205-27.
[http://dx.doi.org/10.1002/em.2860080204] [PMID: 3698943]

[96]   Nakari T, Huhtala S. Comparison of toxicity of congener-153 of PCB, PBB, and PBDE to Daphnia magna. Ecotoxicol Environ Saf 2008; 71(2): 514-8.
[http://dx.doi.org/10.1016/j.ecoenv.2007.10.012] [PMID: 18082264]

[97]   Song YF, Wilke BM, Song XY, Gong P, Zhou QX, Yang GF. Polycyclic aromatic hydrocarbons (PAHs), polychlorinated biphenyls (PCBs) and heavy metals (HMs) as well as their genotoxicity in soil after long-term wastewater irrigation. Chemosphere 2006; 65(10): 1859-68.
[http://dx.doi.org/10.1016/j.chemosphere.2006.03.076] [PMID: 16707147]

[98]   Marabini L, Calò R, Fucile S. Genotoxic effects of polychlorinated biphenyls (PCB 153, 138, 101, 118) in a fish cell line (RTG-2). Toxicol. *in vitro* 2011; 25(5): 1045-52.
[http://dx.doi.org/10.1016/j.tiv.2011.04.004] [PMID: 21504788]

[99]   Babich H, Borenfreund E. *In Vitro* cytotoxicity of polychlorinated biphenyls (PCBs) and toluenes to cultured Bluegill Sunfish BF-2 cells. Aquatic Toxicology and Hazard Assessment: 10th Volume. Pennsylvania: ASTM International 1988.
[http://dx.doi.org/10.1520/STP34059S]

[100]   Gupta BN, Moore JA. Toxicologic assessments of a commercial polybrominated biphenyl mixture in the rat. Am J Vet Res 1979; 40(10): 1458-68.
[PMID: 230756]

[101]   Allen JR, Lambrecht LK, Barsotti DA. Effects of polybrominated biphenyls in nonhuman-primates. J Am Vet Med Assoc 1978; 173(11): 1485-9.

CHAPTER 7

Actinomycete Enabled Remediation Strategies: Potential Tool for Pollutant Removal from Diverse Niches

Anu Kalia[1,*], **Sukhjinder Kaur**[2] and **Madhurama Gangwar**[2]

[1] *EMN Laboratory, Department of Soil Science, College of Agriculture, India*

[2] *Department of Microbiology, College of Basic Sciences and Humanities, Punjab Agricultural University, Ludhiana, Punjab, India*

Abstract: Industrialization, urbanization and misuse of various natural processes have led to the generation of many pollutants in the environment. These include halogenated nitroaromatic compounds, petroleum hydrocarbons, pesticides, phthalate esters and solvents. Remediation of contaminated sites through conventional methods relies on physicochemical treatments thereby posing considerable technical and economic issues or challenges. A promising technology for the efficient removal of pollutants, bioremediation, involves utilization of capabilities of microorganisms in a cost-effective manner. Among microorganisms, actinomycetes have omni-presence in diverse ecological niches and they play a major role as a decomposer community involved in reutilization of substances and degradation of complex polymers as well as in the synthesis of useful bioactive compounds. Therefore, actinomycetes have gained unusual attention as potential candidates for bioremediation. Moreover, prominence of omic studies and nanotechnology has revealed the metabolic regulations and processes that actinomycete utilize to manage the toxicity of pollutants. This manuscript will discuss the research interventions on bioremediation of various pollutants through actinomycetes as remediation agents and their contribution towards environment cleanup.

Keywords: Actinomycetes, *Amycolatopsis tucumanensis*, Bioactive compounds, Bioremediation approaches, Biotransformation, Diverse niche, Extremophilic Actinobacteria, Environmental pollution, Enzymes, Functional genomics, Green revolution, Hydrocarbons, Heavy metals, Marine Actinobacteria, Nano-remediation, Pollutants, Proteomics, Recalcitrants, *Streptomyces rochei*, Xenobiotics.

---

[*] **Corresponding author Anu Kalia:** EMN Laboratory, Department of Soil Science, College of Agriculture, India; Tel: +91-161-2401960; Extn: 288 (O); E-mail: kaliaanu@pau.edu

Ashita Sharma, Manish Kumar, Satwinderjeet Kaur & Avinash Kaur Nagpal (Eds.)

INTRODUCTION

The burgeoning human population has led to rapid urbanization. Besides this, other anthropogenic activities have enhanced the type and concentration of the potential pollutants above the critical levels thereby disturbing the natural diversity of environment. As the human population is growing, agricultural cooperatives have keen interest to meet the present need of food and food products, by application of various insecticides and pesticides to ensure protection of crop from attack of various pests, diseases and rodents [1, 2]. Therefore, the phenomena of green revolution, ear-marked through the replacement of old and conventional agricultural practices by modern, intensive cultivation techniques hailed to meet the goals of higher food productivity. However, these techniques and products of green revolution have aggravated to contamination of the land and water resources particularly due to large-scale agrochemical usage.

The active ingredients of pesticides are toxic compounds which may exhibit residual or recalcitrant effects causing environmental contamination that may result in serious public health problems including severe genetic and immune disorders. The persistence of the pesticide residues in ground and surface water bodies has remained a matter of foremost concern for the environmentalists. Moreover, there is paucity of a practical, economical and on-field effective treatment protocol for remediation of pesticides and/or their residues in polluted water and soil. However, the techniques generally applied as decontamination treatments such as land-filling, reusing, pyrolysis or incineration of the affected soil or sub-soil for remediation of contaminated sites may prompt for the generation of more detrimental pesticide intermediates. Above all, these techniques are costly and sometimes difficult to perform at larger scale, particularly for the land area under crop cultivation [1]. Apart from pesticides, the problem of eco-contamination with a myriad of pollutants has been compounded by irresponsible disposal of waste for instance for the industrial and urban sewage waste [1].

Any unwanted substance introduced into the environment is referred to as a 'contaminant' or 'pollutant'. Deleterious effects of contaminant(s) leading to damage to the ecological niche are manifested as 'pollution', a process through which a resource (natural or man-made) is rendered unfit for usage. Large-scale pollution due to relentless dumping of xenobiotics or other toxic chemical substances derived from various industries and due to conventional agri-practices has grave global ecological consequences. The mobility of the applied chemicals through leaching or seepage and surface run-offs is the potential source of contamination of the soil and water bodies. Besides, the continuous cycling through the processes of volatilization and condensation of pesticides and other organic compounds have even led to their presence in rain, fog and snow [3]. Hence, appropriate management policies, proper remedial plans and sustainable use of resources without changing the natural ecosystem should be the prime goal of modern day sustainable eco-prudent strategies.

A great diversity exists for the possible prevalent contaminants spanning over petroleum hydrocarbons (PHCs), polycyclic aromatic hydrocarbons (PAHs), halogenated hydrocarbons like polychlorinated biphenyls (PCBs) and trichloroethylene (TCE), pesticides, solvents, metals, dyes and salts [4]. To compound the pollution problems the tendency of few of these synthetic chemicals to resist degradation leads to accumulation of the recalcitrant pollutant or its derivative or residue. This further manifests into an elaborate process of bio-magnification and may involve tempering of the non-target factors or disruption of the delicate soil food-web. These toxic compounds or their derivatives need to be remediated from the contaminated niches by application of myriad of remediation strategies. The conventional technologies of pollutant removal include chemical methods such as chemical precipitation, chemical oxidation or reduction, ion exchange, and electrochemical treatment though the physical treatment techniques involve membrane filtration, reverse osmosis and evaporation. The serious pitfalls of these techniques can be attributed to high cost or relatively low to ineffective remediation [5].

Remediation through biological entities can be an effective and environmentally prudent technology for the removal of these compounds from the environment. Among the biological remediation components, microorganisms may be the most effective remediation agents. As these unicellular prokaryotic to eukaryotic forms play a pivotal role in decomposition and nutrient cycling, they are involved in the maintenance and sustainability of any ecosystem. The metabolic versatility, ability to inhabit diverse niches and proficiency to rapidly adapt in response to environmental variations embellish them to interact, both chemically and physically, with diverse environmental pollutants and other recalcitrant substances thereby leading to structural modifications, and transformation of the parent compound(s) to a safer or lesser toxic intermediate or even complete degradation of the target molecules [6]. Therefore, microorganisms can be used to enhance remediation of a variety of contaminants such as heavy metals, hydrocarbons, synthetic polymers like polyethenes, food wastes, greenhouse gases and others. The microorganisms may be indigenous to a contaminated region (the vanguard population) or can be obtained from another place and transferred to the polluted site (exotic cultures). Introduction of exotic microbial populations in a contaminated site for enhanced pollutant degradation is termed as 'Bio-augmentation' [7]. Among the microbes, actinomycetes are members of an important genera of microbes which can effectively disintegrate and bioremediate various types of environmental pollutants.

Actinobacteria as a Potential Tool for Pollutant Removal

These microbes belong to indigenous microbiota of soil of class Actinobacteria, and represent a group of substantially heterogeneous, Gram-positive bacteria, recognized to possess high G+C contents in their DNA [8]. The efficiency of actinobacteria as potential candidates for bioremediation of diverse organic pollutants contaminating the land can be attributed to their metabolic versatility and ability for production of an array of extracellular enzymes to metabolize/ mineralize these compounds. Under greater stress conditions, these bacteria readily form spores which are resistant to desiccation. Furthermore, these bacteria exist in long to short chains of the bacterial cells *i.e.* filamentous growth that favors colonization of the surfaces such as pores in the soil particles, rhizoplane as well as other smooth to rough surfaces or on diverse matrices. These bacteria can be cultivated under several cultural conditions and can be formulated as pure cultures or as combination of two or more cultures *i.e.* consortia [9].

Another interesting feature of this group of microorganisms, especially regarding the degradation of hydrophobic compounds, is their ability to produce surfactant compounds which can either be produced in the cell and incorporated in the membrane such as mycolic acid or can also be produced and excreted out of the cell (extracellular biosurfactants such as glycolipids or trehalose-lipid of *Rhodococcus* sp., lipoprotein in *Arthrobacter* sp.). The biosurfactant mechanisms are important components of natural biogeochemical cycles for metals and metalloids with some processes being of potential application to the treatment of contaminated materials. The metabolic versatility of actinomycetes has rendered them excellent bioremediation potentials arousing immense interest across globe for various specialized applications in diverse niches [10].

Biology of Actinobacteria

Actinobacteria, a separate taxonomic group within domain bacteria include members that belong to order Actinomycetales. At present, 221 genera of these bacteria are categorized with taxon hierarchies of class, order, and family as 6, 19 and 50 respectively even though new taxa are continued to be deciphered. As their name reflects (in Greek, "*atkis*" means ray and "*mykes*" means fungus), these bacteria share some morphological features with fungi such as occurrence of filamentous growth, and production of aerial or substrate mycelium. The taxon displays cosmopolitan distribution, with group members inhabiting diverse niches from water to land ecosystems [11]. These bacteria are mostly free-living chemo-organotrophs known to possess immense physiological diversity as these may exist as aerobes, obligate to facultative anaerobes exhibiting true Gram-positive to variable Gram staining reaction due to presence of cell walls rich in muramic acid. Even the strict high G+C DNA contents in actinobacteria have been turned around to some fresh water inhabiting actinobacteria to possess genomes with low G+C contents [12]. The phylum comprises phenotypically different organisms exhibiting variety of morphologies such as spherical cocci to extremely differentiated filamentous forms having pseudo-hyphal branched structures, 'mycelia' besides great variability for the shape and size of exogenous spores which are beneficial for long-distance dispersal.

The physiological and metabolic versatility of actinobacteria can be extrapolated, for instance, for synthesis of a variety of extracellular catalytic proteins or enzymes and secondary metabolites [11]. Genus *Streptomyces* belonging to order Actinomycetales can be acronymed as 'secondary metabolite factory' producing an array of organic compounds accounting for 80% out of ~45% of bioactive compounds produced by all actinobacteria. These compounds have several pharmaceutical applications particularly their use as antibiotic drugs to curb pathogenic microbial infections to anti-cancer agents and anti-metabolites [13]. In addition, production of biocatalysts for degradation of complex polymers, enzyme inhibitors, immune suppressing compounds, surfactant molecules, phytotoxins, and probiotic metabolites are of substantially greater biotechnological potentials and industrial applications [14].

Extremophilic Actinobacteria

Actinobacteria can inhabit extreme ecological niches like hot sulfur springs characterized by very low pH, alkaline or sodic soils with high pH, extreme low temperature prevailing environments *i.e.* glaciers in Tundra, in deserts or arid lands exhibiting low levels of available moisture, and in nutrient-poor niches such as degraded lands [15]. Therefore, the extraordinary abundance of genera actinobacteria may be considered either extremophile (EP) or as extremotolerant (ET) microbes inhabiting extreme niches by virtue of their physiological versatility probably as a consequence of metabolic adaptability [16]. Rather, co-occurrence of ET/EP actinobacteria in a niche exhibiting two or multiple extreme abiotic conditions may lead to evolution of poly-EP/ poly-ETs adapted to strive under multiple stress conditions [17]. The EP actinobacteria exhibit numerous adaptive strategies like occurrence of diverse trophy modes varying from hetero-, sapro- to chemoautotrophy (antibiosis) and by production of specific biocatalysts to combat the stress.

Mechanism of Adaptation in Extreme Conditions

Thermotolerance in actinobacteria can be attributed to production of thermostable proteins. These thermostable proteins possess higher number of disulfide linkages among the sulfur containing amino acids and exhibit greater occurrence of ionic and hydrophobic interactions. The EP/ET actinobacteria may also exhibit presence of specialized folding proteins, 'chaperones' which help in refolding of the partially denatured proteins due to high temperature stress. Furthermore, the DNA damage caused due to high temperature can be prevented by production of specialized DNA binding proteins [18]. High temperature tolerance in actinobacteria can also be achieved through synthesis of enzyme(s) involved in autotrophy such as carbon monoxide dehydrogenase (CODh) can catalyse oxidation of CO into CO_2. Later, the CO_2 can be routed to the normal C3 cycle through reduction by RuBisCO enzyme for the synthesis of carbohydrates and other components [19]. Likewise, the chemoautotrophic mode of the thermophile genera *Acidithiomicrobium* sp. which inhabits the geothermal vents ensures for utilization of sulfur as an energy source for the growth and survival of this actinobacteria in these hot acidic environment [20].

Actinobacteria particularly thermophilic members are known to be versatile decomposers and can act on diverse biomolecular substrates. This attribute is of immense importance keeping in view the nutrient recycling and replenishment in soil which leads to improvement in soil fertility and productivity. The biodegradation of complex organic substrates demand for production and extracellular secretion of various enzymes involved in hydrolysis and oxidation cascades by actinobacteria. Moreover, these bacteria possess fast colonization rates forming mycelia-like structures ensuring superior establishment on inoculation in the niche to exhibit better bioremediation potential. Additionally, their proficient catalytic bio-machinery degrades and even metabolizes recalcitrant natural as well as synthetic polymeric compounds and their derivatives including polyaromatic hydrocarbons, plastics, pesticides, rubber and many more [18]. Many thermophilic actinobacteria can act on polymers to produce basic monomers besides the functional group carbon atoms. The phenolic compounds and their derivatives are more dreaded compounds keeping in view their genotoxicity potential leading to teratogenic and carcinogenic issues and thus are required to be remediated. Some thermophilic actinomycetes can metabolize such lethal chemicals to non-toxic forms. This transformation happens through production of various biocatalysts including hydroxylases, hydratases, oxidases, dioxygenases, peroxidases, amidases, and lyases.

The probable tolerance mechanisms followed by acido and alkaliphilic actinobacteria to deal with extreme pH conditions involve proton pumps which maintain the optimum H^+ concentrations in the cell to regulate the optimum physiological pH in the cytoplasm [21]. Besides these active pumps, the cell wall structure and chemistry of alkaliphiles helps in stabilization of the cell membrane due to predominance of the negative charges as this decreases the charge density at the cell surface [22]. The poly-EP or poly-ETs further develop alternative adaptation mechanisms like exclusion of excess salt through the sodium/proton antiporters in haloalkaliphiles. Similarly, the haloalkaliphiles also exhibit phenomena of osmoregulation which prevents desiccation of cell cytoplasm through synthesis and accumulation of high amounts of compatible solutes.

Marine Actinobacteria

The actinobacteria inhabiting the marine or estuarine niches can be more aptly utilized for the remediation of various organic xenobiotics and heavy metals as these bacteria exhibit specialized characteristics of bio-film formation through production of extracellular polymeric substances. The benefit of utilizing marine microbes for *in situ* bioremediation is the direct use of organisms in unfavorable conditions without any manipulations at genetic level. Through competition for survival and evolution, marine organisms have formed special mechanism for self-protection as compared with terrestrial organisms [23]. This leads to production of unique and multifarious chemical substances that are being unraveled for different purposes [24]. The high variability for abiotic components like temperature, pH, sea currents, salinity, wind pattern and many more in the marine niches makes it most unfavorable to sustain life forms. Therefore, specialized adaptations exist for the microorganisms inhabiting such extreme and continuously altering environments [25]. Several marine alkali-tolerant and alkaliphilic actinobacteria can degrade hydrocarbon and other pollutants. *Dietzia* sp. has been reported to degrade n-alkane pollutants by production of surfactant compounds [26]. Such bioemulsifiers have potential applications in pharmaceuticals, cleansers, textiles and beauty care products.

Bio-films are associations of microbes packed collectively on animate and inanimate substrates. Bio-films offer the benefits of microbial cell immobilization by bringing them together spatially and in medium substrates resulting in improved cooperative, metabolic activities on the surface of bio-film. Microbial bio-films have extensive applications in many fields including medicine, agriculture and pollutants removal from the environment. A microbial consortia bio-film inhabiting high salt niches have also been observed to remediate several hydrocarbon pollutants [27]. Incidentally, on probing the individual components of the biofilm, alkali-tolerant actinobacteria like *Kocuria flava* and *Dietzia kunjamensis* were among the major constituents of the biofilm [18].

Many studies have shown remediation of xenobiotic compounds and heavy metals by marine bacteria. An example of this was reported by [23], who have evaluated the carbaryl degradation (xenobiotic compound) and heavy metals (Cu and Zn) absorption from the effluent by marine actinomycete *Streptomyces acrimycini* NGP (JX843532), *Streptomyces albogriseolus* NGP (JX843531) and *Streptomyces variabilis* NGP (JX43530), obtained from the coastal region of Tamil Nadu, India. The isolates were able to degrade carbaryl in 30.30, 35.90 and 36.20 percent. Likewise, the Cu and Zn absorption by respective isolates were observed to be 26.67, 57.35, 26.69 and 61.22, 62.33, 66.47 percent as determined by spectroscopic techniques.

Actinomycete Mediated Transformation of Various Compounds

Bioremediation of Hydrocarbon Contaminated Sites

Hydrocarbons and their derivatives like petroleum hydrocarbons (n-alkanes, aliphatics, aromatic compounds like poly-aromatic hydrocarbons or PAHs) and other minor constituents are the organic pollutants of major concern. These compounds reach the soil and water through negligent disposal of the untreated effluents from oil refineries, and breakdown episodes during storage and transportation of the crude oil and its allied products thereby causing pollution [28]. Biodegradability of hydrocarbons and consequently their extent of persistence in particular environment depends on many factors like chemical structure, and number and diversity of microbial populations involved in efficient degradation of the pollutant. Hydrocarbons vary in their susceptibility of attack by microbes. Microbes can easily mineralize aliphatic alkanes with preference for n-alkanes followed by branched alkanes and cyclic alkanes. Similarly, among aromatics the low molecular weight (MW) ones are preferentially attacked followed by high MW aromatics [29].

Among others, polycyclic aromatic hydrocarbons (PAHs) are a group of priority organic pollutants consisting of benzene rings in linear, angular or cluster arrangement. These are of significant concern due to their genotoxic, mutagenic and/or carcinogenic properties posing vast threat to human health by means of bioaccumulation in food-chain [30]. Phenanthrene, pyrene and naphthalene are the frequent soil contaminants. These xenobiotics are mostly derived by partial burning of wood, coal and/or petroleum at respective industrial sites. Majority PAHs are recalcitrant owing to their low bioavailability and extreme resistance to nucleophilic attack [31]. Currently, hydrocarbon fuels mainly diesel contains excess amounts of PAHs. These occur as colorless or white/pale yellow solids with lesser solubility in water, higher melting and boiling points and low vapor pressure [32].

Polycyclic aromatic hydrocarbons have been categorized into two types *i.e.* low-molecular weight (LMW) PAHs (two or three rings like naphthalene and phenanthrene) which are relatively volatile, soluble and more degradable while the high molecular weight (HMW) PAHs (four or more rings like Pyrene) sorb powerfully in soil and sediments and hence, are resistant to degradation by microbes owing to their higher molecular weight and greater hydrophobicity [33]. Degradation of PAH has been well documented in many genera of actinomycetes. Pizzul *et al.* [34] have evaluated the phenanthrene remediating potential of five strains of two actinobacterial genera, *Rhodococcus* and *Gordonia* with or without co-substrate supplementation. They observed efficient degradation of phenanthrene by both the genera only on supplementation by the co-substrate. Furthermore, Teng *et al.* [35] recorded more than 60% and near complete removal of benzopyrene and anthracene PAH compounds respectively which were amended @ 200 mg kg^{-1} in soil within two weeks of incubation by *Rhodococcus* sp.

Streptomyces strain can also degrade multiple PAH compounds. A *Streptomyces rochei* strain PAH-13 obtained from heavy crude oil refinery, India could degrade several PAHs (concentration 100 ppm) like anthracene, fluorene, phenanthrene and pyrene using 0.1% yeast extract as co-substrate within 15 days [36]. Similarly, Bourguignon *et al.* [33] evaluated fifteen different actinobacterial strains to degrade naphthalene, phenanthrene, and pyrene in both minimal salt medium and with glucose supplementation as co-substrate and recorded variable degradation potentials.

Continuing with the studies of PAH degradation Shekhar *et al.* [37], have isolated one hundred and thirty-four indigenous isolates of actinomycete from different sites contaminated with Petrol. Among these isolates, fifty-one actinomycete strains showed growth on hydrocarbon amended mineral salt medium with different concentrations and days of incubation. *Actinomyces naeslundii*, *Actinomyces viscosus*, *Actinomyces israelii*, *Actinomyces meyeri* and *Nocardia formicae* isolated from soil samples of a crude oil spill site exhibited ability to degrade crude oil. *A. viscous* and *A. israel* showed 98% degradation of 10 g L^{-1} crude oil in 12 days and 97% degradation of 30, 50, and 75 g L^{-1} in 16 and 18 days respectively [38]. Therefore, the above studies clearly demonstrate the potential of actinomycetes to grow on hydrocarbon as a substrate and thus can be efficiently utilized for degradation of hydrocarbon contaminants.

Bioremediation of Heavy Metal Contaminated Sites

Heavy metals are significant pollutants imposing toxic effects on microbial assemblages. Heavy metals are released or introduced in the eco-niches due to imprudent effluent or waste disposal by steel manufacturing, chemical processing and electroplating industries [39, 40]. Heavy metals cannot be destroyed though few strategies can render these elements less or non-bioavailable through chelation process by use of metal chelating compounds. The metal chelation can occur by using chemical and/or physical techniques or by shifting the electron valency through coupled oxidation-reduction reactions. The biological remediation strategies particularly microbial and phyto-remediation rely on the above mechanisms for removal of contaminant heavy metal elements [41].

Heavy metals like Cu and Zn are essential for the microbial biological processes, while some metals may induce the protective oxidative damage cascades to improve the survival of several groups of organisms under various stress environments. At higher concentrations, all heavy metals become toxic and redundancy in concentrations of trace metals poses harmful effect to microbes [42]. Physiological and biochemical characteristics of microorganisms can be altered with heavy metals even though microbes remain associated with plant roots. Fundamental mechanisms of toxicity by heavy metals include their role as redox catalysts leading to formation of reactive oxygen species (ROS), ceasing of functioning of vital enzymes, leading to destruction of cellular paradigms involved in regulation of ions and direct influence on DNA and protein adducts formation [43].

Heavy metals like copper may catalyze formation of ROS in Fenton and Haber-Weis reaction. Thus, heavy metals can affect almost all cellular activities and components due to their role as electron carriers [44]. Moreover, heavy metals can disrupt essential functions of enzymes through interaction with substrates by competitive or non-competitive mean, altering configuration and gene expression of enzymes and ROS formation causes generation of carbonyl groups by enzyme oxidation [43]. Ion imbalance can also be caused by heavy metals due to their ability to adhere on the cell membrane/cell wall and enter *via* the membrane ion channels, by self-diffusion or by specialized trans-membrane carrier proteins [45].

Heavy metals may facilitate the efflux of metals by stimulating the microbial antioxidant defensive system, thereby reducing metals within cells [46]. For instance, iron chelating secondary metabolites or siderophores like desferrioxamine E and B, and coelichelin may also chelate other heavy metals to maintain cellular metal homeostasis [47]. Furthermore, the heavy metals can be detoxified through the antioxidant pathways like the Glutathione system. *Actinomyces turicensis* AL36Cd exhibits high cadmium tolerance [48] while *Frankia*, nitrogen fixing actinobacteria of tree genera *Alnus* exhibits tolerance to many heavy metals, Cd, Co, Cu, Cr, Ni, and Zn [49 - 51]. Specifically, for reclamation of the heavy metal contaminated sites, if host plants possessing metal tolerant ability are planted, then population of *Frankia* survives and forms effective symbioses [50]. Likewise, two strains of extremophilic genera, *Acidimicrobium ferrooxidans* can tolerant high Zn concentrations [52].

Albarracín *et al.* [53] observed high tolerance to copper among actinobacterial isolates in qualitative assays using $CuSO_4$ (80 mg L^{-1}). The heavy metal adsorption and accumulation is the other effective mechanism of metal tolerance in actinobacteria. *Amycolatopsis tucumanensis* DSM 45259 strain removed phytoavailable Cu from Cu contaminated soil microcosms through bio-immobili-zation process. Similarly, Daboor *et al.* [54] observed high metal absorption ability of *Streptomyces chromofuscus* K101.

Baz *et al.* [55] investigated heavy metal resistance and bioaccumulation potential of fifty-nine actinomycete strains for five heavy metals. The minimum inhibitory concentration (MIC) was found to be 0.55 for Pb, 0.15 for Cr, and 0.10 mg·mL^{-1} for both Zn and Cu. Among the twenty-seven isolates evaluated, a strong potential for accumulation of Pb was exhibited by two genera, *Streptomyces* and *Amycolatopsis* respectively. Furthermore, a single actinobacterial genera may exhibit bioremediation potential for both heavy metal and organic xenobiotic. Therefore, on evaluation of chromium and lindane pesticide remediation potentials of three *Streptomyces* sp. and *Amycolatopsis tucumanensis* DSM 45259, maximum Cr(VI) removal was displayed by consortium of four strains, while maximum removal of lindane was exhibited by *Streptomyces* sp. M7 isolate [56].

Another study on heavy metal resistance by rare actinomycetes genera was conducted involving screening of forty heavy metal resistant actinobacteria [57]. Thirteen isolates exhibited resistance to multiple heavy metals *i.e.* 140 mM $ZnCl_2$, 7 mM $CuSO_4$, 9.2 mM $CdCl_2$, and 60 mM $NiCl_2$. All these strains belonged to genera *Streptomyces*, *Nonomuraea*, *Saccharothrix*, *Streptosporangium* and *Promicromonospora*. Further, 96.5% reduction in cadmium residual concentration was observed with strain *Promicromonospora* sp. UTMC 2243 [57].

Bioremediation of Xenobiotic Compounds

Actinobacteria possess considerable potential for biotransformation and biodegradation of xenobiotics/ pesticides with widely different chemical structures and chemistries like organochlorine, s-triazine, triazinone, carbamate organophosphates, acetanilide, and sulfonyl ureas. Few xenobiotics can be mineralized by single isolates, but for complete degradation, consortia have to be formulated and applied [23]. Different pesticides can be degraded by various genera of actinobacteria like *Arthrobacter* for 4-chlorophenol, atrazine and monocrotophos and *Streptomyces* sp. for alachlor degradation. Other genera include the micro-symbiotic *Frankia, Nocardia, Mycobacterium, Pseudono-cardia, Rhodococcus,* and *Janibacter*. These microorganisms display different properties with ability to degrade many pesticides, including organochlorines (OC), carbamates (CB), chloroacetanilides, organophosphorus (OP), pyrethroids and urea [9].

Degradation of xenobiotics by Arthrobacter spp.

The genus *Arthrobacter* can degrade various xenobiotics through synthesis of unique bio-catalysts of catabolic pathways encoded by plasmids for xenobiotic detoxification. Members of this genus are omnipresent owing to their versatility in nutrition and ability to tolerate several environmental stresses. Kumar *et al.* [58] documented high endosulfan degrading potential of *Arthrobacter* strain due to oxidation of parent compound(s) to its sulfate form. Likewise, De Paolis *et al.* [59] observed HCH metabolizing actinobacteria, *Arthrobacter fluorescens* and *A. giacomelloi* to metabolize alpha, beta and gamma HCH by formation of probable persistent compounds like pentachlorocyclohexenes and tetrachlorocyclohexenes which were mineralized within 72 hours of incubation. Similarly, Devers-Lamrani *et al.* [60] described the diuron-degrading potentials of *Arthrobacter* sp. BS1, BS2, and SED1. Also, Sagarkar *et al.* [61] reported atrazine and other s-triazine degradation by *Arthrobacter* sp. within 24 hours of incubation by conversion of atrazine to cyanuric acid.

Degradation of xenobiotics by Rhodococcus spp.

Genus *Rhodococcus* consists of different species which are one among the natural vanguard soil microbial populations of both contaminated and fertile environments. Moreover, members of genus *Rhodococcus* are potential bioremediation agents as these actinobaceria can survive extreme conditions, can resist nutrient deprivation or starvation and most interestingly, these bacteria can commence pollutant breakdown even in the presence of easily assimilable carbon sources. The hydrophobic characteristic of the rhodococcal cell wall due to occurrence of mycolic acid is helpful in degradation of less available pesticides.

Kolekar *et al.* [64] characterized atrazine degradation by *Rhodococcus* sp. BCH2, demonstrating its potential to utilize pesticide as source of carbon and nitrogen. Likewise, Verma *et al.* [62] observed the unique endosulfan degrading potentials of *Rhodococcus* MTCC 6716 strain (approximately 93% degradation within 15 days) without production of the toxic endosulfan sulfate intermediate. Later, Verma *et al.* [63] deciphered the endosulfan concentration dependent expression of the genes coding for the enzymes or biocatalytic proteins for the endosulfan degradation in *Rhodococcus*.

Degradation of xenobiotics by Streptomyces spp.

Members of the genus *Streptomyces* have received significant attention as potential bioremediation agents for contaminated environments. Various

Streptomyces strains can grow in high concentration of pollutant compound and possess ability to degrade diverse pesticides spanning over triazines, organo-chlorines, organo-phosphates, pyretheroids and others.

Streptomyces enabled degradation of organochlorine pesticides (OCPs) has been widely probed by several researchers. In this regard, Benimeli *et al.* [65] screened the ability of ninety-three actinobacterial strains against eleven OCPs. Four strains were selected based upon multiple tolerance to pesticides and identified as *Streptomyces* sp. The M7 strain utilized lindane (γ-hexachlorocyclohexane) as the sole C-source in minimal salt medium and released the maximum chloride ions indicating the highest degradation potential and γ-HCH removal within 24 hours of incubation without intracellular accumulation of lindane [66]. Glucose supplementation (0.6 gL^{-1}) in lindane @ 100 μgL^{-1} minimal medium did not curbed lindane removal by *Streptomyces* M7 as the former remained the preferred C-source till first 24 hours followed by lindane indicated by the diauxic growth curve. Furthermore, in the soil microcosm studies this strain exhibited substantial percent γ-HCH removal ranging from 14 to 78% [67]. In this study the effect of M7 inoculation on growth of maize plants in lindane contaminated soils was also evaluated at initial and final stages. A decrease in residual concentration of lindane was observed in different pesticide concentrations. The optimum inoculum of *Streptomyces* sp. M7 for most effective bioremediation was 2 g kg^{-1} soil with 56% removal of lindane. Alternatively, germination and vigor index of maize plants was affected with soil containing 100 μg kg^{-1} lindane concentration whereas improved vigor index and 68% lindane removal was obtained when soil was inoculated with *Streptomyces* sp. M7 [67]. Benimeli *et al.* [68] have also revealed good growth of this strain in lindane as the sole carbon source.

Successively, Cuozzo *et al.* [69] reported induction of dechlorinase enzyme of *Streptomyces* sp. M7 in response to γ-HCH as sole C-source. In another report, nine strains exhibited biodegradation of one or more OCs as evidenced from release of Cl$^-$ ions [70]. Chen *et al.* [71] identified a deltamethrin degrading strain as *S. aureus* HP-S-01 which exhibited detoxification of hydrolysis product by oxidization of 3-phenoxybenzaldehyde to 2-hydroxy-4-methoxy benzophenone. This strain was capable for degradation of other pyrethroids like bifenthrin, cyfluthrin and cypermethrin.

Degradation of xenobiotics by other genera of actinobacteria

Saccharomonospora genus is known for possessing proficient enzymatic machinery for the hydrolysis of several phenolic compounds [18] and thus can also act on organochlorine compounds like pentachlorophenol. Likewise, various species of the genus *Janibacter* can degrade several aromatic and chlorinated compounds including the mono and polychlorinated biphenyls (PCB), PCP and PAHs [72]. The extra advantage of the genus *Janibacter* is its ability to inhabit salt affected soils. Therefore, being a halotolerant, it can also exhibit degradation of high amounts of PCP in contaminated soils (up to 300 mg L^{-1}). The PCP degradation activity of *Janibacter* can be further improved by addition of a surfactant like Tween 80 which improves the low solubility of polychlorinated phenol.

The members of genus *Gordonia* can degrade recalcitrant pollutants as these microbes can grow in variety of ecological niches and can survive under unfavorable environments. The insecticide chlorpyrifos (CPF) degrading strain of *Gordonia* exhibited rapid degradation potentials [73] with complete degradation of even the water soluble intermediate of CPF *i.e.* 3,5,6-trichloro-2-pyridinol (TCP) after 72 hours of incubation.

Bioremediation of other recalcitrants

Many plastic degrading actinobacterial species belonging to the genera *Actinomadura* and *Microbispora* besides the above discussed *Streptomyces* and *Saccharomonospora* strains have been isolated. These actinobacteria can act on and degrade several biodegradable polyesters like polyethylene succinate, poly-caprolactone, poly-hydroxybutyrate, polytetramethylene succinate, poly-L-lactide and terephthalic acid, thereby reducing environmental impacts [74].

The dye degrading actinobacteria have also been reported which were isolated from different agri-farms and waste disposal lands [75]. These actinobacteria can degrade several dyes under aerobic growth conditions and the best degrading strain B2 exhibited 99% similarity with *Streptomyces pactum*.

Future Prospects

Bioremediation by microorganisms through functional genomics and proteomics

Bioremediation through microorganisms is an environment friendly, safe, and effective technology. However, partial or incomplete information on emphatic cellular responses among diverse microbial communities confines the improvement in the mineralization of the pollutant [76]. Therefore, application of functional genomics and proteomics in bioremediation research offers a universal insight into microbial protein composition in inoculated test microbe in response to the target pollutant and its relative bioavailable concentrations [77]. Hence, it will help unravel the functional molecular aspects of microbial remediation offering a deeper insight into the comprehensive microbial metabolic and gene regulatory networks enhancing our understanding of the intricate interdependencies among the genes [77]. The context dependent predominant proteome expression for appropriate growth conditions will help develop rational strategies for maintenance of effective bioremediation. The functional genomics and proteomics studies of the pollutant degrading microbes will also unravel novel genes/proteins involved in pollutant detoxification and biodegradation [76]. This understanding would empower us to edit the requisite genes, proteins and even the metabolic pathways allowing to possibly boost the degradative capabilities of the microbe(s) or may also help in pyrimading or pooling all the downstream structural genes coding for complete mineralization of the pollutant in a single microbial genome [77].

Nanobioremediation (NBR): Actinomycetes derived Nanoparticles for Bioremediation of Toxic Pollutants

Nanobioremediation (NBR) is an emerging strategy aiming at environmental cleanup by removal of pollutants from contaminated niches through nanomaterials synthesized by green/ biological techniques from biomacromolecules in cell free extracts of microbes and/or plants [78 - 81]. The conventional pollutant remediation techniques primarily involve use of chemical and physical remediation strategies, incineration and meager use of bioremediation in specialized niches. These techniques mostly involve *ex situ* remediation which increases the cost [79]. The nano-enabled remediation protocols can be flexible *i.e.* both *in situ* and *ex situ* operations can be fairly dealt with. Moreover, the fabrication of new nano-scale products will aptly replace the existing production processes, through better performance, resulting in environmental and cost savings [79, 81]. The advantage of this technology is its ability to upsurge the efficiency of the conventional microbial and phytoremediation methods such that land soils and even surface and sub-surface water contaminated with diverse pollutants like heavy metals, organic and inorganic compounds can be dealt with and effectively remediated [79]. It may involve the application of different biologically synthesized nanoparticles and other nanomaterials comprised of Au, Ag, TiO_2, Carbon nanomaterials and magnetic NPs like SPIONs which can directly adsorb or immobilize the test contaminant from diverse matrices [79]. The actinobacteria particularly the extremophiles have a special mention as these are unique owing to their physiological and metabolic versatility and have been reported to synthesize various nanomaterials [81]. Meanwhile, another articulation could be immobilization of the potential microbial biocatalysts or enzymes on nanomaterials (nano-immobilization) to enhance the multiplicity of protection, improved functioning and collection for re-use purposes after remediation [79]. The plant mediated remediation of the pollutants can also be improved by better root growth, secretion of secondary metabolites and accumulation of the pollutants when these plants are challenged or treated with nanomaterials [78]. Therefore, the adaptations and alterations of NBR strategies in future will prolong the extent and quality of remediation of multiple contaminated from a single niche. However, the nanomaterial dosages for effective removal of the contaminants with no residual nanotoxic non-target ill-effects on the beneficial soil microbiota have to be screened and tested before the wide spread commercial adaptation of this technology [79, 82].

SUMMARY

Environmental pollutants are a major concern worldwide. Any pollutant finds its way to land, water and atmosphere components of the biosphere with the former two serving as the prime sinks. This leads to cascade of events resulting in alteration of the contaminated niche and thus, affecting its biotic and abiotic components. The soil and aquatic microbial communities have a crucial role for proper functioning of the respective ecosystem. Moreover, these forms are diverse and exhibit complex interdependencies among them which get perturbed in the presence of a contaminant(s). Among these diverse microbial communities, actinomycetes have occupied a prominent role as potential producers of structurally complex and unique metabolites. These bacteria can remove diverse pollutants from environment and action-based remediation strategies can be utilized for remediation of normal to extreme environments. The genetic and metabolic tweaking of the extremophilic actinobacteria can further the remediation potentials for multiple pollutants. Hence, novel chimeric approaches like nanobioremediation should be looked for to improve the ecological and economic benefits of remediation through actinobacterial agents.

CONSENT FOR PUBLICATION

Not applicable.

CONFLICT OF INTEREST

The authors declare no conflict of interest, financial or otherwise.

ACKNOWLEDGEMENTS

The authors wish to thank, the Head, Department of Microbiology and Department of Soil Science, Punjab Agricultural University, Ludhiana, Punjab for providing the necessary infrastructural facilities to carry out the research work on actinomycetes.

REFERENCES

[1]   Sabale SR. Contamination and need of bioremediation of pesticide residues in fresh water aquifers. J Bioremediat Biodegrad 2014; 5: e158.
[http://dx.doi.org/10.4172/2155-6199.1000e158]

[2]   Aktar MW, Sengupta D, Chowdhury A. Impact of pesticides use in agriculture: their benefits and hazards. Interdiscip Toxicol 2009; 2(1): 1-12.
[http://dx.doi.org/10.2478/v10102-009-0001-7] [PMID: 21217838]

[3]   Ramakrishnan B, Megharaj M, Venkateswarlu K, *et al.* The impacts of environmental pollutants on microalgae and cyanobacteria. Crit Rev Environ Sci Technol 2010; 40(8): 699-821.
[http://dx.doi.org/10.1080/10643380802471068]

[4]   Megharaj M, Ramakrishnan B, Venkateswarlu K, Sethunathan N, Naidu R. Bioremediation approaches for organic pollutants: a critical perspective. Environ Int 2011; 37(8): 1362-75.
[http://dx.doi.org/10.1016/j.envint.2011.06.003] [PMID: 21722961]

[5]   Gunatilake SK. Methods of removing heavy metals from industrial wastewater. J Multidisciplinary Engineering Sci Stud 2015; 1(1): 13-8.

[6]   Dadrasnia A, Shahsavari N, Emenike CU. Remediation of Contaminated Sites. Hydrocarbon. Intech 2013; pp. 65-82.
[http://dx.doi.org/10.5772/51591]

[7]   Singh R, Singh P, Sharma R. Microorganism as a tool of bioremediation technology for cleaning environment: A review. Proc Int Acad Ecol Environ Sci 2014; 4: 1-6.

[8]   Ventura M, Canchaya C, Tauch A, *et al.* Genomics of Actinobacteria: tracing the evolutionary history of an ancient phylum. Microbiol Mol Biol Rev 2007; 71(3): 495-548.
[http://dx.doi.org/10.1128/MMBR.00005-07] [PMID: 17804669]

[9]   Alvarez A, Saez JM, Davila Costa JS, *et al.* Actinobacteria: Current research and perspectives for bioremediation of pesticides and heavy metals. Chemosphere 2017; 166: 41-62.
[http://dx.doi.org/10.1016/j.chemosphere.2016.09.070] [PMID: 27684437]

[10]   Kügler JH, Le Roes-Hill M, Syldatk C, Hausmann R. Surfactants tailored by the class Actinobacteria. Front Microbiol 2015; 6: 212.
[PMID: 25852670]

[11]   Goodfellow M, Kampfer P, Busse HJ, *et al.* Bergey's Manual of Systematic Bacteriology. New York: Springer 2012.
[http://dx.doi.org/10.1007/978-0-387-68233-4]

[12]   Ghai R, McMahon KD, Rodriguez-Valera F. Breaking a paradigm: cosmopolitan and abundant freshwater actinobacteria are low GC. Environ Microbiol Rep 2012; 4(1): 29-35.
[http://dx.doi.org/10.1111/j.1758-2229.2011.00274.x] [PMID: 23757226]

[13]   Olano C, Méndez C, Salas JA. Antitumor compounds from marine actinomycetes. Mar Drugs 2009; 7(2): 210-48.
[http://dx.doi.org/10.3390/md7020210] [PMID: 19597582]

[14]   Manivasagan P, Venkatesan J, Sivakumar K, Kim SK. RETRACTED: Marine actinobacterial metabolites: current status and future perspectives. Microbiol Res 2013; 168(6): 311-32.
[http://dx.doi.org/10.1016/j.micres.2013.02.002] [PMID: 23480961]

[15]   Zenova GM, Manucharova NA, Zvyagintsev DG. Extremophilic and extremotolerant actinomycetes in different soil types. Eurasian Soil Sci 2011; 44: 417-36.
[http://dx.doi.org/10.1134/S1064229311040132]

[16]   Bull AT. Actinobacteria of the extremobiosphere.Extremophiles. Berlin: Springer-Verlag 2010; pp. 1203-40.

[17]   Gupta GN, Srivastava S, Khare SK, *et al.* Extremophiles: an overview of microorganism from extreme environment. Int J Agric Environ Biotechnol 2014; 7: 371-80.
[http://dx.doi.org/10.5958/2230-732X.2014.00258.7]

[18]   Shivlata L, Satyanarayana T. Thermophilic and alkaliphilic Actinobacteria: biology and potential applications. Front Microbiol 2015; 6: 1014.
[http://dx.doi.org/10.3389/fmicb.2015.01014] [PMID: 26441937]

[19]   King GM, Weber CF. Distribution, diversity and ecology of aerobic CO-oxidizing bacteria. Nat Rev Microbiol 2007; 5(2): 107-18.
[http://dx.doi.org/10.1038/nrmicro1595] [PMID: 17224920]

[20]   Norris PR, Davis-Belmar CS, Brown CF, Calvo-Bado LA. Autotrophic, sulfur-oxidizing actinobacteria in acidic environments. Extremophiles 2011; 15(2): 155-63.
[http://dx.doi.org/10.1007/s00792-011-0358-3] [PMID: 21308384]

[21]   Kumar L, Awasthi G, Singh B. Extremophiles: a novel source of industrially important enzymes. Biotechnol 2011; 10: 121-35.
[http://dx.doi.org/10.3923/biotech.2011.121.135]

[22]   Wiegel J, Kevbrin VV. Alkalithermophiles. Biochem Soc Trans 2004; 32(Pt 2): 193-8.
[http://dx.doi.org/10.1042/bst0320193] [PMID: 15046570]

[23]   Selvam K, Vishnupriya B. Bioremediation of xenobiotic compound and heavy metals by the novel marine actinobacteria. Int J Pharm Chem Sci 2013; 2: 1589-97.

[24]   Hozzein WN, Abmed MB, Tawab MSA. Efficiency of some actinomycetes isolated in biological treatment and removal of heavy metals from waste water. Afr J Biotechnol 2012; 11: 1163-8.

[25]   Kulkarni P, Bee H. Marine-bacteria Actinomycetes: A case study for their potential in bioremediation. Biotechnol Res J 2015; 1: 171-4.

[26]   Nakano M, Kihara M, Iehata S, Tanaka R, Maeda H, Yoshikawa T. Wax ester-like compounds as biosurfactants produced by Dietzia maris from n-alkane as a sole carbon source. J Basic Microbiol 2011; 51(5): 490-8.
[http://dx.doi.org/10.1002/jobm.201000420] [PMID: 21656811]

[27]   Al-Mailem DM, Eliyas M, Khanafer M, Radwan SS. Biofilms constructed for the removal of hydrocarbon pollutants from hypersaline liquids. Extremophiles 2015; 19(1): 189-96.
[http://dx.doi.org/10.1007/s00792-014-0698-x] [PMID: 25293792]

[28]   Nogales B, Lanfranconi MP, Piña-Villalonga JM, Bosch R. Anthropogenic perturbations in marine microbial communities. FEMS Microbiol Rev 2011; 35(2): 275-98.
[http://dx.doi.org/10.1111/j.1574-6976.2010.00248.x] [PMID: 20738403]

[29]   Chikere CB, Okpokwasili GC, Chikere BO. Monitoring of microbial hydrocarbon remediation in the soil. 3 Biotech 2011; 1(3): 117-38.
[http://dx.doi.org/10.1007/s13205-011-0014-8] [PMID: 22611524]

[30]   Moscoso F, Teijiz I, Deive FJ, Sanromán MA. Efficient PAHs biodegradation by a bacterial consortium at flask and bioreactor scale. Bioresour Technol 2012; 119: 270-6.
[http://dx.doi.org/10.1016/j.biortech.2012.05.095] [PMID: 22738812]

[31]   Cheng M, Zeng G, Huang D, *et al.* Hydroxyl radicals based advanced oxidation processes (AOPs) for remediation of soils contaminated with organic compounds: a review. Chem Eng J 2016; 284: 582-98.
[http://dx.doi.org/10.1016/j.cej.2015.09.001]

[32]   Bisht S, Pandey P, Bhargava B, Sharma S, Kumar V, Sharma KD. Bioremediation of polyaromatic hydrocarbons (PAHs) using rhizosphere technology. Braz J Microbiol 2015; 46(1): 7-21.
[http://dx.doi.org/10.1590/S1517-838246120131354] [PMID: 26221084]

[33]   Bourguignon N, Isaac P, Alvarez H, Amoroso MJ, Ferrero MA. Enhanced polyaromatic hydrocarbon degradation by adapted cultures of actinomycete strains. J Basic Microbiol 2014; 54(12): 1288-94.

[http://dx.doi.org/10.1002/jobm.201400262] [PMID: 25205070]

[34]   Pizzul L, Castillo MDP, Stenström J. Characterization of selected actinomycetes degrading polyaromatic hydrocarbons in liquid culture and spiked soil. World J Microbiol Biotechnol 2006; 22: 745-52.
[http://dx.doi.org/10.1007/s11274-005-9100-6]

[35]   Teng Y, Luo Y, Ping L, Zou D, Li Z, Christie P. Effects of soil amendment with different carbon sources and other factors on the bioremediation of an aged PAH-contaminated soil. Biodegradation 2010; 21(2): 167-78.
[http://dx.doi.org/10.1007/s10532-009-9291-x] [PMID: 19707880]

[36]   Chaudhary P, Sharma R, Singh SB, Nain L. Bioremediation of PAH by *Streptomyces* sp. Bull Environ Contam Toxicol 2011; 86(3): 268-71.
[http://dx.doi.org/10.1007/s00128-011-0211-5] [PMID: 21301805]

[37]   Shekhar SK, Godheja J, Modi DR, *et al.* Growth potential assessment of actinomycetes isolated from petroleum contaminated soil. J Bioremediat Biodegrad 2014; 5: 1-8.

[38]   Olajuyigbe FM, Ehiosun KI. Assessment of crude oil degradation efficiency of newly isolated actinobacteria reveals untapped bioremediation potentials. Bioremediat J 2016; 20: 133-43.
[http://dx.doi.org/10.1080/10889868.2015.1113926]

[39]   Chaturvedi AD, Pal D, Penta S, Kumar A. Ecotoxic heavy metals transformation by bacteria and fungi in aquatic ecosystem. World J Microbiol Biotechnol 2015; 31(10): 1595-603.
[http://dx.doi.org/10.1007/s11274-015-1911-5] [PMID: 26250544]

[40]   Zhou Y, Tang L, Zeng G, *et al.* Current progress in biosensors for heavy metal ions based on DNAzymes/DNA molecules functionalized nanostructures: a review. Sens Actuators B Chem 2016; 223: 280-94.
[http://dx.doi.org/10.1016/j.snb.2015.09.090]

[41]   Liu SH, Zeng GM, Niu Q, *et al.* Bioremediation mechanisms of combined pollution of PAHs and heavy metals by bacteria and fungi: A mini review. Bioresour Technol 2016.
[http://dx.doi.org/10.1016/j.biortech.2016.11.095] [PMID: 27916498]

[42]   Xu P, Zeng GM, Huang DL, *et al.* Use of iron oxide nanomaterials in wastewater treatment: a review. Sci Total Environ 2012; 424: 1-10.
[http://dx.doi.org/10.1016/j.scitotenv.2012.02.023] [PMID: 22391097]

[43]   Gauthier PT, Norwood WP, Prepas EE, Pyle GG. Metal-PAH mixtures in the aquatic environment: a review of co-toxic mechanisms leading to more-than-additive outcomes. Aquat Toxicol 2014; 154: 253-69.
[http://dx.doi.org/10.1016/j.aquatox.2014.05.026] [PMID: 24929353]

[44]   Giner-Lamia J, López-Maury L, Florencio FJ. Global transcriptional profiles of the copper responses in the cyanobacterium *Synechocystis* sp. PCC 6803. PLoS One 2014; 9(9): e108912.
[http://dx.doi.org/10.1371/journal.pone.0108912] [PMID: 25268225]

[45]   Chen S, Yin H, Ye J, *et al.* Influence of co-existed benzo[a]pyrene and copper on the cellular characteristics of *Stenotrophomonas maltophilia* during biodegradation and transformation. Bioresour Technol 2014; 158: 181-7.
[http://dx.doi.org/10.1016/j.biortech.2014.02.020] [PMID: 24603491]

[46]   Zeng GM, Chen AW, Chen GQ, *et al.* Responses of *Phanerochaete chrysosporium* to toxic pollutants: physiological flux, oxidative stress, and detoxification. Environ Sci Technol 2012; 46(14): 7818-25.
[http://dx.doi.org/10.1021/es301006j] [PMID: 22703191]

[47]   Złoch M, Thiem D, Gadzała-Kopciuch R, Hrynkiewicz K. Synthesis of siderophores by plant-associated metallotolerant bacteria under exposure to Cd(2.). Chemosphere 2016; 156: 312-25.
[http://dx.doi.org/10.1016/j.chemosphere.2016.04.130] [PMID: 27183333]

[48]   Oyetibo GO, Ilori MO, Adebusoye SA, Obayori OS, Amund OO. Bacteria with dual resistance to

elevated concentrations of heavy metals and antibiotics in Nigerian contaminated systems. Environ Monit Assess 2010; 168(1-4): 305-14.
[http://dx.doi.org/10.1007/s10661-009-1114-3] [PMID: 19688604]

[49]   Bélanger PA, Beaudin J, Roy S. High-throughput screening of microbial adaptation to environmental stress. J Microbiol Methods 2011; 85(2): 92-7.
[http://dx.doi.org/10.1016/j.mimet.2011.01.028] [PMID: 21315114]

[50]   Wheeler CT, Hughes LT, Oldroyd J, *et al.* Effects of nickel on *Frankia* and its symbiosis with *Alnus glutinosa* (L.) Gaertn. Plant Soil 2001; 231: 81-90.
[http://dx.doi.org/10.1023/A:1010304614992]

[51]   Rehan M, Kluge M, Fränzle S, Kellner H, Ullrich R, Hofrichter M. Degradation of atrazine by *Frankia* alni ACN14a: gene regulation, dealkylation, and dechlorination. Appl Microbiol Biotechnol 2014; 98(13): 6125-35.
[http://dx.doi.org/10.1007/s00253-014-5665-z] [PMID: 24676750]

[52]   Mangold S, Potrykus J, Björn E, Lövgren L, Dopson M. Extreme zinc tolerance in acidophilic microorganisms from the bacterial and archaeal domains. Extremophiles 2013; 17(1): 75-85.
[http://dx.doi.org/10.1007/s00792-012-0495-3] [PMID: 23143658]

[53]   Albarracín VH, Amoroso MJ, Abate CM. Bioaugmentation of copper polluted soil microcosms with *Amycolatopsis tucumanensis* to diminish phytoavailable copper for *Zea mays* plants. Chemosphere 2010; 79(2): 131-7.
[http://dx.doi.org/10.1016/j.chemosphere.2010.01.038] [PMID: 20163821]

[54]   Daboor SM, Haroon AM, Esmael NAE, *et al.* Heavy metal adsorption of *Streptomyces chromofuscus* K101. J Coast Life Med 2014; 2: 431-7.

[55]   Baz SE, Baz M, Barakate M, *et al.* Resistance to and accumulation of heavy metals by actinobacteria isolated from abandoned mining areas. Sci World 2014.
[http://dx.doi.org/10.1155/2015/761834]

[56]   Aparicio JD. SimónSolá MZ, Atjián MC, *et al.* Co-contaminated soils bioremediation by actinobacteria. In: Bioremediation S, America L, Eds. Alvarez A, Polti M. 2014; pp. 79-91.

[57]   Hamedi J, Dehhaghi M, Mohammdipanah F. Isolation of extremely heavy metal resistant strains of rare actinomycetes from high metal content soils in Iran. Int J Environ Res 2015; 9: 475-80.

[58]   Kumar M, Lakshmi CV, Khanna S. Biodegradation and bioremediation of endosulfan contaminated soil. Bioresour Technol 2008; 99(8): 3116-22.
[http://dx.doi.org/10.1016/j.biortech.2007.05.057] [PMID: 17646098]

[59]   De Paolis MR, Lippi D, Guerriero E, Polcaro CM, Donati E. Biodegradation of α-, β-, and γ-hexachlorocyclohexane by Arthrobacter fluorescens and Arthrobacter giacomelloi. Appl Biochem Biotechnol 2013; 170(3): 514-24.
[http://dx.doi.org/10.1007/s12010-013-0147-9] [PMID: 23553101]

[60]   Devers-Lamrani M, Pesce S, Rouard N, Martin-Laurent F. Evidence for cooperative mineralization of diuron by *Arthrobacter* sp. BS2 and *Achromobacter* sp. SP1 isolated from a mixed culture enriched from diuron exposed environments. Chemosphere 2014; 117: 208-15.
[http://dx.doi.org/10.1016/j.chemosphere.2014.06.080] [PMID: 25061887]

[61]   Sagarkar Bhardwaj S. s-triazine degrading bacterial isolate *Arthrobacter* sp. AK-YN10, a candidate for bioaugmentation of atrazine contaminated soil. Appl Microbiol Biotechnol 2015; 99
[http://dx.doi.org/10.1007/s00253-015-6975-5] [PMID: 26403923]

[62]   Verma K, Agrawal N, Farooq M, Misra RB, Hans RK. Endosulfan degradation by a *Rhodococcus* strain isolated from earthworm gut. Ecotoxicol Environ Saf 2006; 64(3): 377-81.
[http://dx.doi.org/10.1016/j.ecoenv.2005.05.014] [PMID: 16029891]

[63]   Verma A, Ali D, Farooq M, Pant AB, Ray RS, Hans RK. Expression and inducibility of endosulfan metabolizing gene in *Rhodococcus* strain isolated from earthworm gut microflora for its application in

bioremediation. Bioresour Technol 2011; 102(3): 2979-84.
[http://dx.doi.org/10.1016/j.biortech.2010.10.005] [PMID: 21035330]

[64]   Kolekar PD, Phugare SS, Jadhav JP. Biodegradation of atrazine by *Rhodococcus* sp. BCH2 to N-isopropylammelide with subsequent assessment of toxicity of biodegraded metabolites. Environ Sci Pollut Res Int 2014; 21(3): 2334-45.
[http://dx.doi.org/10.1007/s11356-013-2151-6] [PMID: 24062064]

[65]   Benimeli CS, Amoroso MJ, Chaile AP, Castro GR. Isolation of four aquatic streptomycetes strains capable of growth on organochlorine pesticides. Bioresour Technol 2003; 89(2): 133-8.
[http://dx.doi.org/10.1016/S0960-8524(03)00061-0] [PMID: 12699931]

[66]   Benimeli CS, Castro GR, Chaile AP, Amoroso MJ. Lindane removal induction by *Streptomyces* sp. M7. J Basic Microbiol 2006; 46(5): 348-57.
[http://dx.doi.org/10.1002/jobm.200510131] [PMID: 17009290]

[67]   Benimeli CS, Fuentes MS, Abate CM, *et al.* Bioremediation of lindane contaminated soil by *Streptomyces* sp. M7 and its effects on *Zea mays* growth. Int Biodet Biodegrad 2008; 61: 233-9.
[http://dx.doi.org/10.1016/j.ibiod.2007.09.001]

[68]   Benimeli CS, Castro GR. Lindane uptake and degradation by aquatic *Streptomyces* sp. strain M7. Int Biodeterior Biodegradation 2007; 59: 148-55.
[http://dx.doi.org/10.1016/j.ibiod.2006.07.014]

[69]   Cuozzo SA, Rollan GG, Abate CM, *et al.* Specific dechlorinase activity in lindane degradation by *Streptomyces* sp. M7. World J Microbiol Biotechnol 2009; 25: 1539-46.
[http://dx.doi.org/10.1007/s11274-009-0039-x]

[70]   Fuentes MS, Benimeli CS, Cuozzo SA, *et al.* Isolation of pesticide degrading actinomycetes from a contaminated site: bacterial growth, removal and dechlorination of organochlorine pesticides. Int Biodet Biodegrad 2010; 64: 434-41.
[http://dx.doi.org/10.1016/j.ibiod.2010.05.001]

[71]   Chen S, Lai K, Li Y, Hu M, Zhang Y, Zeng Y. Biodegradation of deltamethrin and its hydrolysis product 3-phenoxybenzaldehyde by a newly isolated *Streptomyces aureus* strain HP-S-01. Appl Microbiol Biotechnol 2011; 90(4): 1471-83.
[http://dx.doi.org/10.1007/s00253-011-3136-3] [PMID: 21327411]

[72]   Khessairi A, Fhoula I, Jaouani A, *et al.* Pentachlorophenol degradation by *Janibacter* sp., a new actinobacterium isolated from saline sediment of arid land. BioMed Res Int 2014; 2014: 296472.
[http://dx.doi.org/10.1155/2014/296472] [PMID: 25313357]

[73]   Abraham J, Shanker A, Silambarasan S. Role of Gordonia sp JAAS1 in biodegradation of chlorpyrifos and its hydrolysing metabolite 3,5,6-trichloro-2-pyridinol. Lett Appl Microbiol 2013; 57(6): 510-6.
[http://dx.doi.org/10.1111/lam.12141] [PMID: 23909785]

[74]   Tseng M, Hoang KC, Yang MK, Yang SF, Chu WS. Polyester-degrading thermophilic actinomycetes isolated from different environment in Taiwan. Biodegradation 2007; 18(5): 579-83.
[http://dx.doi.org/10.1007/s10532-006-9089-z] [PMID: 17653512]

[75]   Abraham J, Sekhar A, Singh N, *et al.* Evaluation of dye degradation using *Streptomyces pactum* strain JAAS1. Res J Pharm Biol Chem Sci 2016; 7: 2691-700.

[76]   Zhao B, Poh CL. Insights into environmental bioremediation by microorganisms through functional genomics and proteomics. Proteomics 2008; 8(4): 874-81.
[http://dx.doi.org/10.1002/pmic.200701005] [PMID: 18210372]

[77]   Hivrale AU, Pawar PK, Rane NR, *et al.* Application of Genomics and Proteomics in Bioremediation.Toxicity and Waste Management using Bioremediation. Hershey, PA, USA: IGI Global 2016; pp. 97-112.
[http://dx.doi.org/10.4018/978-1-4666-9734-8.ch005]

[78]   Yadav KK, Singh JK, Gupta N, *et al.* A review of nanobioremediation technologies for environmental

cleanup: A novel biological approach. J Mat Environ Sci 2017; 8: 740-57.

[79]   Cecchin I, Reddy KR, Thome A, *et al.* Nanobioremediation: Integration of nanoparticles and bioremediation for sustainable remediation of chlorinated organic contaminants in soils. Intl Biodeter Biodegra 2017; 119: 419-28.
[http://dx.doi.org/10.1016/j.ibiod.2016.09.027]

[80]   Ahmad A, Senapati S, Khan MI, *et al.* Extracellular biosynthesis of monodisperse gold nanoparticles by a novel extremophilic actinomycete, *Thermomonospora* sp. Langmuir 2003; 19: 3550-3.
[http://dx.doi.org/10.1021/la026772l]

[81]   Manivasagan P, Venkatesan J, Sivakumar K, Kim SK. Actinobacteria mediated synthesis of nanoparticles and their biological properties: A review. Crit Rev Microbiol 2016; 42(2): 209-21.
[http://dx.doi.org/10.3109/1040841X.2014.917069] [PMID: 25430521]

[82]   Rizwan M, Sing M, Mitra CK, *et al.* Ecofriendly application of nanomaterials: Nanobioremediation. J Nanoparticles 2014. Article ID 431787.

CHAPTER 8

Role of Phytoconstituents in Modulating Xenobiotic Metabolism

Harsimran Kaur[1], Sandeep Kaur[2], Satwinderjeet Kaur[2] and **Paramjeet Kaur[3,*]**

[1] *Crop Protection Division, Department of Agriculture, Khalsa College, Amritsar, India*

[2] *Department of Botanical and Environmental Sciences, Guru Nanak Dev University, Amritsar, India*

[3] *Department of Botany, Khalsa College, Amritsar, India*

Abstract: Contaminant-induced harmful effects are a global concern. Liver is the main site of xenobiotic metabolism and plays a vital role in averting accumulation of a wide range of compounds by converting them into a form suitable for elimination. Phase I drug metabolizing enzymes, primarily cytochrome P450s carry out bioactivation of carcinogens, thus converting them into electrophilic species which are genotoxic and cytotoxic. These reactive intermediates form protein adducts and induce DNA and RNA damage. Phase II drug metabolizing enzymes, such as glutathione-S-transferases, UDP-glucuronosyl transferases, sulfotransferases and *N*-acetyltransferases detoxify the reactive electrophilic species by conjugating these hydrophobic intermediates to a water-soluble group, thus masking their reactive nature and allowing subsequent excretion. Beneficial effects of natural dietary compounds in detoxification and elimination have been demonstrated in various studies. They have been reported to be effective in inhibiting chemically-induced carcinogenesis. Phytochemicals are known to influence the biotransformation of xenobiotics and may play an important role in reducing their toxicity and carcinogenicity. Indoles, isothiocyanates, allium organosulfur compounds, flavonoids, phenolic acids, terpenoids and psoralens may alter the levels of Phase I and Phase II drug metabolizing enzymes by affecting the transcriptional rates of their genes, the turnover rates of specific mRNAs or enzymes or the enzyme activity by inhibitory or stimulatory actions. Agents that preferentially activate Phase II over Phase I enzymes are considered as promising chemopreventives. Phase I metabolism involves oxidation, reduction or hydrolysis reactions *via* cytochrome P450 enzymes and lead to the conversion of drugs to more polar (water soluble) active metabolites by unmasking or inserting a polar functional group such as -OH, -SH and $-NH_2$. Xenobiotics metabolized *via* Phase I reactions have longer half-lives. Phase II metabolism involves conjugation reactions such as glucuronidation, acetylation and sulfation. Conjugation reactions increase water solubility of drug by adding a polar moiety thus converting them into water soluble inactive metabolites. The present book chapter discusses the interaction between phytochemicals and

* **Corresponding author Paramjeet Kaur:** Department of Botany, Khalsa College, Amritsar-143005, (Punjab) India; Tel: 9478155496; E-mails: paramjeetbot@gmail.com, paramjeetkca@gmail.com

Ashita Sharma, Manish Kumar, Satwinderjeet Kaur & Avinash Kaur Nagpal (Eds.)

xenobiotic metabolizing enzymes in detoxification of harmful exogenous compounds which have been implicated in carcinogenesis.

Keywords: Aflatoxin B1, Aldoketoreductase, Benzo[a]pyrene, Carbon Tetrachloride, Cytochrome P450 Monooxygenase, 7,12-dimethylbenz (a) anthracene,1,2-dimethyl hydrazine, Epoxide Hydrolase, Flavin Containing monooxygenase, Gamma Glutamyl Transferase, Glucose-6-phosphate de-hydrogenase, Glutathione Peroxidase, Glutathione Reductase, Glutathione-S-transferase, Hamster Buccal Pouch, Hepatocarcinogenesis, Malondialdehyde, Nuclear Factor-like 2/Antioxidant Response Element, Phase I & II, Polycyclic Aromatic Hydrocarbon, Xenobiotics.

INTRODUCTION

All organisms are unavoidably exposed to both natural and manmade xenobiotics such as plant alkaloids, toxins from microbes, drugs, pollutants, industrial chemicals and pesticides. Once a xenobiotic is absorbed and distributed in the body, it undergoes different fates. Hydrophilic compounds remain unchanged and are eliminated through urine while lipophilic compounds are excreted unchanged in faeces [1, 2]. A few xenobiotics are retained by the body while some organics and inorganics get stored in fats and bones over a long period of time. However, most of the xenobiotics undergo biotransformation in the presence of xenobiotic metabolizing enzymes (XMEs) resulting in their bioactivation or detoxification [3]. These XMEs include Phase I and Phase II enzymes present not only in the liver but also in extrahepatic organs like the lungs, intestines, colon and skin. Chemical reactions carried out by these enzymes (oxidation, hydrolysis, reduction, condensation and conjugation) add polar or reactive groups to the exogenous parent chemical thus facilitating its excretion. The reactions of Phase I enzymes basically prepare the xenobiotics for Phase II enzyme reactions *i.e.* Phase I enzymes "functionalize" the xenobiotics by uncovering or producing a reactive chemical group upon which Phase II enzymes can act [4 - 6]. Phase II enzyme reactions "conjugate" the transformed xenobiotic metabolites resulting in the formation of inactive excretory products. Lately, Phase III metabolism has been proposed that recognizes the function of membrane transporters in eliminating these conjugated end products of xenobiotics from the body *via* bile or urine [7].

Different factors modify xenobiotic metabolism in humans including induction of XMEs, inhibition of XMEs activity by environmental and pharmacological agents, age, gender, altered hormonal status and diet [8 - 10]. Also, existence of distinct genetic polymorphisms markedly influences the rate of xenobiotic metabolism in humans.

Natural products are defined as molecules or substances *viz.* primary and secondary metabolites which are formed naturally by any organism (plant, animals or microbes). Among natural plant products, plant secondary metabolites *i.e* phytochemicals are of utmost importance as they exhibit an array of bioactive properties. Phytochemicals are the non-nutritive and bioactive components present in the plant based diet that have proved to be beneficial in detoxification and elimination of xenobiotics. These plant derived natural compounds influence the biotransformation of xenobiotics, thereby reducing their toxicity and carcinogenicity. They may alter the levels of Phase I and Phase II drug metabolizing enzymes. Agents that preferentially activate Phase II over Phase I enzymes are considered as promising chemopreventive agents (Table **1**, Fig. **1**).

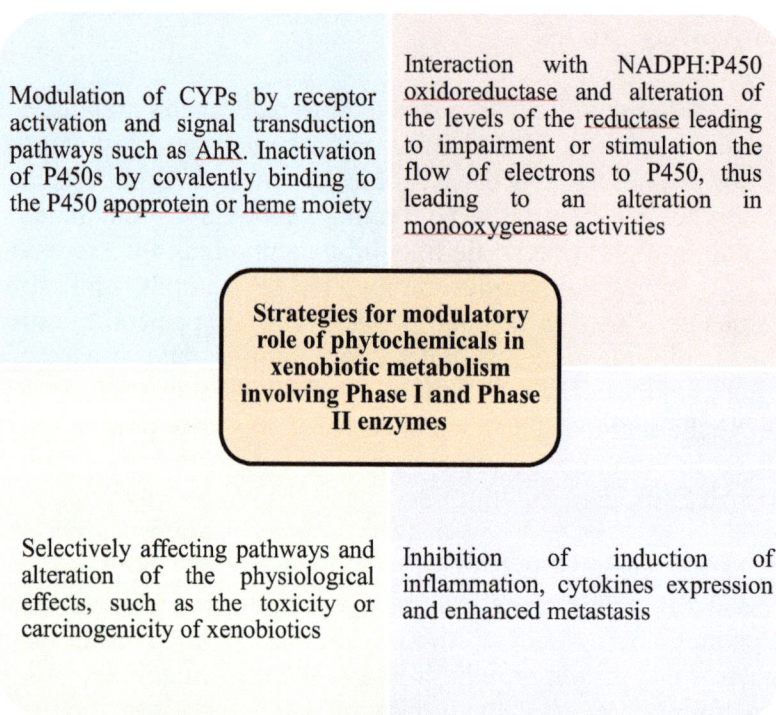

Modulation of CYPs by receptor activation and signal transduction pathways such as AhR. Inactivation of P450s by covalently binding to the P450 apoprotein or heme moiety

Interaction with NADPH:P450 oxidoreductase and alteration of the levels of the reductase leading to impairment or stimulation the flow of electrons to P450, thus leading to an alteration in monooxygenase activities

Strategies for modulatory role of phytochemicals in xenobiotic metabolism involving Phase I and Phase II enzymes

Selectively affecting pathways and alteration of the physiological effects, such as the toxicity or carcinogenicity of xenobiotics

Inhibition of induction of inflammation, cytokines expression and enhanced metastasis

Fig. (1). Strategies for modulation of Phase I and Phase II enzymes.

Table 1. Phytochemicals involved in xenobiotic metabolism.

Compound	Structure	Mutagen	Enzymes	Reference
Phenolics				
Protocatechuic acid	HO—⟨benzene ring⟩—COOH, HO	B[a]P, DMBA	CYP1A1/1A2, 1B1, 2B ↓	[59]

(Table 1) cont.....

Compound	Structure	Mutagen	Enzymes	Reference
Ellagic acid		DMBA	CYP1A1, 1B1 ↓; NAD-(P)H:QR1, GST ↑	[64]
Gallic acid		DMH	CYPb5 ↓; GST, DTD, GGT ↑	[105]
Tannic acid		B[a]P, DMBA	CYP1A1/1A2, 1B1, 2B ↓	[106]
Phloretin		TCDD	CYP1A1 ↓	[107]
Carotenoids				
Canthaxanthin		DMBA	GST, UGT, QR↑	[62, 66]

(Table 1) cont.....

Compound	Structure	Mutagen	Enzymes	Reference
Astaxanthin		Cyclophosphamide DMBA H_2O_2	NAD-(P)H:QR, HO1, GST↑	[63, 64]
Stillbene				
Resveratrol		B[a]P, DMBA, DBP, TCDD, Pyrogallol	CYPs, CYPb5, CYP1A1, CYP1A2 ↓; GST, NAD-(P)H:QR1, GST ↑	[106, 108 - 111]
Diarylheptanoids				
Curcumin		B[a]P, TCDD	CYPs, CYPb5, CYP1A1 ↓, GST ↑	[109, 112, 113]
Flavonoids				
(-)-Epigallocatechin-3-gallate		B[a]P, AFB1	CYPs ↓	[84, 114]
Quercetin		TCDD	CYP1A1 ↓	[84]

(Table 1) cont.....

Compound	Structure	Mutagen	Enzymes	Reference
Genistein		DMBA	CYP1A1, 1B1 ↓	[92]
Biochanin A		CCl$_4$	P450 2E1 ↓; GPx, GST ↑	[93]
Rutin		Thioacetamide, paracetamol, CCl$_4$	CYP 2E ↓; GSH, GPx ↑	[106, 115]

↑upregulation
↓downregulation

ENZYMES INVOLVED IN XENOBIOTIC METABOLISM

Phase I Biotransformation Enzymes

The monooxygenases, cytochrome P450 monooxygenases and flavin containing monooxygenases, hydrolases and reductases make up the majority of the Phase I enzymes along with some less common enzymes [11]. They are accountable for majority of xenobiotic/toxicant metabolism and generate sites for conjugation. Phase I enzymes introduce -OH, -SH, -NH$_2$ and -COOH functional groups into the xenobiotic molecule *via* oxidation, reduction or hydrolysis reactions [12]. The end products of Phase I reactions are not eliminated rapidly and undergo subsequent reactions with endogenous substrates such as glucuronic acid, acetic acid or sulfuric acid that combines with either the newly added or exposed or an already existing functional group to form a highly polar conjugated end product that makes their excretion easy [13, 14].

Cytochrome P450 Monooxygenases

The cytochrome P450 (CYP) superfamily is first line of defense in our body and is the major contributor for metabolism of vast array of xenobiotics including dietary herbs, drugs and carcinogens. These enzymes maintain cellular homeostasis by facilitating elimination of xenobiotics and by participating in the production of endogenous molecules like steroids, bile acids, lipid-soluble vitamins and fatty acids. Approximately 57 genes and 58 pseudogenes are present in humans that encode the CYP and exhibit differences in their patterns of tissue expression and catalytic specificity.

CYP enzymes contain heme molecule that binds noncovalently to the polypeptide chain. Heme binds oxygen at the active site of the enzyme. Oxidation reactions of CYP involve a series of steps. In the first step, substrate binds to the oxidized CYP, followed by single electron reduction in the presence of electrons derived from nicotinamide adenine dinucleotide phosphate (NADPH) *via* NADPH-cytochrome P450. Further steps involve reduction of molecular oxygen after the acceptance of the second electron from NADPH cytochrome b5 reductase, resulting in the release of water and the oxygenated product of the reaction.

CYPs catalyse monooxygenation reactions including epoxidation (aflatoxin), *N*-dealkylation (alachlor), *O*-dealkylation (chlorfenvinphos), *S*-oxidation (phorate) and oxidative desulfuration (parathion) so as to increase the polarity of the substrate molecule [15]. A large number of substrates can be biotransformed by CYP enzymes and this enzyme superfamily is claimed to be the most versatile redox catalysts [16]. The versatility of CYPs in metabolizing xenobiotics is attributed to their large substrate binding site [17]. In case for CYP3A4 the binding site is about 1000 A° long and allows the binding of either multiple small molecules or large molecules [18].

CYP isoforms CYP1A2, CYP2A6, CYP2E1 and CYP3A4 present in human liver microsomes bioactivates acetaminophen to a reactive metabolite *N*-acetyl-*p*-benzoquinone imine (NAPQI) *via* oxidation reaction. NAPQI conjugates with glutathione either spontaneously or enzymatically in the presence of glutathione-S-transferases (GSTs). Non-enzymatic conjugation reaction yields a reduction product, 3-(glutathione-S-yl)-acetaminophen (APAP-GSH) conjugate, an oxidation product, glutathione disulfide (GSSG) and free acetaminophen while the enzymatic reaction yields APAP-GSH and free acetaminophen. APAP-GSH is finally excreted in the urine as mercapturic acid and cysteine conjugates [19].

Flavin containing Monoo xygenases

Like cytochrome P450, microsomes also contain flavin containing monoo-xygenases (FMOs). These enzymes are flavoproteins containing a single prosthetic group, flavin adenine dinucleotide (FAD). These microsomal enzymes are dependent upon molecular oxygen and NADPH for catalyzing oxidative metabolism of a variety of xenobiotics, including nucleophilic sulfur, nitrogen, phosphorous and selenium heteroatoms. FMOs are active towards a wide range of substrates such as phosphines, sulfides, hydrazines, iodide, boron-containing compounds, selenides and primary, secondary and tertiary amines. FMOs are responsible for approximate 6% of all the Phase I biotransformation reactions. These enzymes convert lipophilic xenobiotics to oxygenated polar moieties, thus offering a proficient detoxification system [20]. Mammalian FMO gene family is comprised of five enzymes (FMO1 to FMO5) containing about 550 amino acid residues in each enzyme. The highly conserved glycine-rich region (residues 4-32) region of each enzyme binds noncovalently to 1 mole of FAD near the active site which is adjacent to another highly conserved glycine-rich region (residues 186-213) that binds NADPH [21]. Besides, six pseudogenes (FMO6P-FMO11P) have also been identified [22].

The mechanism of FMO catalysis begins with the reduction of FAD moiety by NADPH to $FADH_2$. $NADP^+$, the oxidized cofactor, remains bound to the enzyme. $FADH_2$ then binds to oxygen generating peroxide, 4a-hydroperoxyflavin of FAD. In the final step water is released from 4a-hydroxyflavin (which restores FAD to its resting oxidized state) along with the release of $NADP^+$. This final step is rate-limiting and it occurs only after substrate oxygenation [23].

In humans, FMOs plays a major role in the biotransformation of several xenobiotics (cocaine, nicotine, methamphetamine, tyramine), drugs (benzydamine, cimetidine, olanzapine, clozapine, guanethidine, methimazole, sulindac sulfide, tamoxifen) and endogenous substrates (trimethylamine, cysteamine). Binding of the xenobiotic to the enzyme occurs only after the reduction of flavin by NADPH and reoxidation by molecular oxygen. FMO_3, is predominantly responsible for converting (*S*)-nicotine to (*S*)-nicotine-*N*-1'-oxide. The reaction proceeds stereospecifically with the formation of *trans* isomer only and this is the only isomer secreted in the urine of cigarette smokers.

Oxygenation of substrates by FMOs will not inactivate the enzyme, although some of the end products are strong electrophiles. However, oxygenation products of reactions catalyzed by FMOs and the oxygenation products of the same substrate by CYPs can inactivate the CYPs, a process known as mechanism-based inhibition, metabolism-dependent inhibition and suicide inactivation. For example, the spironolactone thiol undergoes *S*-oxygenation in the presence of FMO generating an electrophilic sulfenic acid which inactivates CYPs and also binds covalently to various proteins.

Reductases

Carbonyl Reductases

Aldehyde or ketone-based xenobiotics are catalyzed by carbonyl reductases to their corresponding alcohols thus making them susceptible for Phase II biotransformation (glucuronidation or sulfonation). Aldehydes are reduced to primary alcohols while ketones are reduced to secondary alcohols. Besides, carbonyl reduction also metabolizes endogenous chemicals like steroid hormones, prostaglandins and biogenic amines. Carbonyl reductases are NADPH-dependent monomeric enzymes found in blood and the cytosol of liver, brain, kidney and other tissues. These enzymes are classified into two superfamilies: aldo-keto reductases (AKRs) and short-chain reductases (SDRs). The AKR enzyme family includes aldehyde reductases, aldose reductases and some hydroxysteroid dehydrogenases (HSDs), whereas the SDR enzyme family includes carbonyl reductases along with a few other HSDs. Polycyclic aromatic hydrocarbons (PAHs) *viz.* quinones and menadione are excellent substrates for human carbonyl reductases [24].

Quinone Reductase

Reduction of quinones to hydroquinones is carried out by a cytosolic flavoprotein NADPH-quinone oxidoreductase (NADPH:QR) also called DT-diaphorase. Formation of stable hydroquinones involves two-electron reduction of the quinone moiety with a stoichiometric NADPH oxidation. Hydroquinones are easily conjugated and eliminated from the body [25]. Although carbonyl reductases also catalyze a similar two electron reduction but reduction *via* NADPH:QR generates non toxic metabolites. Besides quinones, NADPH:QR can also reduce various toxic chemicals including azo dyes, quinone epoxides, quinoneimines and *C*-nitroso derivatives of arylamines.

Hydrolases

Epoxide Hydrolases

Epoxide hydrolases (EHs) are a group of enzymes that form a link between Phase I and Phase II xenobiotic metabolism. The group of EHs is divided into two superfamilies, the microsomal EH (mEH) and the soluble cytosolic EH (sEH). In humans, both the superfamilies are subjected to polymorphism. A number of alleles have been described for these enzymes that have moderate effects on the enzymes' half life [26, 27] but do not appear to affect enzyme activity.

Both mEH and sEH hydrolyze a wide spectrum of compounds but exhibit different substrate specificity, for example mEH hydrolyzes epoxides on cyclic systems, while sEH has little activity towards these compounds. The two enzymes can also be distinguished on the basis of their stereoselective hydration, particularly of stilbene-oxide; mEH hydrolyzes the *cis*-isomer (pH 9.0), while sEH hydrolyzes the *trans*-isomer (pH 7.4). Epoxide hydrolase is an inducible enzyme and its induction is associated with CYPs induction and several CYPs inducers such as *trans*-stilbene oxide and phenobarbital. The levels of epoxide hydrolase in liver microsomes are enhanced by antioxidants like butylated hydroxytoluene (BHT), ethoxyquin *etc*. Alcohols, imidazoles and ketones are also known to stimulate mEH activity *in vitro*.

The catalytic site of EH is comprised of three amino acid residues that form a catalytic triad. In mEH the triad consists of Asp_{226} which functions as the nucleophile, both Glu_{404} and Glu_{376} as the acid and His_{431} as the base. In sEH, the corresponding residues are Asp_{333}, Asp_{495} and His_{503} respectively. The nucleophile Asp_{226} attacks the oxirane ring at the carbon atom, thus initiating enzymatic activity and this leads to the formation of α-hydroxyesterenzyme intermediate, with a negative charge on the oxygen atom which is stabilized by an assumed oxyanion hole. The His_{431} residue abstracts proton from water, thereby activating the same. The activated water attacks the Cγ atom of Asp_{226} which results in ester bond hydrolysis of the acyl- enzyme intermediate. The active enzyme is thus restored and this leads to the formation of a vicinal diol having *trans*-configuration [28]. The second step involves cleavage of the ester bond present in the acyl-enzyme intermediate.

These enzymes cleave genotoxic alkene epoxides and oxiranes to their corresponding diols by *trans* addition of water. Epoxides are reactive electrophiles that have the ability to attack nucleophilic centers present in proteins and nucleic acids, leading to formation of adducts that results in cellular transformation and toxicity. Epoxide hydrolases deactivates potentially toxic derivatives of Phase I metabolism and results in the formation of a hydrophilic metabolite [29]. Most of the oxiranes and epoxides are metabolic intermediates formed during oxidation of aromatic and unsaturated aliphatic xenobiotics catalyzed by cytochrome P450. Benzo[a]pyrene is oxidized to B[a]p-4,5-oxide by CYPs and this reactive intermediate is highly mutagenic, but in mammalian cells it is rapidly hydrolysed into an inactivate B[a]p-4,5-diol by epoxide hydrolase [30]. In general, epoxide hydrolases and CYPs have similar cellular localization. mEH also influence metabolism of drugs like carbamazepine, an antiepileptic drug. This prodrug is converted to carbamazepine-10,11-epoxide, which is the pharmacologically active derivative, by CYP. This metabolite is then hydrolyzed to a dihydrodiol in the presence of mEH that results in the deactivation of the drug.

Esterases

Amides, carboxylic acid esters and thioesters are hydrolysed by carboxylesterases present in different tissues and serum and also by two other esterases present in the blood, the dimeric acetylcholinesterase situated on erythrocyte membranes and pseudocholinesterase located in the serum. Within the cells carboxylesterases are located in the cytosol and the endoplasmic reticulum and are involved in the detoxification/metabolic activation of different drugs and environmental toxicants. In humans primarily two types of carboxylesterases isozymes are found, hCE1 and hCE2 with the highest quantities found in the liver.

Carboxylesterases in the presence of an alcohol can catalyze transesterification of xenobiotics like cocaine (a methyl ester) into ethylcocaine (corresponding ethyl ester). They also catalyze the bioactivation of prodrugs into their respective free acids. For example, the cancer chemotherapeutic agent irinotecan, which is a prodrug and an analog of camptothecin, is bioactivated by the intracellular and plasma carboxylesterases into a potent topoisomerase inhibitor SN-38, which retards the growth of the tumor. SN-38 is later subjected to glucuronidation which results in the loss of its biological activity and eliminates SN-38 through the bile [31].

The catalytic mechanism of carboxylesterases also involves a catalytic triad comprised of a nucleophile (serine), an acid (glutamate) and a base (histidine) at the enzymes active site similar to the epoxide hydrolases. In the triad, both the enzymes have glutamate/aspartate as the acid and histidine as the base but have a different type of amino acids as the nucleophile. This results in marked differences in their catalytic activity with carboxylesterases primarily being involved in the hydrolysis of amides and esters while epoxide hydrolases hydrolyze epoxides and oxides. Even the catalytic mechanism for both the enzyme is different. In case of carboxylesterases, the substrate's carbonyl carbon atom is initially attacked by the nucleophile (Ser_{203}) resulting in the formation of α-hydroxyesterenzyme intermediate that is further attacked by the activated water resulting in the release of an alcoholic product. Thus, in case of carboxylesterase, the oxygen derived from the activated water molecule gets added into the product. In case of epoxide hydrolase, the oxygen is derived from the nucleophile Asp_{226}.

Phase II Biotransformation Enzymes

The biotransformation reactions of Phase II metabolism include glucuronidation, sulfonation, methylation, acetylation and conjugation with glutathione and amino acids. The cofactors involved in these reactions react with the functional groups present either on the xenobiotic or exposed/introduced during Phase I biotransformation. Phase II biotransformation reactions are largely involved with increasing the hydrophilicity of the xenobiotic thereby promoting the excretion of the foreign chemicals.

It is not always necessary that Phase II metabolic reactions be preceded by Phase I biotransformation reactions. For example, morphine, codeine and heroin are converted to morphine-3-glucuronide. In case of morphine, direct conjugation with glucuronic acid leads to the formation of this metabolite. However, in the other cases glucuronic acid conjugation follows Phase I biotransformation: *O*-demethylation in case of codeine and hydrolysis (deacetylation) in case of heroin. Similarly, acetaminophen can be directly sulfated and glucuronidated, while phenacetin must go through Phase I metabolism where it is *O*-deethylated to acetaminophen before undergoing Phase II biotransformation [21].

Uridine Diphosphate Glucuronosyl Transferases

Lipophilic xenobiotics are primarily excreted from body after conjugation with glucuronide in the presence of a cofactor, uridine diphosphate-glucuronic acid (UDP-glucuronic acid) and the enzyme UDP-glucuronosyl transferases (UGTs). These enzymes are located in the endoplasmic reticulum of tissues of liver, kidney, skin, brain, intestine, spleen and nasal mucosa.

The electron-rich nucleophilic heteroatom of the xenobiotic (*O, N* or *S*) are primarily the sites of glucuronidation. Therefore, glucuronidation substrates contain functional groups such as phenols and aliphatic alcohols (that form *O*-glucuronide ethers), carboxylic acids (that form *O*-glucuronide esters), aromatic and aliphatic amines (that form *N*-glucuronides) and free sulfhydryl groups (that form *S*-glucuronides). In humans, tertiary amines including cyclobenzaprine, tripelennamine and imipramine are substrates for *N*-glucuronidation and leads to the production of positively charged quaternary glucuronides. Xenobiotics like sulfinpyrazone, phenylbutazone and feprazone that contain carbon atoms form *C*-glucuronides. Glucuronide conjugates of xenobiotics are hydrophilic in nature and are removed from the body *via* urine or bile. The size of the parent compound/ Phase I metabolite determines the route of excretion. Smaller size glucuronides are secreted through urine while the larger ones are excreted in bile.

Acetaminophen, an analgesic drug when taken at a therapeutic dose undergoes glucuronidation by transfer of glucuronosyl group from UDP-glucuronic acid, thus making the drug more water soluble. The isoform UGT1A6 is active at low concentrations of acetaminophen while UGT1A9 and UGT1A1 are active at the toxic doses of the drug. UGT1A9 catalyze a broad range of pharmacologically relevant concentrations of acetaminophen [19].

Glutathione-S-Transferases

Glutathione-S-transferases (GSTs) are the most important enzymes of Phase II detoxification process involved with the metabolism of xenobiotics. They also play a significant role in cellular protection from oxidative stress. The GSTs catalyze conjugation between endogenous tripeptide glutathione (GSH) and the xenobiotic compounds resulting in the formation of thioether conjugates. These conjugates are formed by nucleophilic attack of glutathione thiolate anion (GS^-) upon the xenobiotics electrophilic carbon atom. Glutathione conjugation of xenobiotics with electrophilic heteroatoms (*O*, *S* and *N*) is also of common occurrence. GSTs are located in most tissues with high concentrations present in the liver, kidney, intestine, testis and lung. Within the cell they are localized in the cytoplasm (95%) and endoplasmic reticulum (5%). GSTs can act upon substrates that are either hydrophobic or contain an electrophilic atom or can react with glutathione nonenzymatically at some measurable rate. The mechanism of glutathione-S-transferase activity involves GSH deprotonation to GS^- by the active site tyrosinate ($Tyr-O^-$), which acts as a general base catalyst [32]. Glutathione conjugates thus formed can be excreted directly in the bile or may undergo further metabolism in the kidney to mercapturic acids and excreted in urine.

A major role played by GSTs is in the detoxification of epoxides like B[a]P-7,-diol-9,10-epoxide and AFB1-exo-8,9-epoxide (AFBO) derived from Phase I biotransformation of alpha-beta unsaturated ketones and PAHs. AFBO, the reactive intermediate of aflatoxin biotransformation, conjugates with glutathione which further undergoes sequential metabolism in liver and kidneys and ultimately excreted in urine as aflatoxin-*N*-acetylcysteine, a mercapturic acid [33]. Pesticides like halogenated hydrocarbon insecticides, organophosphate insecticides and *S*-triazine herbicides and reactive drug metabolites like *N*-acetyl-*p*-benzoquinone, acetaminophen metabolite are also conjugated with glutathione resulting in their detoxification and elimination.

Sulfotransferases

A number of xenobiotics as well as endogenous substrates that undergo *O*-glucuronidation can also undergo sulfate conjugation/sulfonation *e.g.* acetaminophen. Sulfate conjugation involves conjugation of the xenobiotic with the sulfonate group (SO_3^-) from 3'-phosphoadenosine-5'-phosphosulfate (PAPS), an endogenous universal donor, to hydroxyl, amino or sulfhydryl groups of the substrate resulting in the formation of sulfonates, sulfamates and thiosulfates, respectively. The reaction is catalyzed by cytosolic enzymes, sulfotransferases (SULT), found mainly in the liver, kidney, lungs, intestinal tract, platelets and brain. Sulfonation generally produces a highly polar sulfuric acid ester. In humans two broad classes of SULTs are present. The membrane bound class of SULTs is present in the Golgi apparatus and conjugates SO_3^- group with endogenous proteins, peptides, carbohydrates and lipids. The soluble SULTs are present in the cytosol and mediate metabolism of bile acids, steroids, neurotransmitters and xenobiotics. Therefore, the cytosolic SULTs are relevant to xenobiotic metabolism, disposition and potential bioactivation. Human cytosolic SULT super- family comprises of 13 enzymes that differ strongly not only in their tissue distribution but also in their substrate specificity [34].

The substrates for sulfonation are generally phenols and aliphatic alcohols, that are often Phase I biotransformation products. Besides, certain aliphatic amines (despramine), aromatic amines (2-aminonaphthalene, aniline) aromatic hydroxylamines (*N*-hydroxy-2-aminonaphthalene), aromatic hydroxyamide (*N*-hydroxy-2-acetylaminofluorene) and *N*-oxide (minoxidil) can also be sulfated without prior Phase I biotransformations [35]. SULT isoforms SULT1A1, SULT2A1 and SULT1A3/4 transfers a sulfo group from PAPS to acetaminophen thus making it more polar and facilitates its elimination.

Methyltransferases

One of the most prevelant reaction in nature are methylation reactions and are involved mainly in the metabolism of endogenous substances. Methylation reaction are different from other Phase II conjugation reactions since they decrease the water solubility of the xenobiotics and masks the functional groups that otherwise may be conjugated in the presence of other Phase II enzymes. Variety of xenobiotics containing *N*, *O* and *S* atoms can be conjugated by a nonspecific group of methyltransferases (METs) with different degrees of specificity. *N*- and *O*-methyltransferases are located mainly in the liver and lungs, while the *S*-methyltransferases are present mainly in colonic and cecal mucosa. The cofactor required by methyl-transferases is *S*-adenosylmethionine (SAM) which is produced from ATP and L-methionine *via* L-methionine adenosyl-transferas.

N-methylation is important biotransformation step for xenobiotics with primary, secondary and tertiary amino groups. Nicotine, a pyridine containing compound, undergo methylation resulting in the formation of polar quaternary amines. However, *N*-methylation may lead to bioacitvation of certain xenobiotics with enhanced toxicity. 4-phenyl-1,2,3,4-tetrahydropyridine is converted to 1-methyl-4-phenyltetrahydropyridine which further undergoes oxidation to form a reactive intermediate *N*-methyl-phenylpyridine, which is implicated in the development of Parkinson's disease [36]. *O*-methylation reactions are known to be associated with metabolism of carboxy groups of proteins, catechols and hydroxyindoles [37]. *S*-methylation biotransforms sulfhydryl containing xenobiotics like anti-hypertensive, antidiuretic, antirheumatic and immunosuppresive drugs. In humans, thiol methyltrans-ferase (TMET) and thiopurine methyltransferase (TPMET) catalyze *S*-methylation. TMET is microsomal and preferentially methylates aliphatic sulfhydryl compounds like D-penicillamine, captopril and disulfiram derivatives while TPMT is cytoplasmic and methylates heterocyclic and aromatic compounds like 6-mercaptopurine, azathioprine and 6-thioguanine [38]. Certain metals can also be methylated like arsenic. Inorganic mercury can be dimethylated while inorganic selenium can undergo trimethylation.

N-acetyltransferases

Xenobiotics containing hydrazines and aromatic amines are *N*-acetylated by *N*-acetyltransferases (NATs) present in the cytosol of liver and other tissues. In humans these reactions are catalyzed by 2 isoforms NAT1 and NAT2. NAT1 is expressed in most of the tissues whereas NAT2 is expressed exclusively in liver and intestine [39]. The NAT enzyme requires a co-factor, acetyl coenzymeA (acetyl CoA) that provides the acetyl group for conjugation with aromatic hydrazines and aromatic amines leading to the formation of aromatic hydrazide and aromatic amides metabolites, respectively. Substrates for NAT1 include *p*-aminosalicylic acid, *p*-aminobenzoic acid, antibiotics sulfanilamide and sulfa-methoxazole, caffeine and 2-aminofluorene [40]. Human NAT2 is a xenobiotic metabolising enzyme that detoxifies drugs like isoniazid, hydralazine, endralazine, procainamide, aminoglutethimide, nitrazepamand and dapsone [40, 41]. Both NAT1 and NAT2 exhibit polymorphism.

N-acetylation reaction proceeds in two sequential steps. In the first step, acetyl group of acetyl-CoA gets transferred to the active site cysteine residue within the NAT enzyme with the release of coenzyme A. The second step involves the transfer of the acetyl group from the acylated enzyme to the amino group of the substrate, thereby regenerating the enzyme. Several drugs after their Phase I biotransformation undergo *N*-acetylation. For example, caffeine is converted to paraxanthine after *N3*-demethylation by CYP1A2 which is further converted to 1-methylxanthine following *N*-demethylation. This intermediate is then *N*-acetylated by NAT2 into 5-acetylamino-6-formylamino-3-methyluracil [21].

Amino Acid Conjugation

Xenobiotics containing carboxylic group or an aromatic hydroxylamine may conjugate with amino acids *via* two different pathways. The first pathway involves the conjugation between the carboxylic acid group of xenobiotics with the amino group of amino acids like glutamine, glycine and taurine. In this case the xenobiotic gets activated by conjugation with CoA, resulting in the production of acyl-CoA thioether which forms an amide linkage with the amino group of the amino acid. The second pathway involves conjugation of the aromatic hydroxylamine xenobiotics with the carboxylic acid group of amino acids such as serine and proline. Here the amino acid is activated by aminoacyl-tRN--synthetase, which then reacts with an aromatic hydroxylamine yielding reactive *N*-hydroxyesters [42]. Many xenobiotics like food preservatives, herbicides and drugs such as simvastatin, acetylsalicylic acid and valproic acid undergo amino acid conjugation. Besides many xenobiotics are metabolized to carboxylic acids which further conjugate with amino acids.

The steric hindrance around the carboxylic acid group and the substituents of either the aromatic ring or the aliphatic side chain determines the ability of the xenobiotic to undergo amino acid conjugation. Amino acid conjugation is also a detoxification route for endogenous acids. The enzymes for amino acid conjugation of xenobiotic carboxylic acids are primarily located in the mitochondria of liver and kidney whereas conjugation of bile acids is extramitochondrial and their enzymes are located in the peroxisomes and endoplasmic reticulum.

γ-Glutamyl Transpeptidase

γ-glutamyl transpeptidase (γ-GT) is a membrane-bound heterodimeric glycoprotein composed of a large (51 kD) and a small subunit (22 kD) located on the luminal surface of the proximal tubules of kidney, in the intestine and in the bile ducts of the liver. The main role of γ-GT is to hydrolyze the GSH-conjugates/GSH at the γ-linkage between the α-amino group of cysteine present in GSH and the γ carboxyl group of glutamate leaving the cysteinyl-glycine peptide prone to cleavage by aminopeptidases finally hydrolyzing GSH to amino acids for renal reabsorption. Hydrolysis by γ-GT is accounted as the first step in the conversion of GSH-conjugates of xenobiotics to cysteinyl-glycine, that ultimately produce mercapturic acids which are eliminated *via* urine or bile [43].

Besides detoxifying the xenobiotics, Phase I and Phase II biotransformation metabolic reactions may convert many foreign chemicals into reactive intermediates thus implicating in their toxicity. For decades Phase I reactions are known to be associated with the formation of reactive metabolites exhibiting mutagenic and carcinogenic activity. The CYP3A4 metabolizes aflatoxin (AFB1) to its reactive metabolic intermediate AFBO that can react with guanine to form a mutagenic DNA adduct, aflatoxin-*N*7-guanine. The DNA adducts are moderately resistant to DNA repair mechanisms, resulting in gene mutations that leads to the development of cancer. Besides, the reactive AFBO effects the length of the telomere during cell division, the mitotic phase, growth phase (G1 and G2) and DNA synthesis (S phase) in the cell cycle, leading to carcinogenesis [33]. Benzo[a]pyrene and other polycyclic aromatic hydrocarbons are metabolized into their diol-epoxides by human CYP1 enzymes (CYP1A1, 1A2 and 1B1) along with epoxide hydrolases and show high reactivity towards DNA [30]. Examples of Phase II bioactivation of xenobiotics includes glucuronidation of *N*-hydroxylated reactive metabolites of aromatic amine containing xenobiotics that leads to the formation of unstable *N*-glucuronides. The glucuronides quickly decompose to nitrenium ion containing highly reactive intermediates that have the capacity to damage DNA. Bladder cancer, idiosyncratic hypersensitivity and gastroenteropathy toxicities are known to be implicated by reactive glucuronide conjugates produced after Phase II metabolism [44].

Sometimes the end product of xenobiotic metabolism released from one organ is nontoxic but is later converted to toxic metabolites in other organs. Bromobenzene exposure results in renal necrosis and is caused by the formation of glutathione conjugate, 2-bromo-3-hydroquinone in the liver. This nontoxic conjugate is released in the blood and is taken up by the kidney where it is further metabolized to benzothiazine, a toxic derivative [45, 46]. A brief outline of all the enzymes involved in xenobiotic metabolism is shown in Fig. (**2**).

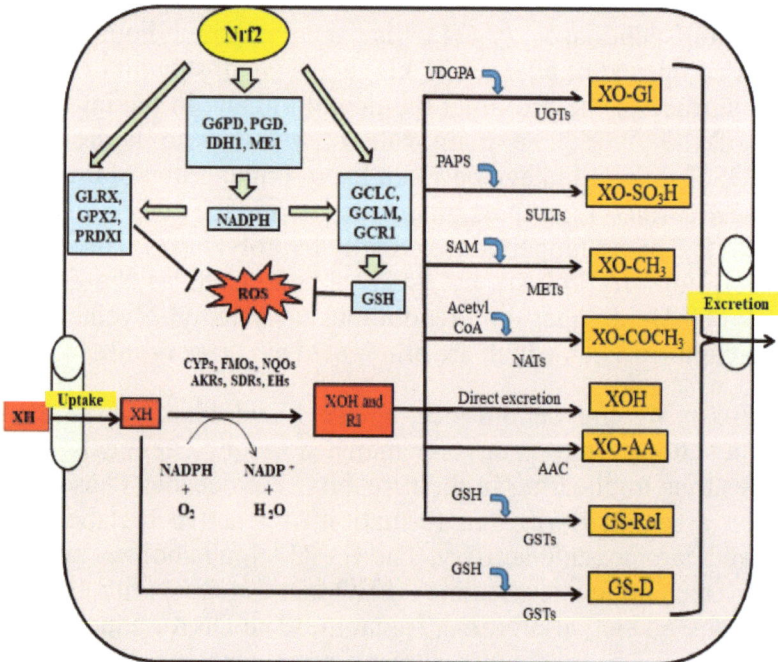

Fig. (2). Enzymes involved in xenobiotic metabolism and their regulation by Nrf2.

Phytochemicals in Xenobiotic Metabolism

Phenolics

Phenolics are antioxidant compounds which include monophenolics, diphenolics, and polyphenolics. Phenolics have received considerable attention due to promising antioxidant, radical scavenging, chemopreventive and antiapoptotic potential [47, 48]. The chemoprotective effect of polyphenols against chemically induced cancers have been attributed to a decrease in the metabolic rates of carcinogens by Phase I enzymes, due to direct inhibition of catalytic activity and modulation of gene expression. The level of expression of Phase II enzymes, augment the ability of the target tissues to detoxify the reactive intermediates [49, 50].

Protocatechuic acid, ellagitanins, catechin gallates, tannic acid and phloretin are promising phenolics which have shown chemopreventive potential. Proto-catechuic acid showed protective effects against hepatocarcinogenesis (HC) induced by diethylnitrosamine in male F344 rats at a dose of 500 and 1000 ppm. The inhibitory effects during the initiation and post-initiation phases were attributed to alteration in hepatic ornithine decarboxylase activity [51].

Ellagitannins are esters of glucose with hexahydroxydiphenic acid, when hydrolyzed, they yield ellagic acid (EA), the dilactone of hexahydroxydiphenic acid. Dietary EA decreases the occurrence of N-2-fluorenylacetamide-induced HC in Fisher 344 rats. P450 reductase activity decreased by up to 28% and the hepatic Phase II enzymes GST, NADPH:QR and UGT activities increased by up to 26, 17 and 75% respectively upon EA treatment. *In vitro* studies confirmed that concentration of 100 μM was effective in inhibiting P450 2E1, 1A1 and 2B1 activities by 87, 55 and 18% respectively, but did not affect 3A1/2 activity [52]. The oral administration of EA to N-nitrosodiethylamine induced HC in rats as a protective compound, showed significant increases in tested enzymes *viz.* glutathione peroxidase (GPx), γ-GT, GSH, AST, ALT, ALP, GST, G6PD and direct and total bilirubin thus exerting protective effects [53].

Catechin gallates have shown promising results as chemoprotective agents. They are inducers of Phase II enzymes. HepG2 cells and primary cultures of human hepatocytes indicated a complexity of actions of individual components *versus* complex mixtures [54]. Gallic acid (GA) was found to be effective in DMH induced rats. DMH treatment led to a decrease in the activities of Phase II enzymes and an increase in the activities of Phase I enzymes. GA supplementation increased the activities of Phase II enzymes and decreased the activities of Phase I enzymes, in addition to the decreased tumor incidence [55]. GA administration in HC bearing rats showed significant downregulation in the gene expression levels of hepatic γ-GT, heat shock protein gp96 and amelioration of the destabilized liver tissue architecture caused by *N*-nitrosodiethylamine intoxication [56].

Potential ameliorative effects of Tannic acid (TA) were investigated in CCl_4^- intoxicated mice. TA pre-treatment (25 or 50 g/kg/day) modulated the activities of AST, ALT and MDH along with upregulation of SOD, CAT and GPx [57]. Investigations have revealed that proanthocyanidins treatment can potentially inhibit expression of CYP2E1 in liver, which prevents free radical formation from CCl_4 in a steatosis model, likely through exerting antioxidant actions and inhibiting the free radical-generating CYP2E1 enzyme [58]. TA also decreased the mutagenicity/carcinogenicity of several amine derivatives and polycyclic aromatic hydrocarbons in rodents. The activities of murine cytochrome P450 and Phase II enzymes at a dose range of 20-80 mg/kg showed a significant inhibition of CYP2E1, NADPH:QR and GST in female Swiss mice [59].

AFB1 is a potent carcinogen which gets activated to ultimate carcinogenic intermediate, AFBO. Phloretin was shown to have strong chemopreventive effect against AFB1 through its inhibitory effect on CYP1A2, CYP3A4 and its inductive effect on GST activity (GSTA3, GSTA4, GSTM1, GSTP1 and GSTT1) in AML 12 cells. Induction of the GST isozyme genes was mediated through Nrf2/ARE pathway [60]. Similar protective effects of phloretin were observed against 7,12-dimethylbenz(a)anthracene (DMBA) induced buccal pouch carcinogenesis in male golden Syrian hamsters. Phloretin (40 mg/kg b. wt) normalized the neoplastic changes, decreased the levels of lipid by products, retained the antioxidants and restored Phase I and II enzymes [61].

Carotenoids

The two 4-oxocarotenoids, Canthaxanthin and Astaxanthin (ASTX) are substantial inducers of liver P4501A1 and 1A2 in the rats fed on 300 mg/kg diet for 15 days. Canthaxanthin increased the liver content of P450, the activities of NADH- and NADPH-cytochrome c reductase and produced a substantial increase of some P450-dependent activities. Phase II UGT and QR activities were also increased. ASTX induced the same pattern of enzymes activities as Canthaxanthin, but to a lesser extent. These two oxocarotenoids form a new class of inducers of P4501A which are ligands of the AH receptor [62]. ASTX pre- and post-treatment at a dose 25mg/kg increased the level of Nrf2 and Phase-II enzymes, *i.e.* NADPH:QR and HO-1 against cyclophosphamide-induced oxidative stress. DNA damage, cell death and induction of GST-P foci in rat liver were found to be mediated through Nrf2-ARE pathway [63]. ASTX was studied for its xenobiotic-metabolizing and antioxidant enzymes in the DMBA-induced

HBP carcinogenesis model. It was effective in reducing the expression of CYP1A1, CYP1B1 and cytochrome P450 isoforms coupled with upregulation of the Phase II detoxification enzymes GST and NADPH:QR through the activation of Nrf2/Keap-1 signalling [64]. ASTX reduced H_2O_2-induced cell viability loss in ARPE-19 cells. Phase II enzymes NADPH:QR1, HO-1, glutamate-cysteine ligase modifier subunit and glutamate-cysteine ligase catalytic subunit were upregulated *via* Nrf2-mediated PI3K/Akt pathway [65].

Canthaxanthin (300 mg/kg diet) increased the liver content of Cytochrome P-450, NADH-cytochrome c reductase and of some P450-dependent enzymes (ethoxy-, methoxy-, pentoxy- and benzoxyresorufin *O*-dealkylases) as well as Phase II enzymes *viz*. UGT1, UGT2 and GST in rats. The results were similar to that of a classical inducer, 3-methylcholanthrene [66]. Lycopene has shown protective activities against DMBA and Polycylic aromatic hydrocarbons induced toxicity in MCF-7 cell line. CYP1 enzymes were assessed in a recombinant protein. Lycopene inhibited recombinant CYP1A1 and CYP1B1 in the µM range. Phase I enzyme inhibition and Phase II enzyme induction were proposed to be the underlying chemoprotective mechanisms of lycopene [67]. Lutein, a carotenoid, acts as an antioxidant in retinal cells and its treatment of a neuronal cell line PC12D, induced Phase II enzyme expression thereby reducing intracellular ROS levels *via* Nrf2 independent pathway [68].

Stilbenes

Resveratrol (RES) is a polyphenol with antioxidant, anti-cancer and anti-inflammatory properties. RES was efficacious against B[a]P and DMBA induced toxicity in female BALB/C mice. It significantly increased (129-174%) the activity of NADPH:QR1 in comparison with B[a]P-treated animals and decreased NADPH:QR1 activity in comparison with DMBA-treated group. In a similar study RES did not affect the GST activity but induced UGT by approximately 100-150% and NADPH:QR to a lesser extent [69, 70]. RES reduced Pyrogallol-mediated increase in alanine aminotransaminase (ALT), aspartate amino-transferase (AST), bilirubin, lipid peroxidation and mRNA expression and catalytic activity of CYP2E1 and CYP1A2 in Swiss albino mice. Pyrogallol-mediated decrease in GST, GPx and glutathione reductase activities and GSH content was significantly attenuated in RES co-treated animals [71]. The hepato-protective effects of RES against acetaminophen induced toxicity were due to significant inhibition of CYP2E1, CYP3A11, and CYP1A2 activities that caused its activation into toxic metabolite NAPQI [72]. RES attenuated CCl_4-induced liver fibrosis in mice *via* the Akt/NF-κB pathways. It upregulated the activity of serum AST, ALT, TNF-α, α-SMA and collagen I [73].

Diarylheptanoids

Curcumin (CUR) has been widely used as a spice and coloring agent in foods. Chemopreventive effects of CUR on murine HC have been widely reported. C3H/HeN mice injected i.p. with DEN and fed on 0.2% CUR showed a 62% reduction in incidence of HC [74]. CUR at a concentration of 100 mg/kg/day was effective against Nitrosodiethylamine (DENA) initiated and Phenobarbital promoted HC in Wistar rats [75]. The researchers further confirmed that CUR prevented the drop in hepatic glutathione antioxidant defense, decreased lipid peroxidation and minimized the histological alterations induced by DENA/PB [76, 77]. Glutathione-S-transferase-P antibodies confirmed that the expression levels of these hepato-carcinogenic markers were reduced by treatment with CUR in DEN-induced HC in rat model [78]. CUR attenuated the DEN induced severe histological and immune histochemical changes in liver tissues of a male Sprague Dawley rats by modulating the activity of ALT, AST, ALP, γ-GT and total bilirubin level. The hepatocarcinoma incidences were 100.0% and 36.7% in the DEN-alone and DEN-CUR groups, respectively [79]. DENA induced male albino rats treated with CUR *via* intra-gastric intubation at doses of 300, 200 and 100 mg/kg b.wt. respectively for 20 weeks showed significantly reduced serum levels of AFP, ALT activity and MDA content. It increased the gene expression and activities of GPx, GRD, CAT and SOD [80].

Flavonoids

(-)-Epigallocatechin-3-gallate (EGCG) is the most abundant catechin found in tea with various biological activities. EGCG (0.5 μM/L) was found to be efficacious in inducing Phase II enzymes GST, and NADPH:QR at mRNA and protein levels in immortalized human bronchial epithelial cells and lung adenocarcinoma cells (A549) [81]. ECGC inhibited CYP isoforms with properties similar to those of green tea extracts and produced competitive inhibitions against CYP2B6 and CYP2C8. In human intestinal microsomes, EGCG moderately inhibited CYP3A activity in a noncompetitive manner with an IC_{50} value of 31.1 μM [82]. EGCG (3.2 mg/g diet for 2 weeks) mitigated hepatotoxicity and decreased mRNA expression of glutathione reductase in humans. Dietary pretreatment prevented the decreased mRNA expression and increased glutathione peroxidase GPx2, GPx3, GPx5, and GPx7 expression. EGCG (750 mg/kg, i.g., once daily for 3 days) increased plasma AST by 80-fold, decreased both reduced and total hepatic glutathione by 59% and 33% respectively and increased hepatic levels of phosphorylated histone 2AX [83]. Male Sprague-Dawley rats fed with EGCG (0.54 %, w/w) and injected with acetaminophen (1 g/kg body weight) i.p. showed lower activity of plasma ALT, AST, CYP3A, CYP2E1, UGT and sulfo-transferase. Reduced acetaminophen-glucuronate and acetaminophen-glutathione contents in plasma and liver were also reported [84]. EGCG incubated with rat liver microsomes at 1-100μM for 30 min showed an inhibition of CYP activities, except for CYP2E1 whose activity was unaffected. EGCG effects were partially abolished in the presence of 1mM glutathione, suggesting they are particularly relevant to the *in vivo* conditions when glutathione is depleted by toxicant insults [85]. On the contrary, EGCG was able to evoke hepatotoxicity at a non lethal high dose of 75 mg/kg i.p. and the nuclear distribution of Nrf2 was significantly increased resulting in the upregulation of gens for HO1, NADPH:QR, GST and those involved in glutathione and thioredoxin systems. At the maximum tolerated dose of 45 mg/kg, i.p., repeated EGCG treatments did not alter the major antioxidant defense. At a lethal dose (200 mg/kg, i.p.), a single EGCG treatment triggered hepatotoxicity by suppression of major antioxidant enzymes, and the Nrf2 rescue pathway was found to be vital for counteracting EGCG toxicity [86].

A pre-treatment with the flavonoid antioxidant quercetin (QC) before DENA initiation, significantly prevented development of γ-GT positive lesions and reduced lipid peroxidation in rats [87]. Therapeutic efficacy of nanoencapsulated QC was evaluated in combating DEN-induced HC in rats. Elevated levels of conjugated diene in DEN-treated rats were lowered significantly by nanoparticulated QC along with upregulation of GST activity and downregulation of cytochrome c expression in the liver [88]. CUR and QC treatments at a dose of 60 mg/kg and 40 mg/kg respectively to B[a]P induced mice were able to significantly decrease the levels of lipid peroxidation, ROS generation, SOD and GST [89].

Genistein pretreatment at a dose of 5mg/kg b.w./day showed hepatoprotective potential against d-galactosamine induced inflammation and hepatotoxicity in male Wistar rats. It significantly decreased serum ALT, AST along with suppression of NF-κB activation. It was able to maintain the redox potential and strengthen the antioxidant defense system of the cell [90]. Germinated and fermented soybean extract with 0.14 μg/mg Genistin significantly inhibited t-BHP-induced ROS production in HepG2 cells and rat liver; upregulated the mRNA levels of antioxidant enzymes including CAT, SOD, GR GPx and improved the lever marker enzymes [91]. On the other hand, higher dose of Genistein (125, 250, 500 and 1000 mg/kg) produced several undesirable effects by affecting multiple cellular pathways in Swiss mice. It significantly elevated ALT, AST, and alkaline phosphatase (ALP) and upregulated Cyp4a14, Sult1e1, Gadd45g, Cidec and Myc genes at 500 and 1000 mg/kg [92].

Biochanin A, phytoestrogenic isoflavone found in red clover showed hepatoprotective activity against CCl$_4$-induced hepatotoxicity model in rats by successively normalizing many parameters such as cytochrome CYP2E1, CAT,

SOD, GPx, GST, GR and lipid peroxidation [93]. Administration of Biochanin A (20 mg/kg b.w./day) and selenium (3 mg/kg·bw/day) resulted in a significant reversal of hepatic and oxidative stress markers in arsenic-intoxicated rats. These biochemical perturbations were supported by histopathological observations of the liver [94].

Rutin, a well-known polyphenolic natural flavonoid has shown protective effects against thioacetamide, paracetamol and CCl_4 at concentrations ranging from 10-20 mg/kg as evidenced by normalization of liver marker enzymes and hepatic architecture [95, 96]. Similar results were seen in case of ethanol induced hepatotoxicity in male albino rats wherein 100 mg/kg rutin significantly decreased the levels of liver marker enzymes, lipid peroxidation and elevated the activities of liver SOD, CAT, GSH, GPx, vitamins C and E when compared to untreated ethanol supplemented rats [97]. In Sprague-Dawley male rats, rutin at a dose of 50 and 70 mg/kg, increased the activity of endogenous liver antioxidant enzymes CAT, SOD, GPx, GST, GR and decreased the amount of lipid peroxidation and CYP2E1 expression induced with CCl_4 [98]. In a similar study involving male Wistar albino rats rutin modulated the activities of ALT, AST, GPx, GST and CAT in CCl_4 induced animals [99]. The authors further confirmed that CCl_4 administration caused alteration in expression of IL-6/STAT3 pathway genes, resulting in hepatotoxicity whereas rutin exerted protective effects by reversing these changes [100]. Hypercholesterolemia-induced hepatotoxicity in rat was abridged by rutin as evidenced through reduction of ALT, AST and increased expression of GPx, GR, GSTα, sulfiredoxin-1 and glutamate-cysteine ligase [101]. Rutin was effective against methotrexate induced hepatotoxicity in rats. Cotreatment groups showed lower histological injury compared to Methotrexate group. MDA and ALT levels were increased, while SOD and GPx were decreased in methotrexate group compared with cotreatment group [102].

Transcriptional Regulation of Phase I and Phase II Enzymes

Nrf2 is a transcription factor that is responsible for the regulation of redox status of the cell. Its activity is up regulated during the process of inflammation, changes in redox status, stimulation by growth factors *etc* thereby conferring adaptations to various forms of stress. It is known to control the expression of Phase I and II enzymes as well as multidrug resistance-associated protein (MRP) transporters, MRP2, MRP3, MRP4, and MRP5 [103]. Phase II enzymes are under the control of ARE which resembles the binding sequence of NF-E2 and induction of GST subunits and NADPH:QR1 is considerably reduced in Nrf2 null mice. Therefore, it was concluded that Nrf2 is indispensable for the transcriptional stimulation of Phase II enzymes [104]. Glutathione homeostasis is the endogenous antioxidant machinery which is also under the control of Nrf2. ARE is an indispensable constituent of upstream regulatory sequences present on genes for majority of Phase II detoxification enzymes, including the glutamate cysteine ligase catalytic subunit [104] (Fig. **3**).

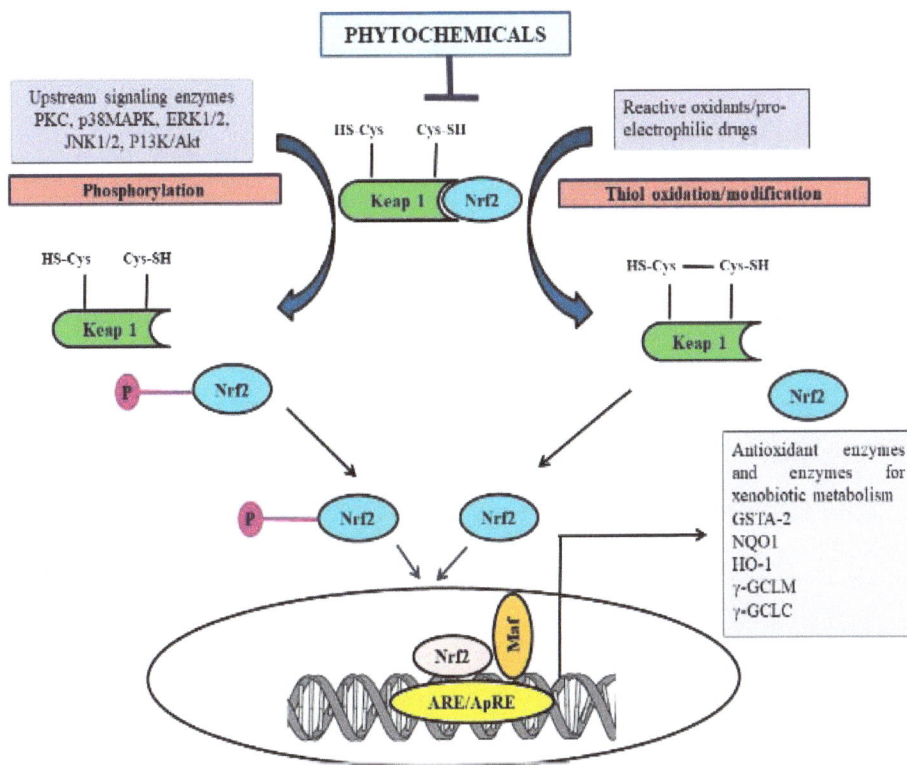

Fig. (3). Signal transduction pathways involving Nrf2-ARE activation. Nrf2 dissociation from Keap1 is brought about by activation of protein kinases such as PI3K, PKC, JNK, p38MAPK and ERK which phosphorylate Nrf2 or reactive oxidants that interact with cysteine residues present in Keap1. Following dissociation Nrf2 translocates to the nucleus and interacts with its coactivator CBP/p300. It forms heterodimers with Maf and bind with ARE thereby trigerring the transcription of enzymes involved in Phase II metabolism and detoxification of xenobiotics.

SUMMARY

Exposure to environmental pollutants and contaminants is a serious concern as it has severe health implications. Their adverse effects are significant and unavoidable. These are detoxified from the body by Phase I and Phase II enzymes. Accumulation and activation of toxins leading to carcinogenesis is a global concern because cancer is one of the leading causes of mortality and morbidity. Intake of phytoconstituents favoring detoxification of xenobiotics may have significant influence on modulation of carcinogenesis. They induce an upregulation of metabolic pathways involved in xenobiotic transformation and elimination. The chemopreventive potential of these compounds is mainly under the control of Nrf2 signaling pathway. Continuous research is revealing the crucial role and effects of compounds of plant origin on toxin metabolism. Furthermore, they are especially safe and associated with low toxicity, making them remarkable candidates as chemopreventive agents.

CONSENT FOR PUBLICATION

Not applicable.

CONFLICT OF INTEREST

The authors declare no conflict of interest, financial or otherwise.

ACKNOWLEDGEMENTS

Declared none.

REFERENCES

[1]   Sandermann H Jr. Higher plant metabolism of xenobiotics: the 'green liver' concept. Pharmacogenetics 1994; 4(5): 225-41.
[http://dx.doi.org/10.1097/00008571-199410000-00001] [PMID: 7894495]

[2]   Anzenbacher P, Anzenbacherová E. Cytochromes P450 and metabolism of xenobiotics. Cell Mol Life Sci 2001; 58(5-6): 737-47.
[http://dx.doi.org/10.1007/PL00000897] [PMID: 11437235]

[3]   Croom E. Metabolism of xenobiotics of human environments. Prog Mol Biol Transl Sci 2012; 112: 31-88.
[http://dx.doi.org/10.1016/B978-0-12-415813-9.00003-9] [PMID: 22974737]

[4]   Xu C, Li CYT, Kong ANT. Induction of phase I, II and III drug metabolism/transport by xenobiotics. Arch Pharm Res 2005; 28(3): 249-68.
[http://dx.doi.org/10.1007/BF02977789] [PMID: 15832810]

[5]   Yu S, Kong AN. Targeting carcinogen metabolism by dietary cancer preventive compounds. Curr Cancer Drug Targets 2007; 7(5): 416-24.
[http://dx.doi.org/10.2174/156800907781386669] [PMID: 17691900]

[6]   Wen X, Donepudi AC, Thomas PE, Slitt AL, King RS, Aleksunes LM. Regulation of hepatic phase II metabolism in pregnant mice. J Pharmacol Exp Ther 2013; 344(1): 244-52.
[http://dx.doi.org/10.1124/jpet.112.199034] [PMID: 23055538]

[7]   Encarnación-Medina J, Rodríguez-Cotto RI, Bloom-Oquendo J, Ortiz-Martínez MG, Duconge J, Jiménez-Vélez B. Selective ATP-binding cassette subfamily C gene expression and proinflammatory mediators released by BEAS-2B after PM2. 5, budesonide, and cotreated exposures. Mediators Inflamm 2017; 2017: 6827194.
[http://dx.doi.org/10.1155/2017/6827194] [PMID: 28900313]

[8]   Frei B, Higdon JV. Antioxidant activity of tea polyphenols *in vivo*: evidence from animal studies. J

Nutr 2003; 133(10): 3275S-84S.
[http://dx.doi.org/10.1093/jn/133.10.3275S] [PMID: 14519826]

[9]   Issa AY, Volate SR, Wargovich MJ. The role of phytochemicals in inhibition of cancer and inflammation: New directions and perspectives. J Food Compos Anal 2006; 19: 405-19.
[http://dx.doi.org/10.1016/j.jfca.2006.02.009]

[10]   Jana S, Mandlekar S. Role of phase II drug metabolizing enzymes in cancer chemoprevention. Curr Drug Metab 2009; 10(6): 595-616.
[http://dx.doi.org/10.2174/138920009789375379] [PMID: 19702535]

[11]   Lerapetritou MG, Georgopoulos PG, Roth CM, Androulakis LP. Tissue-level modeling of xenobiotic metabolism in liver: An emerging tool for enabling clinical translational research. Clin Transl Sci 2009; 2(3): 228-37.
[http://dx.doi.org/10.1111/j.1752-8062.2009.00092.x] [PMID: 20443896]

[12]   Wang P, Heber D, Henning SM. Quercetin increased the antiproliferative activity of green tea polyphenol (-)-epigallocatechin gallate in prostate cancer cells. Nutr Cancer 2012; 64(4): 580-7.
[http://dx.doi.org/10.1080/01635581.2012.661514] [PMID: 22452782]

[13]   Zamek-Gliszczynski MJ, Hoffmaster KA, Nezasa K, Tallman MN, Brouwer KL. Integration of hepatic drug transporters and phase II metabolizing enzymes: mechanisms of hepatic excretion of sulfate, glucuronide, and glutathione metabolites. Eur J Pharm Sci 2006; 27(5): 447-86.
[http://dx.doi.org/10.1016/j.ejps.2005.12.007] [PMID: 16472997]

[14]   Abass K, Turpeinen M, Rautio A, Hakkola J, Pelkonen O. Metabolism of pesticides by human cytochrome P450 enzymes *in vitro*–a survey In Insecticides-advances in integrated pest management In Tech. 2012; pp. 165-94.

[15]   Kulkarni AP, Hodgson E. The metabolism of insecticides: the role of monooxygenase enzymes. Annu Rev Pharmacol Toxicol 1984; 24: 19-42.
[http://dx.doi.org/10.1146/annurev.pa.24.040184.000315] [PMID: 6375545]

[16]   Isin EM, Guengerich FP. Complex reactions catalyzed by cytochrome P450 enzymes. Biochim Biophys Acta 2007; 1770(3): 314-29.
[http://dx.doi.org/10.1016/j.bbagen.2006.07.003] [PMID: 17239540]

[17]   Johnson EF, Stout CD. Structural diversity of human xenobiotic-metabolizing cytochrome P450 monooxygenases. Biochem Biophys Res Commun 2005; 338(1): 331-6.
[http://dx.doi.org/10.1016/j.bbrc.2005.08.190] [PMID: 16157296]

[18]   Yano JK, Wester MR, Schoch GA, Griffin KJ, Stout CD, Johnson EF. The structure of human microsomal cytochrome P450 3A4 determined by X-ray crystallography to 2.05-A resolution. J Biol Chem 2004; 279(37): 38091-4.
[http://dx.doi.org/10.1074/jbc.C400293200] [PMID: 15258162]

[19]   Mazaleuskaya LL, Sangkuhl K, Thorn CF, FitzGerald GA, Altman RB, Klein TE. PharmGKB summary: pathways of acetaminophen metabolism at the therapeutic *versus* toxic doses. Pharmacogenet Genomics 2015; 25(8): 416-26.
[http://dx.doi.org/10.1097/FPC.0000000000000150] [PMID: 26049587]

[20]   Başaran R, Benay CAN. Flavin containing monooxygenases and metabolism of xenobiotics. Turk J Pharm Sci 2017; 14: 90-4.
[http://dx.doi.org/10.4274/tjps.30592]

[21]   Parkinson A, Ogilvie BW. Biotransformation of xenobiotics.Casarett and Doull's Toxicology. New York: McGraw-Hill 2001; pp. 113-86.

[22]   Cashman JR, Zhang J. Human flavin-containing monooxygenases. Annu Rev Pharmacol Toxicol 2006; 46: 65-100.
[http://dx.doi.org/10.1146/annurev.pharmtox.46.120604.141043] [PMID: 16402899]

[23]   Ziegler DM. An overview of the mechanism, substrate specificities, and structure of FMOs. Drug

Metab Rev 2002; 34(3): 503-11.
[http://dx.doi.org/10.1081/DMR-120005650] [PMID: 12214662]

[24]   Wermuth B, Platts KL, Seidel A, Oesch F. Carbonyl reductase provides the enzymatic basis of quinone detoxication in man. Biochem Pharmacol 1986; 35(8): 1277-82.
[http://dx.doi.org/10.1016/0006-2952(86)90271-6] [PMID: 3083821]

[25]   Deller S, Macheroux P, Sollner S. Flavin-dependent quinone reductases. Cell Mol Life Sci 2008; 65(1): 141-60.
[http://dx.doi.org/10.1007/s00018-007-7300-y] [PMID: 17938860]

[26]   Sandberg M, Hassett C, Adman ET, Meijer J, Omiecinski CJ. Identification and functional characterization of human soluble epoxide hydrolase genetic polymorphisms. J Biol Chem 2000; 275(37): 28873-81.
[http://dx.doi.org/10.1074/jbc.M001153200] [PMID: 10862610]

[27]   Hosagrahara VP, Rettie AE, Hassett C, Omiecinski CJ. Functional analysis of human microsomal epoxide hydrolase genetic variants. Chem Biol Interact 2004; 150(2): 149-59.
[http://dx.doi.org/10.1016/j.cbi.2004.07.004] [PMID: 15535985]

[28]   Armstrong RN. Kinetic and chemical mechanism of epoxide hydrolase. Drug Metab Rev 1999; 31(1): 71-86.
[http://dx.doi.org/10.1081/DMR-100101908] [PMID: 10065366]

[29]   Decker M, Arand M, Cronin A. Mammalian epoxide hydrolases in xenobiotic metabolism and signalling. Arch Toxicol 2009; 83(4): 297-318.
[http://dx.doi.org/10.1007/s00204-009-0416-0] [PMID: 19340413]

[30]   Shimada T. Xenobiotic-metabolizing enzymes involved in activation and detoxification of carcinogenic polycyclic aromatic hydrocarbons. Drug Metab Pharmacokinet 2006; 21(4): 257-76.
[http://dx.doi.org/10.2133/dmpk.21.257] [PMID: 16946553]

[31]   Kobayashi K, Bouscarel B, Matsuzaki Y, Ceryak S, Kudoh S, Fromm H. pH-dependent uptake of irinotecan and its active metabolite, SN-38, by intestinal cells. Int J Cancer 1999; 83(4): 491-6.
[http://dx.doi.org/10.1002/(SICI)1097-0215(19991112)83:4<491::AID-IJC10>3.0.CO;2-M] [PMID: 10508485]

[32]   Dirr H, Reinemer P, Huber R. X-ray crystal structures of cytosolic glutathione S-transferases. Implications for protein architecture, substrate recognition and catalytic function. Eur J Biochem 1994; 220(3): 645-61.
[http://dx.doi.org/10.1111/j.1432-1033.1994.tb18666.x] [PMID: 8143720]

[33]   Bbosa GS, Kitya D, Odda J, Ogwal-Okeng J. Aflatoxins metabolism, effects on epigenetic mechanisms and their role in carcinogenesis. Health 2013; 5: 14-34.
[http://dx.doi.org/10.4236/health.2013.510A1003] [PMID: 25002916]

[34]   Gamage N, Barnett A, Hempel N, *et al.* Human sulfotransferases and their role in chemical metabolism. Toxicol Sci 2006; 90(1): 5-22.
[http://dx.doi.org/10.1093/toxsci/kfj061] [PMID: 16322073]

[35]   Grillo MP. Bioactivation by phase-II-enzyme-catalyzed conjugation of xenobiotics Encyclopedia of Drug Metabolism and Interactions. John Wiley & Sons 2011.

[36]   Sanchez RI, Kauffman FC. Regulation of xenobiotic metabolism in the liver. Comprehensive Toxicology 2010; 9: 109-28.
[http://dx.doi.org/10.1016/B978-0-08-046884-6.01005-8]

[37]   Jancova P, Anzenbacher P, Anzenbacherova E. Phase II drug metabolizing enzymes. Biomed Pap Med Fac Univ Palacky Olomouc Czech Repub 2010; 154(2): 103-16.
[http://dx.doi.org/10.5507/bp.2010.017] [PMID: 20668491]

[38]   Weinshilboum RM, Otterness DM, Szumlanski CL. Methylation pharmacogenetics: catechol O-methyltransferase, thiopurine methyltransferase, and histamine N-methyltransferase. Annu Rev

Pharmacol Toxicol 1999; 39: 19-52.
[http://dx.doi.org/10.1146/annurev.pharmtox.39.1.19] [PMID: 10331075]

[39]   Debiec-Rychter M, Land SJ, King CM. Histological localization of acetyltransferases in human tissue.
        Cancer Lett 1999; 143(2): 99-102.
        [http://dx.doi.org/10.1016/S0304-3835(99)00135-4] [PMID: 10503885]

[40]   Ginsberg G, Smolenski S, Neafsey P, *et al.* The influence of genetic polymorphisms on population
        variability in six xenobiotic-metabolizing enzymes. J Toxicol Environ Health B Crit Rev 2009; 12(5-
        6): 307-33.
        [http://dx.doi.org/10.1080/10937400903158318] [PMID: 20183525]

[41]   Butcher NJ, Boukouvala S, Sim E, Minchin RF. Pharmacogenetics of the arylamine N-
        acetyltransferases. Pharmacogenomics J 2002; 2(1): 30-42.
        [http://dx.doi.org/10.1038/sj.tpj.6500053] [PMID: 11990379]

[42]   Kato R, Yamazoe Y. Metabolic activation of N-hydroxylated metabolites of carcinogenic and
        mutagenic arylamines and arylamides by esterification. Drug Metab Rev 1994; 26(1-2): 413-29.
        [http://dx.doi.org/10.3109/03602539409029806] [PMID: 8082577]

[43]   Fan PW, Zhang D, Halladay JS, Driscoll JP, Khojasteh SC. Going beyond common drug metabolizing
        enzymes: case studies of biotransformation involving aldehyde oxidase, γ-glutamyl transpeptidase,
        cathepsin B, flavin-containing monooxygenase, and ADP-ribosyltransferase. Drug Metab Dispos
        2016; 44(8): 1253-61.
        [http://dx.doi.org/10.1124/dmd.116.070169] [PMID: 27117704]

[44]   Sallustio BC. Glucuronidation-dependent toxicity and bioactivation. Advances in Molecular
        Toxicology 2008; 2: 57-86.
        [http://dx.doi.org/10.1016/S1872-0854(07)02003-6]

[45]   Monks TJ, Anders MW, Dekant W, Stevens JL, Lau SS, van Bladeren PJ. Glutathione conjugate
        mediated toxicities. Toxicol Appl Pharmacol 1990; 106(1): 1-19. a
        [http://dx.doi.org/10.1016/0041-008X(90)90100-9] [PMID: 2251674]

[46]   Monks TJ, Highet RJ, Lau SS. Oxidative cyclization, 1,4-benzothiazine formation and dimerization of
        2-bromo-3-(glutathion-S-yl)hydroquinone. Mol Pharmacol 1990; 38(1): 121-7. b
        [PMID: 1973524]

[47]   Jomova K, Vondrakova D, Lawson M, Valko M. Metals, oxidative stress and neurodegenerative
        disorders. Mol Cell Biochem 2010; 345(1-2): 91-104.
        [http://dx.doi.org/10.1007/s11010-010-0563-x] [PMID: 20730621]

[48]   Saw CL, Cintrón M, Wu TY, *et al.* Pharmacodynamics of dietary phytochemical indoles I3C and
        DIM: Induction of Nrf2-mediated phase II drug metabolizing and antioxidant genes and synergism
        with isothiocyanates. Biopharm Drug Dispos 2011; 32(5): 289-300.
        [http://dx.doi.org/10.1002/bdd.759] [PMID: 21656528]

[49]   Shu L, Khor TO, Lee JH, *et al.* Epigenetic CpG demethylation of the promoter and reactivation of the
        expression of Neurog1 by curcumin in prostate LNCaP cells. AAPS J 2011; 13(4): 606-14.
        [http://dx.doi.org/10.1208/s12248-011-9300-y] [PMID: 21938566]

[50]   Tan AC, Konczak I, Sze DMY, Ramzan I. Molecular pathways for cancer chemoprevention by dietary
        phytochemicals. Nutr Cancer 2011; 63(4): 495-505.
        [http://dx.doi.org/10.1080/01635581.2011.538953] [PMID: 21500099]

[51]   Tanaka T, Kojima T, Kawamori T, Yoshimi N, Mori H. Chemoprevention of diethylnitrosamine-
        induced hepatocarcinogenesis by a simple phenolic acid protocatechuic acid in rats. Cancer Res 1993;
        53(12): 2775-9.
        [PMID: 8504418]

[52]   Ahn D, Putt D, Kresty L, Stoner GD, Fromm D, Hollenberg PF. The effects of dietary ellagic acid on
        rat hepatic and esophageal mucosal cytochromes P450 and phase II enzymes. Carcinogenesis 1996;

17(4): 821-8.
[http://dx.doi.org/10.1093/carcin/17.4.821] [PMID: 8625497]

[53] Hussein RH, Khalifa FK. The protective role of ellagitannins flavonoids pretreatment against N-nitrosodiethylamine induced-hepatocellular carcinoma. Saudi J Biol Sci 2014; 21(6): 589-96.
[http://dx.doi.org/10.1016/j.sjbs.2014.03.004] [PMID: 25473368]

[54] Ow YY, Stupans I. Gallic acid and gallic acid derivatives: effects on drug metabolizing enzymes. Curr Drug Metab 2003; 4(3): 241-8.
[http://dx.doi.org/10.2174/1389200033489479] [PMID: 12769668]

[55] Giftson JS, Jayanthi S, Nalini N. Chemopreventive efficacy of gallic acid, an antioxidant and anticarcinogenic polyphenol, against 1,2-dimethyl hydrazine induced rat colon carcinogenesis. Invest New Drugs 2010; 28(3): 251-9.
[http://dx.doi.org/10.1007/s10637-009-9241-9] [PMID: 19300909]

[56] Aglan HA, Ahmed HH, El-Toumy SA, Mahmoud NS. Gallic acid against hepatocellular carcinoma: An integrated scheme of the potential mechanisms of action from *in vivo* study. Tumour Biol 2017; 39(6): 1010428317699127.
[http://dx.doi.org/10.1177/1010428317699127] [PMID: 28618930]

[57] Chu X, Wang H, Jiang YM, et al. Ameliorative effects of tannic acid on carbon tetrachloride-induced liver fibrosis *in vivo* and *in vitro*. J Pharmacol Sci 2016; 130(1): 15-23.
[http://dx.doi.org/10.1016/j.jphs.2015.12.002] [PMID: 26810570]

[58] Dai N, Zou Y, Zhu L, Wang HF, Dai MG. Antioxidant properties of proanthocyanidins attenuate carbon tetrachloride (CCl_4)-induced steatosis and liver injury in rats *via* CYP2E1 regulation. J Med Food 2014; 17(6): 663-9.
[http://dx.doi.org/10.1089/jmf.2013.2834] [PMID: 24712752]

[59] Krajka-Kuźniak V, Baer-Dubowska W. The effects of tannic acid on cytochrome P450 and phase II enzymes in mouse liver and kidney. Toxicol Lett 2003; 143(2): 209-16.
[http://dx.doi.org/10.1016/S0378-4274(03)00177-2] [PMID: 12749824]

[60] Gao SS, Chen XY, Zhu RZ, Choi BM, Kim SJ, Kim BR. Dual effects of phloretin on aflatoxin B1 metabolism: activation and detoxification of aflatoxin B1. Biofactors 2012; 38(1): 34-43.
[http://dx.doi.org/10.1002/biof.190] [PMID: 22253071]

[61] Anand MAV, Suresh K. Biochemical profiling and chemopreventive activity of phloretin on 7,12-Dimethylbenz (a) anthracene induced oral carcinogenesis in male golden Syrian hamsters. Toxicol Int 2014; 21(2): 179-85.
[http://dx.doi.org/10.4103/0971-6580.139805] [PMID: 25253928]

[62] Gradelet S, Astorg P, Leclerc J, Chevalier J, Vernevaut MF, Siess MH. Effects of canthaxanthin, astaxanthin, lycopene and lutein on liver xenobiotic-metabolizing enzymes in the rat. Xenobiotica 1996; 26(1): 49-63.
[http://dx.doi.org/10.3109/00498259609046688] [PMID: 8851821]

[63] Tripathi DN, Jena GB. Astaxanthin intervention ameliorates cyclophosphamide-induced oxidative stress, DNA damage and early hepatocarcinogenesis in rat: role of Nrf2, p53, p38 and phase-II enzymes. Mutat Res 2010; 696(1): 69-80.
[http://dx.doi.org/10.1016/j.mrgentox.2009.12.014] [PMID: 20038455]

[64] Kavitha K, Thiyagarajan P, Rathna Nandhini J, Mishra R, Nagini S. Chemopreventive effects of diverse dietary phytochemicals against DMBA-induced hamster buccal pouch carcinogenesis *via* the induction of Nrf2-mediated cytoprotective antioxidant, detoxification, and DNA repair enzymes. Biochimie 2013; 95(8): 1629-39.
[http://dx.doi.org/10.1016/j.biochi.2013.05.004] [PMID: 23707664]

[65] Li Z, Dong X, Liu H, et al. Astaxanthin protects ARPE-19 cells from oxidative stress *via* upregulation of Nrf2-regulated phase II enzymes through activation of PI3K/Akt. Mol Vis 2013; 19: 1656-66.
[PMID: 23901249]

[66] Astorg P, Gradelet S, Leclerc J, Canivenc MC, Siess MH. Effects of β-carotene and canthaxanthin on liver xenobiotic-metabolizing enzymes in the rat. Food Chem Toxicol 1994; 32(8): 735-42.
[http://dx.doi.org/10.1016/S0278-6915(09)80006-9] [PMID: 8070738]

[67] Wang H, Leung LK. The carotenoid lycopene differentially regulates phase I and II enzymes in dimethylbenz[a]anthracene-induced MCF-7 cells. Nutrition 2010; 26(11-12): 1181-7.
[http://dx.doi.org/10.1016/j.nut.2009.11.013] [PMID: 20400267]

[68] Miyake S, Kobayashi S, Tsubota K, Ozawa Y. Phase II enzyme induction by a carotenoid, lutein, in a PC12D neuronal cell line. Biochem Biophys Res Commun 2014; 446(2): 535-40.
[http://dx.doi.org/10.1016/j.bbrc.2014.02.135] [PMID: 24613837]

[69] Szaefer H, Cichocki M, Brauze D, Baer-Dubowska W. Alteration in phase I and II enzyme activities and polycyclic aromatic hydrocarbons-DNA adduct formation by plant phenolics in mouse epidermis. Nutr Cancer 2004; 48(1): 70-7.
[http://dx.doi.org/10.1207/s15327914nc4801_10] [PMID: 15203380]

[70] Szaefer H, Krajka-Kuźniak V, Baer-Dubowska W. The effect of initiating doses of benzo[a]pyrene and 7,12-dimethylbenz[a]anthracene on the expression of PAH activating enzymes and its modulation by plant phenols. Toxicology 2008; 251(1-3): 28-34.
[http://dx.doi.org/10.1016/j.tox.2008.07.047] [PMID: 18694800]

[71] Upadhyay G, Singh AK, Kumar A, Prakash O, Singh MP. Resveratrol modulates pyrogallol-induced changes in hepatic toxicity markers, xenobiotic metabolizing enzymes and oxidative stress. Eur J Pharmacol 2008; 596(1-3): 146-52.
[http://dx.doi.org/10.1016/j.ejphar.2008.08.019] [PMID: 18789925]

[72] Wang Y, Jiang Y, Fan X, *et al.* Hepato-protective effect of resveratrol against acetaminophen-induced liver injury is associated with inhibition of CYP-mediated bioactivation and regulation of SIRT1-p53 signaling pathways. Toxicol Lett 2015; 236(2): 82-9.
[http://dx.doi.org/10.1016/j.toxlet.2015.05.001] [PMID: 25956474]

[73] Zhang H, Sun Q, Xu T, *et al.* Resveratrol attenuates the progress of liver fibrosis *via* the Akt/nuclear factor-κB pathways. Mol Med Rep 2016; 13(1): 224-30.
[http://dx.doi.org/10.3892/mmr.2015.4497] [PMID: 26530037]

[74] Chuang SE, Kuo ML, Hsu CH, *et al.* Curcumin-containing diet inhibits diethylnitrosamine-induced murine hepatocarcinogenesis. Carcinogenesis 2000; 21(2): 331-5.
[http://dx.doi.org/10.1093/carcin/21.2.331] [PMID: 10657978]

[75] Sreepriya M, Bali G. Chemopreventive effects of embelin and curcumin against N-nitrosod--ethylamine/ phenobarbital- induced hepatocarcinogenesis in Wistar rats. Fitoterapia 2005; 76(6): 549-55.
[http://dx.doi.org/10.1016/j.fitote.2005.04.014] [PMID: 16009505]

[76] Sreepriya M, Bali G. Effects of administration of Embelin and Curcumin on lipid peroxidation, hepatic glutathione antioxidant defense and hematopoietic system during N-nitrosodi-ethylamine/ Phenobarbital- induced hepatocarcinogenesis in Wistar rats. Mol Cell Biochem 2006; 284(1-2): 49-55.
[http://dx.doi.org/10.1007/s11010-005-9012-7] [PMID: 16477385]

[77] Mann CD, Neal CP, Garcea G, Manson MM, Dennison AR, Berry DP. Phytochemicals as potential chemopreventive and chemotherapeutic agents in hepatocarcinogenesis. Eur J Cancer Prev 2009; 18(1): 13-25.
[http://dx.doi.org/10.1097/CEJ.0b013e3282f0c090] [PMID: 19077560]

[78] Fujise Y, Okano JI, Nagahara T, Abe R, Imamoto R, Murawaki Y. Preventive effect of caffeine and curcumin on hepatocarcinogenesis in diethylnitrosamine-induced rats Corrigendum in/103892/ijo 20163577 Int J Oncol 2012; 40: 1779-88.

[79] Zhao JA, Peng L, Geng CZ, *et al.* Preventive effect of hydrazinocurcumin on carcinogenesis of diethylnitrosamine-induced hepatocarcinoma in male SD rats. Asian Pac J Cancer Prev 2014; 15(5):

2115-21.
[http://dx.doi.org/10.7314/APJCP.2014.15.5.2115] [PMID: 24716943]

[80] Kadasa NM, Abdallah H, Afifi M, Gowayed S. Hepatoprotective effects of curcumin against diethyl nitrosamine induced hepatotoxicity in albino rats. Asian Pac J Cancer Prev 2015; 16(1): 103-8.
[http://dx.doi.org/10.7314/APJCP.2015.16.1.103] [PMID: 25640336]

[81] Tan XL, Shi M, Tang H, Han W, Spivack SD. Candidate dietary phytochemicals modulate expression of phase II enzymes GSTP1 and NQO1 in human lung cells. J Nutr 2010; 140(8): 1404-10.
[http://dx.doi.org/10.3945/jn.110.121905] [PMID: 20554899]

[82] Misaka S, Kawabe K, Onoue S, et al. Effects of green tea catechins on cytochrome P450 2B6, 2C8, 2C19, 2D6 and 3A activities in human liver and intestinal microsomes. Drug Metab Pharmacokinet 2013; 28(3): 244-9.
[http://dx.doi.org/10.2133/dmpk.DMPK-12-RG-101] [PMID: 23268924]

[83] James KD, Forester SC, Lambert JD. Dietary pretreatment with green tea polyphenol, (-)- epigallocatechin-3-gallate reduces the bioavailability and hepatotoxicity of subsequent oral bolus doses of (-)-epigallocatechin-3-gallate. Food Chem Toxicol 2015; 76: 103-8.
[http://dx.doi.org/10.1016/j.fct.2014.12.009] [PMID: 25528115]

[84] Yao HT, Yang YC, Chang CH, Yang HT, Yin MC. Protective effects of (-)-epigallocatechin-3-gallate against acetaminophen-induced liver injury in rats). Biomedicine (Taipei) 2015; 5(3): 15.
[http://dx.doi.org/10.7603/s40681-015-0015-8] [PMID: 26264479]

[85] Weng Z, Greenhaw J, Salminen WF, Shi Q. Mechanisms for epigallocatechin gallate induced inhibition of drug metabolizing enzymes in rat liver microsomes. Toxicol Lett 2012; 214(3): 328-38.
[http://dx.doi.org/10.1016/j.toxlet.2012.09.011] [PMID: 23010222]

[86] Wang D, Wang Y, Wan X, Yang CS, Zhang J. Green tea polyphenol (-)-epigallocatechin-3-gallate triggered hepatotoxicity in mice: responses of major antioxidant enzymes and the Nrf2 rescue pathway. Toxicol Appl Pharmacol 2015; 283(1): 65-74.
[http://dx.doi.org/10.1016/j.taap.2014.12.018] [PMID: 25585349]

[87] Sánchez-Pérez Y, Carrasco-Legleu C, García-Cuellar C, et al. Oxidative stress in carcinogenesis. Correlation between lipid peroxidation and induction of preneoplastic lesions in rat hepatocarcinogenesis. Cancer Lett 2005; 217(1): 25-32.
[http://dx.doi.org/10.1016/j.canlet.2004.07.019] [PMID: 15596293]

[88] Ghosh A, Ghosh D, Sarkar S, Mandal AK, Thakur Choudhury S, Das N. Anticarcinogenic activity of nanoencapsulated quercetin in combating diethylnitrosamine-induced hepatocarcinoma in rats. Eur J Cancer Prev 2012; 21(1): 32-41.
[http://dx.doi.org/10.1097/CEJ.0b013e32834a7e2b] [PMID: 21968689]

[89] Liu Y, Wu YM, Yu Y, et al. Curcumin and resveratrol in combination modulate drug-metabolizing enzymes as well as antioxidant indices during lung carcinogenesis in mice. Hum Exp Toxicol 2015; 34(6): 620-7.
[http://dx.doi.org/10.1177/0960327114551396] [PMID: 25632966]

[90] Ganai AA, Khan AA, Malik ZA, Farooqi H. Genistein modulates the expression of NF-κB and MAPK (p-38 and ERK1/2), thereby attenuating d-Galactosamine induced fulminant hepatic failure in Wistar rats. Toxicol Appl Pharmacol 2015; 283(2): 139-46.
[http://dx.doi.org/10.1016/j.taap.2015.01.012] [PMID: 25620059]

[91] Kim EY, Hong KB, Suh HJ, Choi HS. Protective effects of germinated and fermented soybean extract against tert-butyl hydroperoxide-induced hepatotoxicity in HepG2 cells and in rats. Food Funct 2015; 6(11): 3512-21.
[http://dx.doi.org/10.1039/C5FO00785B] [PMID: 26299642]

[92] Singh P, Sharma S, Rath SK. Genistein induces deleterious effects during its acute exposure in Swiss mice. BioMed Res Int 2014; 2014: 619617.
[http://dx.doi.org/10.1155/2014/619617] [PMID: 24967385]

[93]   Breikaa RM, Algandaby MM, El-Demerdash E, Abdel-Naim AB. Biochanin A protects against acute carbon tetrachloride-induced hepatotoxicity in rats. Biosci Biotechnol Biochem 2013; 77(5): 909-16.
[http://dx.doi.org/10.1271/bbb.120675] [PMID: 23649249]

[94]   Jalaludeen AM, Ha WT, Lee R, *et al.* Biochanin a ameliorates arsenic-induced hepato-and hematotoxicity in rats. Molecules 2016; 21(1): 69-83.
[http://dx.doi.org/10.3390/molecules21010069] [PMID: 26760991]

[95]   Janbaz KH, Saeed SA, Gilani AH. Protective effect of rutin on paracetamol- and CCl4-induced hepatotoxicity in rodents. Fitoterapia 2002; 73(7-8): 557-63.
[http://dx.doi.org/10.1016/S0367-326X(02)00217-4] [PMID: 12490212]

[96]   Zargar S, Wani TA, Alamro AA, Ganaie MA. Amelioration of thioacetamide-induced liver toxicity in Wistar rats by rutin. Int J Immunopathol Pharmacol 2017; 30(3): 207-14.
[http://dx.doi.org/10.1177/0394632017714175] [PMID: 28590141]

[97]   Shenbagam M, Nalini N. Dose response effect of rutin a dietary antioxidant on alcohol-induced prooxidant and antioxidant imbalance - a histopathologic study. Fundam Clin Pharmacol 2011; 25(4): 493-502.
[http://dx.doi.org/10.1111/j.1472-8206.2010.00861.x] [PMID: 20727014]

[98]   Khan RA, Khan MR, Sahreen S. CCl4-induced hepatotoxicity: protective effect of rutin on p53, CYP2E1 and the antioxidative status in rat. BMC Complement Altern Med 2012; 12: 178.
[http://dx.doi.org/10.1186/1472-6882-12-178] [PMID: 23043521]

[99]   Hafez MM, Al-Shabanah OA, Al-Harbi NO, *et al.* Association between paraoxonases gene expression and oxidative stress in hepatotoxicity induced by CCl4. Oxid Med Cell Longev 2014; 2014: 893212.
[http://dx.doi.org/10.1155/2014/893212] [PMID: 25478064]

[100] Al-Rejaie SS, Aleisa AM, Sayed-Ahmed MM, *et al.* Protective effect of rutin on the antioxidant genes expression in hypercholestrolemic male Westar rat. BMC Complement Altern Med 2013; 13: 136-45.
[http://dx.doi.org/10.1186/1472-6882-13-136] [PMID: 23773725]

[101] Erdogan E, Ilgaz Y, Gurgor PN, Oztas Y, Topal T, Oztas E. Rutin ameliorates methotrexate induced hepatic injury in rats. Acta Cir Bras 2015; 30(11): 778-84.
[http://dx.doi.org/10.1590/S0102-865020150110000009] [PMID: 26647798]

[102] Hayes JD, Dinkova-Kostova AT. The Nrf2 regulatory network provides an interface between redox and intermediary metabolism. Trends Biochem Sci 2014; 39(4): 199-218.
[http://dx.doi.org/10.1016/j.tibs.2014.02.002] [PMID: 24647116]

[103] Itoh K, Chiba T, Takahashi S, *et al.* An Nrf2/small Maf heterodimer mediates the induction of phase II detoxifying enzyme genes through antioxidant response elements. Biochem Biophys Res Commun 1997; 236(2): 313-22.
[http://dx.doi.org/10.1006/bbrc.1997.6943] [PMID: 9240432]

[104] Li M, Chiu JF, Kelsen A, Lu SC, Fukagawa NK. Identification and characterization of an Nrf2-mediated ARE upstream of the rat glutamate cysteine ligase catalytic subunit gene (GCLC). J Cell Biochem 2009; 107(5): 944-54.
[http://dx.doi.org/10.1002/jcb.22197] [PMID: 19459163]

[105] Giftson Senapathy J, Jayanthi S, Viswanathan P, Umadevi P, Nalini N. Effect of gallic acid on xenobiotic metabolizing enzymes in 1,2-dimethyl hydrazine induced colon carcinogenesis in Wistar rats--a chemopreventive approach. Food Chem Toxicol 2011; 49(4): 887-92.
[http://dx.doi.org/10.1016/j.fct.2010.12.012] [PMID: 21172399]

[106] Krajka-Kuźniak V, Szaefer H, Baer-Dubowska W. Modulation of 3-methylcholanthrene-induced rat hepatic and renal cytochrome P450 and phase II enzymes by plant phenols: protocatechuic and tannic acids. Toxicol Lett 2004; 152(2): 117-26.
[PMID: 15302093]

[107] Pohl C, Will F, Dietrich H, Schrenk D. Cytochrome P450 1A1 expression and activity in Caco-2 cells:

modulation by apple juice extract and certain apple polyphenols. J Agric Food Chem 2006; 54(26): 10262-8.
[http://dx.doi.org/10.1021/jf061791c] [PMID: 17177569]

[108] Tsuji PA, Walle T. Benzo[a]pyrene-induced cytochrome P450 1A and DNA binding in cultured trout hepatocytes - inhibition by plant polyphenols. Chem Biol Interact 2007; 169(1): 25-31.
[http://dx.doi.org/10.1016/j.cbi.2007.05.001] [PMID: 17583686]

[109] Liu Y, Wu YM, Zhang PY. Protective effects of curcumin and quercetin during benzo(a)pyrene induced lung carcinogenesis in mice. Eur Rev Med Pharmacol Sci 2015; 19(9): 1736-43.
[PMID: 26004618]

[110] Leung HY, Yung LH, Shi G, Lu AL, Leung LK. The red wine polyphenol resveratrol reduces polycyclic aromatic hydrocarbon-induced DNA damage in MCF-10A cells. Br J Nutr 2009; 102(10): 1462-8.
[http://dx.doi.org/10.1017/S0007114509990481] [PMID: 19811694]

[111] Russell GK, Gupta RC, Vadhanam MV. Effect of phytochemical intervention on dibenzo [a, l] pyrene-induced DNA adduct formation Mutat Res/Fund Mol Mech Mutagen 2015; 774: 25-32.

[112] Choi H, Chun YS, Shin YJ, Ye SK, Kim MS, Park JW. Curcumin attenuates cytochrome P450 induction in response to 2,3,7,8-tetrachlorodibenzo-p-dioxin by ROS-dependently degrading AhR and ARNT. Cancer Sci 2008; 99(12): 2518-24.
[http://dx.doi.org/10.1111/j.1349-7006.2008.00984.x] [PMID: 19018768]

[113] Osawa T. Nephroprotective and hepatoprotective effects of curcuminoids. Adv Exp Med Biol 2007; 595: 407-23.
[http://dx.doi.org/10.1007/978-0-387-46401-5_18] [PMID: 17569222]

[114] Muto S, Fujita KI, Yamazaki Y, Kamataki T. Inhibition by green tea catechins of metabolic activation of procarcinogens by human cytochrome P450 Mutat Res/Fund Mol Mech Mutagen 2001; 479: 197-206.
[http://dx.doi.org/10.1016/S0027-5107(01)00204-4]

[115] Donnez D, Jeandet P, Clément C, Courot E. Bioproduction of resveratrol and stilbene derivatives by plant cells and microorganisms. Trends Biotechnol 2009; 27(12): 706-13.
[http://dx.doi.org/10.1016/j.tibtech.2009.09.005] [PMID: 19875185]

Environmental Contaminants and Natural Products, 2019, 193-209

CHAPTER 9

Potential Health Benefits of Nutraceuticals for Human Health

Sandeep Kaur[1]**, Manish Kumar**[2]**, Kritika Pandit**[1]**, Ajay Kumar**[1] **and Satwinderjeet Kaur**[1,*]

[1] *Department of Botanical and Environmental Sciences, Guru Nanak Dev University, Amritsar-143005, Punjab, India*

[2] *Department of Biology, S.D. College, Barnala, Punjab-148101, India*

Abstract: Nutraceuticals or functional food plays a key role in curbing the inception and etiology of several degenerative diseases and acts as a medicinal food with least side effects in comparison to other therapeutic agents and thereby successfully implemented in modern drug discovery. Global nutraceutical market was expected to rise by 67 billion US dollars approximately in 2016 which indicated surging demand for functional foods worldwide. The major nutraceuticals market includes supplements/medications, foods and functional beverages that provide mental acuity, energy and immunity of the individuals. This review focuses on the novel strategies that implemented the use of nutraceuticals for preventing and treating the diseases/disorders by avoiding unhealthy food and promoting the consumption of foods that go beyond basic nutrition. In the present review, we demonstrate the role of nutraceuticals in modulating certain physiological and pathophysiological processes which are involved in the development of human acute and chronic diseases.

Keywords: Acquired Immuno Deficiency Syndrome, Chronic Diseases, Chronic Mild Stress, Cancer, Flowers, Fruits, Functional Foods, Human Health, Human Immunodeficiency Virus, High Fat Diet, Medicinal Plants, Multi-Drug Resistant, Natural Sources, Nutraceuticals, Phytoconstituents, Protective Factors, Therapeutic Agents.

INTRODUCTION

The term "Nutraceutical" is a combination of two words 'nutrition' and 'pharmaceutical' which indicate a strong relationship between the nutrients and their use in therapeutic drugs [1, 2]. The Nutraceuticals industry of India is expected to rise by 2019-2020 at 20% to USD 6.1 billion due to awareness about

---

* **Corresponding author Satwinderjeet Kaur:** Department of Botanical and Environmental Sciences, Guru Nanak Dev University, Amritsar-143005, (Punjab) India; Tel: +91-9988195508; E-mail: sjkaur2011@gmail.com

the changing life style, health and fitness. On the basis of their chemical grouping and natural source, nutraceuticals are divided into three categories *viz.* nutrients, dietary supplements, herbals and dietary fiber. Out of these, dietary supplements were considered as the rapidly growing segment of industry with an increase of 19.5% per year in comparison to 11.6% per year of natural/herbal products [3]. Nutraceuticals contain phytochemical constituents that have long term health promoting and medicinal qualities. In nutraceuticals, the benefits to health may be arising from its long-term use as food and also play a nutritional role in the diet *i.e.* chemoprevention [4]. In contrast, medicinal plants exhibit their medicinal actions without serving a nutritive role in the diet and are useful in response to specific health issues over short and long-term intervals [5]. The bioactive phytochemicals derived from plants have become a significant source of nutraceutical ingredients. Out of these phytochemicals, several groups of polyphenols (flavanones, isoflavones, anthocyanins, proanthocyanidins, ellagic acid and resveratrol) are currently found useful in the nutraceutical industry. Nutraceuticals as dietary supplements provide an incredible health and medicinal value for the prevention and treatment of various ailments. It is also a food and food product that is considered as a basic part of normal diet and also found beneficial beyond the effects of traditional nutrition. The famous quote "Let food be thy medicine and medicine be thy food" by Hippocrates was found to be best suited for nutraceutical. Dr. Stephen L. DeFelice, founder and chairperson of the Foundation of Innovation Medicine (FIM), Crawford, New Jersey coined and defined the term "Nutraceutical" in 1989 [6]. However, later on Health Canada has modified its meaning and also defined it as an isolated and purified product obtained from foods and then sold in the market in the form of medicine but not as a food. For the last few decades, the interest in the field of nutraceuticals has increased considerably due to their affordable bioavailability, safety and other therapeutic effects [7]. The processed, formulated and fortified foods consumed as a dietary supplementation improve the health of an individual. Mostly, nutraceuticals are considered as components of food derived from raw materials of herbals and botanical products, which are useful in the treatment and prevention of various degenerative and chronic diseases. Till date, a renaissance has been created by exploiting and exploring the multitude of phytochemicals present in nutritive and non-nutritive products for their ability to defend against diseases [8].

The quantitative and qualitative analysis of the nutraceutical or functional food are crucial for the prevention of over dosing or adverse effects on the human health. Nutraceuticals as dietary supplement are a food product that is orally consumed and contain dietary ingredients which intend to supplement the diet. These products included fruits, vegetables, medicinal plants, herbs, minerals, vitamins, nuts, crops and probiotics that play an essential role in modulating the various physiological and pathophysiological processes of the body (Fig. **1**) [9]. The designing or manufacturing of nutraceutical or functional food helps the consumers to take them in their natural state instead of taking them as dietary supplements in the form of tablets and capsules, which lead to several chronic and acute diseases due to their intolerable side effects. Although, myriads of biologically active constituents are known but the conjugation of various inactive and active components affects the biological activity of different nutraceutical products [10]. These products helps in maintaining and restoring the normal body functioning of an individual in disease state and eventually step by step reduce the abnormalities.

Several studies have illustrated the beneficial effects of nutraceutical and functional food in promoting human health. This chapter will highlight the role of different nutraceuticals in the delay, reversal and prevention of innumerable degenerative and chronic diseases.

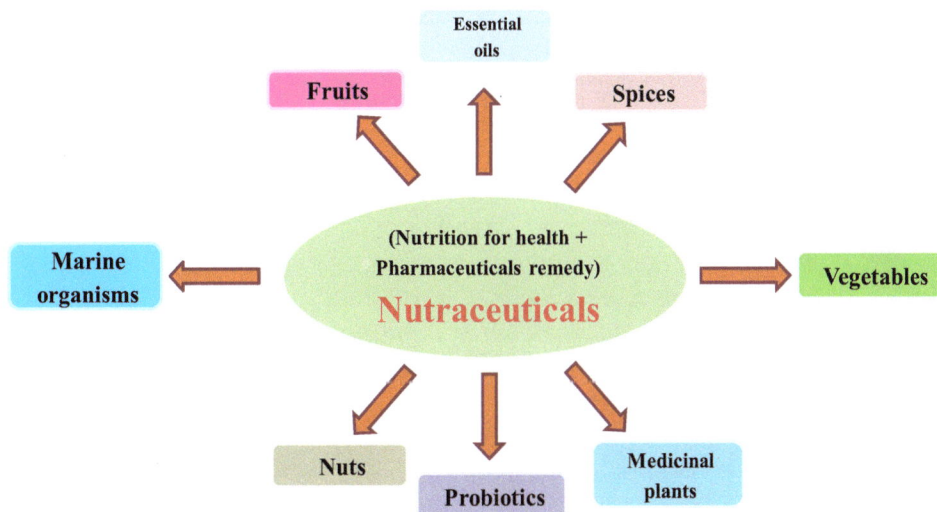

Fig. (1). Representing various nutraceuticals involved in providing health benefits.

Fruits and Vegetables

According to the WHO and FAO reports on the prevention and control of non-communicable diseases, Third Global Forum was held at Rio de Janeiro, Brazil in November, 2003, in which more emphasis was given on the awareness about the adequate consumption of fruits and vegetables. Various research findings have highlighted the importance of fruits and vegetables in promoting human health due to their affordability, bioavailability and safety. The consumption of polyphenol rich fruit, Açai (*Euterpe oleracea*) was found beneficial in improving vascular function, oxidative stress and reducing the risk of cardiovascular diseases in overweight healthy men [11]. In a study by Sciullo and coworkers [12], it was found that the phytochemicals derived from food *viz.* zerumbone, nobiletin, auraptene and (±)-13-hydroxy-10-oxo-trans-11-octadecenoic acid (13-HOA) exhibited strong potential in suppressing the toxicity of environmental toxicants 2,3,7,8-tetrachlorodibenzo-p-dioxin (TCDD) and p,p'-DDT (DDT) in U937 macrophages. It was reported that fruits of *Ziziphus jujuba*, known as jujube or Chinese date contain flavonoid, cAMP and jujuboside that possess neuroprotective properties by providing protection to neuronal cells against neurotoxin stress by increasing the expression level of neurotrophic factors, stimulating the differentiation of neurons to promote memory and learning [13]. The study found that after the consumption of unfiltered apple juice, significant difference in polyphenol pharmacokinetics was observed due to the variation in the rate of metabolism among both male and female individuals [14]. *Punica granatum* L. (Punicaceae), pomegranate vinegar showed potent effects in maintaining the lipid profile and antioxidant status as compared to apple [*Malus domestica* Borkh. (Rosaceae)] and prickly pear [*Opuntia ficus-indica* (L.) Mill. (Cactaceae)] vinegars in hyperlipidemic induced high fat diet fed male Wistar rats [15]. Fu *et al.* [16] evaluated the antioxidant and antibacterial activity of 80% methanol extract of three tropical fruits viz. guava, persimmon and sweetsop against twelve pathogenic strains (*Staphylococcus epidermidis*, *Bacillus cereus*, *Escherichia coli*, *Monilia albican*, *Staphylococcus aureus*, *Shigella flexneri*, *Salmonella typhimurium* and *Pseudomonas aeruginosa*) and 4 multidrug-resistant strains (ESBLs-producing *Escherichia coli*, methicillin-resistant *Staphylococcus aureus*, multidrug-resistant *Acinetobacter baumannii* and carbapenems-resistant *Pseudomonas aeruginosa*). It was observed that these fruit extracts act as natural antioxidant and anti-bacterial agents.

Paturi *et al.* [17] investigated the blueberries and broccoli for their protection against inflammatory bowel disease (IBD) in mdr1a(-/-) mice (IBD mouse model) and found that these dietary supplements were effective in altering the metabolism and composition of colon morphology and cecal microbiota. The consumption of Chinese cabbage helps to reduce the development and progression of stomach cancer [18]. Tse and Eslick [19], reported that isothiocyanates, a major component of Broccoli, cruciferous vegetable showed chemoprotective potential against colon cancer. Consumption of dietary broccoli was found effective in reducing the initiation and progression of hepatocellular carcinoma which was induced by Western diet and diethylnitrosamine in male B6C3F1 mice [20]. Heber *et al.* [21] evaluated that the consumption of broccoli sprout extract enriched with sulforaphane precursor, glucoraphanin showed effectiveness in preventing the asthma and allergic diseases by reducing the impact of diesel exhaust particle (DEP) suspensions or particulate pollution in humans. It was reported [22] that consumption of broccoli, a brassicaceous vegetable regulates the xenobiotic metabolising enzymes, cellular signal transduction, epigenetic mechanisms and thereby prevents the risk of chronic diseases and cancer due to the presence of glucosinolates and other plant pigments in them. Garlic, carrots, cabbage as well as broccoli, when consumed as raw vegetables were effective in reducing the risk of stomach cancer (Fig. **2**) [23].

Lutein, carotenoid belonging to family xanthophyll, is abundantly found in green leafy vegetables such as kale, spinach and in animal fat, egg yolk and human eye retinal macula showed a protective effect against a neurotoxin, 1-Methyl- 4-phenyl- 1, 2, 3, 6 -tetrahydropyridine induced mitochondrial dysfunction, oxidative stress and loss of nigral dopaminergic neurons in Male C57BL/6 mice [24]. The results obtained exhibited an enhancement in striatal dopamine level. Lycopene, a carotenoid, consumed as tomato juice was found effective in reducing the risk of coronary heart disease by interfering with enzymes involved in the synthesis of cholesterol and increasing the degradation of low density lipoprotein [25]. Bhardwaj and Kumar [26], evaluated the neuroprotective efficacy of lycopene against kindling epilepsy induced by pentylenetetrazol in Laca mice by mitigating the kindling score and restoring mitochondrial functions. Lycopene exhibited an inhibitory effect against the nephrotoxicity and oxidative stress induced by cyclosporine in adult male Sprague-Dawley rats [27]. It was evaluated that drinking carrot juice promote cardiovascular health by reducing the production of malondialdehyde and increasing the antioxidant status of plasma [28]. Li *et al.* [29] reported the beneficial role of urosolic acid, natural pentacyclic triterpenoid compound, present in cranberry, apple and olive in protecting against nonalcoholic fatty liver disease in db/db mice (a type 2 diabetic mouse model) by increasing β-oxidation of lipid and inhibiting the hepatic endoplasmic reticulum stress (Fig. **2**).

Edible Flowers

Edible flowers are receiving considerable attention as a natural coloring agent with food-health relationship and have originated from a wide range of plants. In ancient Greece and Rome [30], Asia [31] and medieval France, Europe [32] edible flowers are also used as a food. Recently, many studies have elucidated that the nutritive quality of flowers play a crucial role for their consumption in the food sector. Flower is an essential part of plant which contain a range of natural antioxidants *viz.* anthocyanins, phenolic acids, phenolic and flavonoid compounds [33]. Edible flowers are used for a variety of purposes like garnishing and seasoning the dishes and also as an important ingredient in salads, entrees, soups, desserts and drinks. These flowers act as a source of certain antioxidative compounds, have nutritional value and medicinal properties, including the flowers of species like violet, rose, monks cress, prunus, jasmine and flower of Jamaica [34 - 36]. From the dietary point of view, the tea prepared from edible flowers contains no caffeine and other stimulant drugs *viz.* alkaloid xanthenes caffeine and

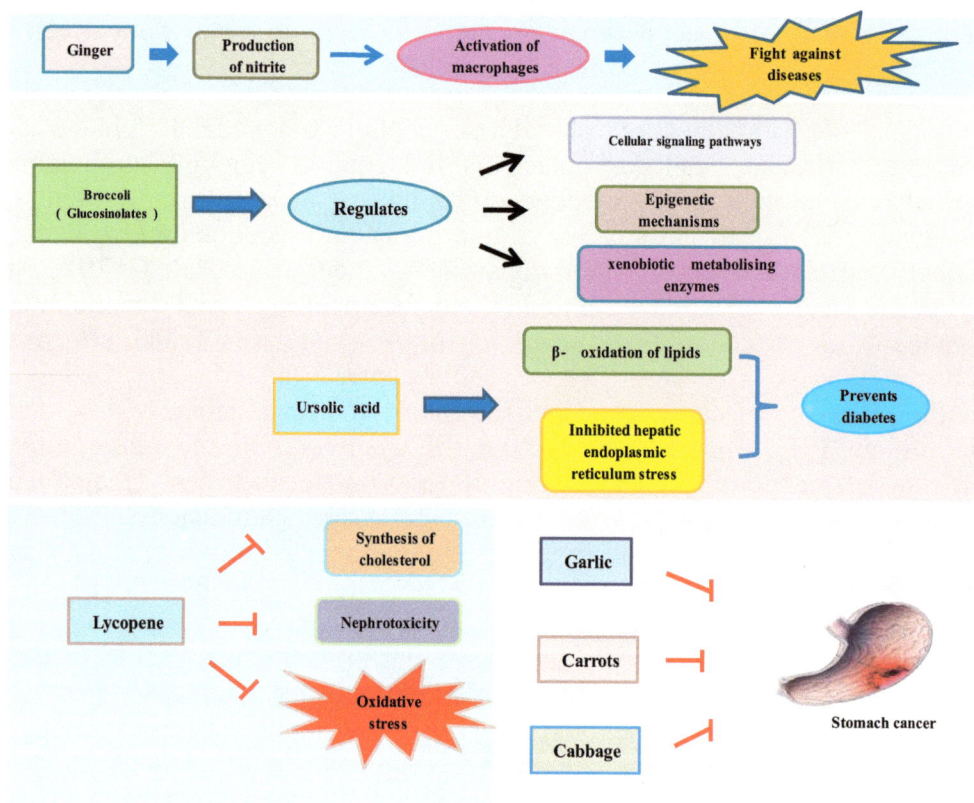

Fig. (2). Nutraceuticals ameliorating the deleterious effect of different diseases by regulating and inhibiting the production of reactive species.

theobromine as compared to the other types of teas [37]. Several studies have found that not all edible flowers are safe to be involved in human diets, some flowers contain toxic substances like hemaglutinnins, cyanogenic glycosides, oxalic acid or alkaloids, which affects their nutritional value and causes several damages to the consumers [30, 38 - 40]. The commonly used edible flowers include 15 species (*Hibiscus rosasinensis, Chrysanthemum morifolium, Hibiscus sabdariffa, Rosa rugose, Paeonia suffruticosa, Rosa chinensis, Lonicera japonica, Osmanthus fragrans, Magnolia denudate, Hemerocallis fulva, Chrysanthemum indicum, Bauhinia variegata, Opuntia ficus indica, Jasminum sambac, Matricaria chamomilla* and *Nelumbo nucifera*) from 10 different families [41].

Diplotaxis simplex (Viv.) Spreng. (Brassicaceae) ethanolic flowers extract was evaluated for its antihyperglycemic potential against Alloxan induced hyperglycemia by inhibiting the activity of α-glucosidase which is a key enzyme for type 2 diabetes mellitus [42]. The study had investigated [43] the antioxidative and oxidative stress suppressing potential of 12 Chinese edible flowers aqueous extracts. The results obtained demonstrated that *Lonicera japonica* Thunb., *Rosa rugosa* Thunb., *Carthamus tinctorius* L., *Chrysanthemum indicum* L., *Rosmarinus officinalis* L., *Magnolia officinalis* Rehd. et Wils., and *Chrysanthemum morifolium* Ramat. Flower extracts showed reduction in the level of oxidative stress by increasing glutathione peroxidase, serum superoxide dismutase level and reducing the concentration of malondialdehyde in hyperlipidemia rats. It was found that ethanolic chamomile (*Matricaria recutita*) flowers extract showed a potent type 2 anti-diabetic effect in the insulin-resistant high-fat diet (HFD)-fed C57BL/6 mice. The data obtained revealed a reduction in glucose intolerance, insulin resistance, non-esterified fatty acids (NEFA), plasma triacylglycerol and LDL/VLDL cholesterol and *in vitro* activation of nuclear receptor peroxisome proliferator-activated receptor gamma (PPARγ) and its isotypes [44].

In a study, Qiu *et al.* [45] reported that the flavonoids rich extract of *Abelmoschus manihot* (L.) Medic flower (an edible hibiscus) mitigates the oxidative stress induced by d-Galactose in mouse model by activating the Nrf2-mediated antioxidant response pathway. The data obtained was accomplished by the elevated mRNA expression of SOD, GPx and CAT and decreased interleukin-1 beta (IL-1β) and tumor necrosis factor-alpha (TNF-α).

Medicinal Plants

Till date, much of the scientific research has focused on the beneficial effects of the plant origin-based Nutraceuticals that have appreciable amount of bioactive phytoconstituents in them. For the last few decades, Nutraceuticals derived from plants have received considerable attention as they are found to be safe, non-toxic and have nutritional as well as therapeutic effects on human health [46]. Several recent evidences have reported the role of medicinal plants as Nutraceuticals/functional foods for curing various acute and chronic degenerative diseases. In a report, it was shown that nutraceuticals including bioactive peptides, plant polyphenols, oligosaccharides, vitamins, carotenoids, and polyunsaturated fatty acids alleviate certain skin disorders by protecting against UV radiation ageing in several human trials [47]. Kim *et al.* [48] evaluated the anti-depressant effects of the aqueous extracts of an oriental traditional medicine, Chaihu-Shuga-San (CSS) in both the forced swimming test (FST) and the chronic mild stress (CMS) experimental animal models. The data obtained indicated that *Rhizoma Cyperi*, a herbal constituent of CSS showed anti-depression activity in FST but not in CMS model. Several studies have reported the use of herbal medicine as a complementary treatment for curing asthma [49]. *Terminalia arjuna* bark was considered as a popular cardiotonic substance in Indian pharmacopoeia and was found to show a decreased level of HDL, LDL, cholesterol and triglycerides in patients as compared to the placebo and vitamin E treated group [50]. Ahmed *et al.* [51] reported the nephro-protective and anti-hyperglycemic effects of ethanolic extracts and fractions of *Solanum trilobatum* leaf (200 mg/kg bw) against the alloxan induced diabetes in rats.

Lin *et al.* [52] evaluated that the extract of *Xanthium strumarium* prevents complete Freund's Adjuvant induced rheumatoid arthritis in male Wistar rats. The data obtained revealed reduction in arthritic score and paw swelling, enhanced body weight loss, decreased in thymus index and suppression of TNF-α and IL-1β in the serum in contrast to remarkable increase in IL-10 cytokine level. *Terminalia macroptera* (Combretaceae) showed a range of applications in the treatment of infections, tuberculosis, hepatitis and healing of wounds, sores, cough and other diseases [53]. Study had reported the use of plant-based drugs for the treatment and prevention of diabetes and hyperlipidemia. These medicinal plant fights against diseases without posing any serious side effects and at low cost [54]. Wang *et al.* [55] investigated the anti-fatigue activity of acidic and neutral Ginseng polysaccharide, which is an active component of a Chinese herb, *Panax ginseng* C. A. Meyer (ginseng) using a forced swim test in mice. Results obtained demonstrated potent anti-fatigue activity of acidic polysaccharide. It was evaluated that a herb, *Rumex japonicus* Houtt. (RJH) was found effective in treating the picryl chloride induced atopic dermatitis like skin lesions in the animal model, NC/Nga mice. The treatment of RJH showed significant reduction in the number of *Staphylococcus aureus*, scratching behavior, IL-4 and IgE level [56]. Campbell *et al.* [57] assessed 71 medicinal herbs for their anticancer activity against five breast cancer cell lines (MCF7, SK-BR-3, BT-474, MDA-MB-231 and MCNeuA). Out of various herbs, *Rheum palmatum* exhibited a potent antiproliferative and apoptosis inducing potential.

Ampelopsis brevipedunculata (Maxim.) Trautv. (Vitaceae) aqueous fraction exhibited a potent hepatoprotective potential by reducing the carbon tetrachloride induced development and progression of hepatic alterations in mice [58]. Limaye *et al.* [59], evaluated the detoxifying and protective potential of Selenium and Curcumin against Aflatoxin B1 (AFB1) which is the most hazardous and carcinogenic mycotoxin to animals and human beings. Several inorganic and organic compounds derived from plants possesses anticancer, antimutagenic and other related immunomodulating characteristics [60]. Huang and coworkers [61], investigated that *Boswellia serrata*, gum resin exudates methanol extract, Boswellin (containing a natural components triterpenoids, beta-boswellic and its related derivatives) when applied to the back of mice significantly decreased its epidermal proliferation, skin inflammation and formation of tumor induced by 12-O-tetradecanoylphorbol-13-acetate and initiated by 7,12-dimethylbenz[a]-anthracene (DMBA).

Gadekar and coauthors [62], reported the use of medicinal plant derived drugs for the treatment and prevention of peptic ulcer caused by the presence of acid and other peptic activity of gastric juices and also by the infection of *Helicobacter pylori*. Al-Howiriny *et al.* [63] investigated the hepatoprotective effects of *Commiphora opobalsamum* "Balessan" ethanolic extract against the carbon tetrachloride: liquid paraffin (1: 1) induced hepatic toxicity. The data obtained indicated a decrease in the level of serum transaminase, alkaline phosphatase, bilirubin, serum glutamate pyruvate transaminase and serum glutamate oxaloacetate transaminase on treatment with extract. McKay and Blumberg [64], reported the beneficial effect of Peppermint (*Mentha piperita* L.) and its oil therapy in curing various diseases and disorders like irritable bowel syndrome, a relaxation effect on gastrointestinal (GI) tissue, immunomodulating actions, anesthetic effects in the central and peripheral nervous system and other chemopreventive properties. Its oil is not recommended to the patients with hiatal hernia, GI reflux or kidney stones.

Voukeng *et al.* [65] evaluated the antibacterial efficacy of six medicinal plants *viz.* *Ageratum conyzoides*, *Alstonia boonei*, *Croton macrostachys*, *Catharanthus roseus*, *Paullinia pinnata* and *Cassia obtusifolia* against a panel of 36 multi-drug resistant (MDR) Gram- positive and Gram- negative bacteria. The data obtained had demonstrated that these plants possess phytoconstituents like polyphenols, saponins, sterols, alkaloids, flavonoids and triterpenes which might contribute in protecting against the MDR bacterial infections. Study had investigated the anti-inflammatory activity of the *Carissa carandas* methanol extract against hind paw edema induced by carrageenan in rats. Zhao and Agarwal [66], investigated that the administration of silymarin, (flavonoid antioxidant, isolated from milk thistle) in different tissues of stomach, liver, skin, lung and small bowel effectively induces the activities of phase II enzymes *viz.* glutathione S-transferase and quinone reductase and has also been found to reduce the growth and progression of cancer in different models. Singh *et al.* [67] reported the chemopreventive efficacy of Silibinin, a flavanone from milk thistle in protecting against the progression and angiogenesis of lung tumors induced by urethane in mice.

Tarkang *et al.* [68] reported that a polyherbal product, Nefang is composed of *Psidium guajava*, *Ocimum gratissimum*, *Cymbopogon citratus*, *Carica papaya*, *Citrus sinensis* (leaves) and *Mangifera indica* (bark and leaf), which were found to provide an effective therapy against *Plasmodium falciparum* malaria. It was investigated that the leaf part of three medicinal plant *viz. Clausena anisata*, *Artemisia afra* and *Haemanthus albiflos* provides a most practical strategy for the prevention and treatment of tuberculosis in Africa [69]. Recently, the antifungal potential of three traditionally used medicinal plants *viz. Glycyrrhiza glabra* (bark), *Ficus religiosa* (stem), and *Plantago* (husk) was examined against oral Candida albicans (ATCC 66027) using Kirby-Bauer disc diffusion method [70]. The results obtained demonstrated that *Glycyrrhiza glabra* showed a potent antifungal activity with highest value of Mean zone of inhibition as compared to other plants and synthetic antifungal agents. Luo *et al.* [71] reported the anti-tumor and anti-metastasis effects of the aqueous extract of a medicinal mushroom, *Coriolus versicolor* in mouse mammary carcinoma 4T1 cells and in 4T1-tumor bearing mouse model.

It was reported that the oral consumption of medicinal plants was found effective for the prevention and treatment of osteoarthritis [72]. Li and coworkers [73], evaluated that several herbal extracts, formulations and medicines belonging to 24 genera and 18 families, such as *Acanthopanax*, *Alpinia* and *Astragalus etc.* were found effective in the treatment of Parkinson's disease (PD). The anti-Parkinsonian activities of these herbal products were tested in both *in vitro* and *in vivo* PD model. The methanol extract in comparison to aqueous extract of three widely used medicinal plants *viz. Euonymus laxiflorus*, *Gardenia jasminoides* and *Rubia lanceolata* was found effective in reducing the uric acid level (hyperuricemia) of serum in rats and *in vitro* xanthine oxidase activity [74]. Dangarembizi *et al.* [75] reported the presence of bioactive compounds *viz.* alkaloids, flavonoids, terpenoids, tannins and other active proteins in *Ficus thonningii*, a wild fig and found that these compounds might contribute to its antioxidant, antimicrobial, antihelmintic, antidiarrhoeal, anti-inflammatory and analgesic properties. *Zingiber officinalis* Roscoe (ginger), water extract was investigated for its antileishmanial activity due to its non-toxicity for human red blood cells and stimulation of nitrite production essential for the activation of macrophages to fight against the disease caused by *L. amazonensis* [76].

Bhalang *et al.* [77] evaluated that the oral aphthous ulceration can be safely and effectively treated with acemannan, a polysaccharide extracted from Aloe vera. Lamorde and coworkers [78], have documented the use of medicinal plants belonging to *Aloe* spp., *Sarcocephalus latifolius*, *Erythrina abyssinica*, *Psorospermum febrifugum*, *Warburgia salutaris* and *Mangifera indica* for the treatment of HIV/AIDS and related infections. It was reported that aloe species contain biaoactive compounds *viz.* p-coumaroyl, Cinnamoyl, caffeoyl aloesin, feruloyl and aloesin-related compounds, which in normal human melanocyte cell lysates inhibited the dihydroxyphenylalanine oxidase activities of tyrosinase and tyrosine hydroxylase and thus for cosmetic application act as a positive pigment-altering agent. Avijgan *et al.* [79] investigated the effectiveness of *Aloe vera* gel for the treatment of chronic ulcers in patients. Vemu *et al.* [80] reported the protective effects of a combination of emu oil and *Aloe vera* against the ulceration induced by indomethacin in Wistar albino rats. Study had showed that the topically applied *Aloe ferox* extract was found to be more effective than *Aloe vera* in providing protection against 2, 4-dinitrochlorobenzene induced chronic atopic dermatitis by inhibiting the cutaneous inflammatory response as well as serum IgE levels in mouse models [81].

SUMMARY

Nutraceutical and functional foods play a crucial role in improving the metabolic and other physiological functioning of the body which indeed is beneficial in overcoming the risk factors of various degenerative and life-threatening cancer related diseases. Thus, from the above referred evidences it becomes clear that the consumption of vegetables, fruits, edible flowers, medicinal herbs and plants helps in reducing or preventing the risk factors of several degenerative diseases and thereby found beneficial in promoting human health by regulating the immune system and other physiological and pathophysiological pathways of body.

CONSENT FOR PUBLICATION

Not applicable.

CONFLICT OF INTEREST

The authors declare no conflict of interest, financial or otherwise.

ACKNOWLEDGEMENTS

This work was supported by the DST-PURSE (SR/S9/Z-23/2010/20(C); Dated 02/6/2011), Programme, Department of Science and Technology (DST) and Fund for Improvement of S & T Infrastructure (FIST) programme of DST and DRS-SAP (II) Programme of University Grants Commission, New Delhi (India).

REFERENCES

[1]   Kalra EK. Nutraceutical-definition and introduction. AAPS J 2003; 5: 2-3.

[2]   Ratnaparkhi PK, Karode NP, Patil KB, *et al.* Nutraceuticals-its current scenario and challenges in dietary supplements. WJPPS 2015; 4(07): 460-74.

[3]   Chauhan B, Kumar G, Kalam N, Ansari SH. Current concepts and prospects of herbal nutraceutical: A review. J Adv Pharm Technol Res 2013; 4(1): 4-8.
[http://dx.doi.org/10.4103/2231-4040.107494] [PMID: 23662276]

[4]   Korver O. Functional foods: the food industry and functional foods: Some European perspec-

tives.Functional Foods for Disease Prevention: II Medicinal Plants and Other Foods. Washington, DC: American Chemical Society 1998; pp. 22-5.
[http://dx.doi.org/10.1021/bk-1998-0702.ch004]

[5]   Briskin DP. Medicinal plants and phytomedicines. Linking plant biochemistry and physiology to human health. Plant Physiol 2000; 124(2): 507-14.
[http://dx.doi.org/10.1104/pp.124.2.507] [PMID: 11027701]

[6]   Bhowmik D, Gopinath H, Kumar BP, Duraivel S, Kumar KPS. Nutraceutical –A bright scope and opportunity of Indian healthcare market. Pharma InnovJ 2013; 1(11): 29-41.

[7]   Rajasekaran A, Sivagnanam G, Xavier R. Nutraceuticals as therapeutic agents: A Review. J Pharma Sci Technol 2008; p. 1.

[8]   Chintale AG, Kadam VS, Sakhare RS, Birajdar GO, Nalwad DN. Role of Nutraceuticals in various Diseases: A Comprehensive Review. IJRPC 2013; 3(2): 290-9.

[9]   Gupta SK, Yadav SK, Mali Patil SM. Nutraceutical–a bright scope and opportunity of Indian healthcare market. IJRDPL 2013; 2(4): 478-81.

[10]   Shahidi F. Nutraceuticals, functional foods and dietary supplements in health and disease. Yao Wu Shi Pin Fen Xi 2012; 20(1): 226-30.

[11]   Alqurashi RM, Galante LA, Rowland IR, Spencer JP, Commane DM. Consumption of a flavonoid-rich açai meal is associated with acute improvements in vascular function and a reduction in total oxidative status in healthy overweight men. Am J Clin Nutr 2016; 104(5): 1227-35.
[http://dx.doi.org/10.3945/ajcn.115.128728] [PMID: 27680990]

[12]   Chen J, Liu X, Li Z, *et al.* A Review of Dietary *Ziziphusjujuba* Fruit (Jujube): Developing health food supplements for brain protection. Evid Based Complement Alternat Med 2017; 2017: 3019568.
[http://dx.doi.org/10.1155/2017/3019568] [PMID: 28680447]

[13]   Wruss J, Lanzerstorfer P, Huemer S, *et al.* Differences in pharmacokinetics of apple polyphenols after standardized oral consumption of unprocessed apple juice. Nutr J 2015; 14: 32.
[http://dx.doi.org/10.1186/s12937-015-0018-z] [PMID: 25890155]

[14]   Sciullo EM, Vogel CF, Wu D, Murakami A, Ohigashi H, Matsumura F. Effects of selected food phytochemicals in reducing the toxic actions of TCDD and p,p′-DDT in U937 macrophages. Arch Toxicol 2010; 84(12): 957-66.
[http://dx.doi.org/10.1007/s00204-010-0592-y] [PMID: 20865247]

[15]   Bouazza A, Bitam A, Amiali M, Bounihi A, Yargui L, Koceir EA. Effect of fruit vinegars on liver damage and oxidative stress in high-fat-fed rats. Pharm Biol 2016; 54(2): 260-5.
[http://dx.doi.org/10.3109/13880209.2015.1031910] [PMID: 25853952]

[16]   Fu L, Lu W, Zhou X. Phenolic compounds and *In vitro* antibacterial and antioxidant activities of three tropic fruits: Persimmon, Guava, and Sweetsop. BioMed Res Int 2016; 2016: 4287461.
[http://dx.doi.org/10.1155/2016/4287461] [PMID: 27648444]

[17]   Paturi G, Mandimika T, Butts CA, *et al.* Influence of dietary blueberry and broccoli on cecal microbiota activity and colon morphology in mdr1a(-/-) mice, a model of inflammatory bowel diseases. Nutrition 2012; 28(3): 324-30.
[http://dx.doi.org/10.1016/j.nut.2011.07.018] [PMID: 22113065]

[18]   Hu JF, Zhang SF, Jia EM, *et al.* Diet and cancer of the stomach: a case-control study in China. Int J Cancer 1988; 41(3): 331-5.
[http://dx.doi.org/10.1002/ijc.2910410302] [PMID: 3346096]

[19]   Tse G, Eslick GD. Cruciferous vegetables and risk of colorectal neoplasms: a systematic review and meta-analysis. Nutr Cancer 2014; 66(1): 128-39.
[http://dx.doi.org/10.1080/01635581.2014.852686] [PMID: 24341734]

[20]   Chen YJ, Wallig MA, Jeffery EH. Dietary Broccoli Lessens Development of Fatty Liver and Liver

Cancer in Mice Given Diethylnitrosamine and Fed a Western or Control Diet. J Nutr 2016; 146(3): 542-50.
[http://dx.doi.org/10.3945/jn.115.228148] [PMID: 26865652]

[21]   Heber D, Li Z, Garcia-Lloret M, *et al.* Sulforaphane-rich broccoli sprout extract attenuates nasal allergic response to diesel exhaust particles. Food Funct 2014; 5(1): 35-41.
[http://dx.doi.org/10.1039/C3FO60277J] [PMID: 24287881]

[22]   Ferguson LR, Schlothauer RC. The potential role of nutritional genomics tools in validating high health foods for cancer control: broccoli as example. Mol Nutr Food Res 2012; 56(1): 126-46.
[http://dx.doi.org/10.1002/mnfr.201100507] [PMID: 22147677]

[23]   Zickute J, Strumylaite L, Dregval L, Petrauskiene J, Dudzevicius J, Stratilatovas E. [Vegetables and fruits and risk of stomach cancer]. Medicina (Kaunas) 2005; 41(9): 733-40.
[PMID: 16227704]

[24]   Nataraj J, Manivasagam T, Thenmozhi AJ, Essa MM. Lutein protects dopaminergic neurons against MPTP-induced apoptotic death and motor dysfunction by ameliorating mitochondrial disruption and oxidative stress. Nutr Neurosci 2016; 19(6): 237-46.
[http://dx.doi.org/10.1179/1476830515Y.0000000010] [PMID: 25730317]

[25]   Rao AV. Lycopene, tomatoes, and the prevention of coronary heart disease. Exp Biol Med (Maywood) 2002; 227(10): 908-13.
[http://dx.doi.org/10.1177/153537020222701011] [PMID: 12424333]

[26]   Bhardwaj M, Kumar A. Neuroprotective effect of Lycopene against PTZ-induced kindling seizures in mice: Possible behavioural, biochemical and mitochondrial Dysfunction. Phytother Res 2016; 30(2): 306-13.
[http://dx.doi.org/10.1002/ptr.5533] [PMID: 26633078]

[27]   Ateşşahin A, Ceribaşi AO, Yilmaz S. Lycopene, a carotenoid, attenuates cyclosporine-induced renal dysfunction and oxidative stress in rats. Basic Clin Pharmacol Toxicol 2007; 100(6): 372-6.
[http://dx.doi.org/10.1111/j.1742-7843.2007.00060.x] [PMID: 17516989]

[28]   Potter AS, Foroudi S, Stamatikos A, Patil BS, Deyhim F. Drinking carrot juice increases total antioxidant status and decreases lipid peroxidation in adults. Nutr J 2011; 10: 96.
[http://dx.doi.org/10.1186/1475-2891-10-96] [PMID: 21943297]

[29]   Li JS, Wang WJ, Sun Y, Zhang YH, Zheng L. Ursolic acid inhibits the development of nonalcoholic fatty liver disease by attenuating endoplasmic reticulum stress. Food Funct 2015; 6(5): 1643-51.
[http://dx.doi.org/10.1039/C5FO00083A] [PMID: 25892149]

[30]   Melillo L. Diuretic plants in the paintings of Pompeii. Am J Nephrol 1994; 14(4-6): 423-5.
[http://dx.doi.org/10.1159/000168758] [PMID: 7847479]

[31]   Cichewicz RH, Nair MG. Isolation and characterization of stelladerol, a new antioxidant naphthalene glycoside, and other antioxidant glycosides from edible daylily (*hemerocallis*) flowers. J Agric Food Chem 2002; 50(1): 87-91.
[http://dx.doi.org/10.1021/jf010914k] [PMID: 11754548]

[32]   Kopec K. Jedlekvety pro zpestrenijidelnicku. VyzivaPotraviny 2004; 59: 151-2.

[33]   Ngoitaku C, Kwannate P, Riangwong K. Total phenolic content and antioxidant activities of edible flower tea products from Thailand. IFRJ 2016; 23(5): 2286-90.

[34]   Wongwattanasathien O, Kangsadalampai K, Tongyonk L. Antimutagenicity of some flowers grown in Thailand. Food Chem Toxicol 2010; 48(4): 1045-51.
[http://dx.doi.org/10.1016/j.fct.2010.01.018] [PMID: 20100534]

[35]   Xiong L, Yang J, Jiang Y, *et al.* Phenolic compounds and antioxidant capacities of 10 common edible flowers from China. J Food Sci 2014; 79(4): C517-25.
[http://dx.doi.org/10.1111/1750-3841.12404] [PMID: 24621197]

[36] Navarro-González I, González-Barrio R, García-Valverde V, Bautista-Ortín AB, Periago MJ. Nutritional composition and antioxidant capacity in edible flowers: characterisation of phenolic compounds by HPLC-DAD-ESI/MS^n. Int J Mol Sci 2014; 16(1): 805-22.
[http://dx.doi.org/10.3390/ijms16010805] [PMID: 25561232]

[37] Al-Howiriny TA, Al-Sohaibani MO, Al-Said MS, Al-Yahya MA, El-Tahir KH, Rafatullah S. Hepatoprotective properties of *Commiphora opobalsamum* ("Balessan"), a traditional medicinal plant of Saudi Arabia. Drugs Exp Clin Res 2004; 30(5-6): 213-20.
[PMID: 15702514]

[38] Mlcek J, Rop O. Fresh edible flowers of ornamental plants are a new source of nutraceutical foods. Trends Food Sci Technol 2011; 22: 561-9.
[http://dx.doi.org/10.1016/j.tifs.2011.04.006]

[39] Sotelo A, López-García S, Basurto-Peña F. Content of nutrient and antinutrient in edible flowers of wild plants in Mexico. Plant Foods Hum Nutr 2007; 62(3): 133-8.
[http://dx.doi.org/10.1007/s11130-007-0053-9] [PMID: 17768684]

[40] Lara-Cortés E, Osorio-Díaz P, Jiménez-Aparicio A, Bautista-Baños S. Contenido nutricional, propiedades funcionales y conservación de flores comestibles. Revisión. Arch Latinoam Nutr 2013; 63(3): 197-208.
[PMID: 25362819]

[41] Lu B, Li M, Yin R. Phytochemical content, health benefits, and toxicology of common edible flowers: A Review (2000–2015). Crit Rev Food Sci Nutr 2016; 56 (Suppl. 1): S130-48.
[http://dx.doi.org/10.1080/10408398.2015.1078276] [PMID: 26462418]

[42] Jdir H, Kolsi RBA, Zouari S, Hamden K, Zouari N, Fakhfakh N. The cruciferous *Diplotaxis simplex*: Phytochemistry analysis and its protective effect on liver and kidney toxicities, and lipid profile disorders in alloxan-induced diabetic rats. Lipids Health Dis 2017; 16(1): 100.
[http://dx.doi.org/10.1186/s12944-017-0492-8] [PMID: 28558824]

[43] Wang F, Miao M, Xia H, Yang LG, Wang SK, Sun GJ. Antioxidant activities of aqueous extracts from 12 Chinese edible flowers *in vitro* and *in vivo*. Food Nutr Res 2016; 61(1): 1265324.
[http://dx.doi.org/10.1080/16546628.2017.1265324] [PMID: 28326000]

[44] Weidner C, Wowro SJ, Rousseau M, *et al.* Antidiabetic effects of chamomile flowers extract in obese mice through transcriptional stimulation of nutrient sensors of the peroxisome proliferator-activated receptor (PPAR) family. PLoS One 2013; 8(11): e80335.
[http://dx.doi.org/10.1371/journal.pone.0080335] [PMID: 24265809]

[45] Qiu Y, Ai PF, Song JJ, Liu C, Li ZW. Total flavonoid extract from *Abelmoschusmanihot* (L.) medic flowers attenuates d-galactose-induced oxidative stress in mouse liver through the Nrf2 Pathway. J Med Food 2017; 20(6): 557-67.
[http://dx.doi.org/10.1089/jmf.2016.3870] [PMID: 28472605]

[46] Pandey N, Meena RP, Rai SK, Pandey-Rai S. Medicinal plants derived nutraceuticals: a re-emerging health aid. Int J Pharma Bio Sci 2011; 2(4): 419-41.

[47] Kim SH, Han J, Seog DH, *et al.* Antidepressant effect of Chaihu-Shugan-San extract and its constituents in rat models of depression. Life Sci 2005; 76(11): 1297-306.
[http://dx.doi.org/10.1016/j.lfs.2004.10.022] [PMID: 15642599]

[48] Szelenyi I, Brune K. Herbal remedies for asthma treatment: between myth and reality. Drugs Today (Barc) 2002; 38(4): 265-303.
[http://dx.doi.org/10.1358/dot.2002.38.4.668337] [PMID: 12532195]

[49] Gupta R, Singhal S, Goyle A, Sharma VN. Antioxidant and hypocholesterolaemic effects of *Terminalia arjuna* tree-bark powder: a randomised placebo-controlled trial. J Assoc Physicians India 2001; 49: 231-5.
[PMID: 11225136]

[50]   Pérez-Sánchez A, Barrajón-Catalán E, Herranz-López M, Micol V. Nutraceuticals for Skin Care: A Comprehensive Review of Human Clinical Studies. Nutrients 2018; 10(4): E403.
[http://dx.doi.org/10.3390/nu10040403] [PMID: 29587342]

[51]   Ahmed KZ, Sidhra SZ, Ponmurugan P, Kumar BS. Ameliorative potential of *Solanum trilobatum* leaf extract and fractions on lipid profile and oxidative stress in experimental diabetes. Pak J Pharm Sci 2016; 29(5): 1578.
[PMID: 27731814]

[52]   Lin B, Zhao Y, Han P, *et al.* Anti-arthritic activity of *Xanthium strumarium* L. extract on complete Freund's adjuvant induced arthritis in rats. J Ethnopharmacol ;2014 :(1)155 248-55.
[http://dx.doi.org/10.1016/j.jep.2014.05.023] [PMID: 24862493]

[53]   Pham AT, Dvergsnes C, Togola A, *et al. Terminalia macroptera*, its current medicinal use and future perspectives. J Ethnopharmacol 2011; 137(3): 1486-91.
[http://dx.doi.org/10.1016/j.jep.2011.08.029] [PMID: 21884779]

[54]   Parikh NH, Parikh PK, Kothari C. Indigenous plant medicines for health care: treatment of Diabetes mellitus and hyperlipidemia. Chin J Nat Med 2014; 12(5): 335-44.
[http://dx.doi.org/10.1016/S1875-5364(14)60041-8] [PMID: 24856756]

[55]   Wang J, Li S, Fan Y, *et al.* Anti-fatigue activity of the water-soluble polysaccharides isolated from Panax ginseng C. A. Meyer. J Ethnopharmacol 2010; 130(2): 421-3.
[http://dx.doi.org/10.1016/j.jep.2010.05.027] [PMID: 20580802]

[56]   Lee HS, Kim SK, Han JB, *et al.* Inhibitory effects of Rumex japonicus Houtt. on the development of atopic dermatitis-like skin lesions in NC/Nga mice. Br J Dermatol 2006; 155(1): 33-8.
[http://dx.doi.org/10.1111/j.1365-2133.2006.07303.x] [PMID: 16792749]

[57]   Campbell MJ, Hamilton B, Shoemaker M, Tagliaferri M, Cohen I, Tripathy D. Antiproliferative activity of Chinese medicinal herbs on breast cancer cells *in vitro*. Anticancer Res 2002; 22(6C): 3843-52.
[PMID: 12553004]

[58]   Yabe N, Matsui H. *Ampelopsis brevipedunculata* (Vitaceae) extract inhibits a progression of carbon tetrachloride-induced hepatic injury in the mice. Phytomedicine 2000; 7(6): 493-8.
[http://dx.doi.org/10.1016/S0944-7113(00)80035-5] [PMID: 11194178]

[59]   Shliankevich MA, Sergeev AV, Golubeva ZF. [The role of plant compounds in the prevention of oncologic diseases]. Vopr Pitan 1993; (4): 26-32.
[PMID: 8073690]

[60]   Huang MT, Badmaev V, Ding Y, Liu Y, Xie JG, Ho CT. Anti-tumor and anti-carcinogenic activities of triterpenoid, beta-boswellic acid. Biofactors 2000; 13(1-4): 225-30.
[http://dx.doi.org/10.1002/biof.5520130135] [PMID: 11237186]

[61]   Gadekar R, Singour PK, Chaurasiya PK, Pawar RS, Patil UK. A potential of some medicinal plants as an antiulcer agents. Pharmacogn Rev 2010; 4(8): 136-46.
[http://dx.doi.org/10.4103/0973-7847.70906] [PMID: 22228953]

[62]   Al-Howiriny TA, Al-Sohaibani MO, Al-Said MS, Al-Yahya MA, El-Tahir KH, Rafatullah S. Hepatoprotective properties of Commiphora opobalsamum ("Balessan"), a traditional medicinal plant of Saudi Arabia. Drugs Exp Clin Res 2004; 30(5-6): 213-20.
[PMID: 15702514]

[63]   Limaye A, Yu RC, Chou CC, Liu JR, Cheng KC. Protective and Detoxifying Effects Conferred by Dietary Selenium and Curcumin against AFB1-Mediated Toxicity in Livestock: A Review Toxins (Basel) 2018; 10
[http://dx.doi.org/10.3390/toxins10010025]

[64]   McKay DL, Blumberg JB. A review of the bioactivity and potential health benefits of peppermint tea (*Mentha piperita* L.). Phytother Res 2006; 20(8): 619-33.

[http://dx.doi.org/10.1002/ptr.1936] [PMID: 16767798]

[65] Voukeng IK, Beng VP, Kuete V. Antibacterial activity of six medicinal Cameroonian plants against Gram-positive and Gram-negative multidrug resistant phenotypes. BMC Complement Altern Med 2016; 16(1): 388.
[http://dx.doi.org/10.1186/s12906-016-1371-y] [PMID: 27724917]

[66] Zhao J, Agarwal R. Tissue distribution of silibinin, the major active constituent of silymarin, in mice and its association with enhancement of phase II enzymes: implications in cancer chemoprevention. Carcinogenesis 1999; 20(11): 2101-8.
[http://dx.doi.org/10.1093/carcin/20.11.2101] [PMID: 10545412]

[67] Singh RP, Deep G, Chittezhath M, et al. Effect of silibinin on the growth and progression of primary lung tumors in mice. J Natl Cancer Inst 2006; 98(12): 846-55.
[http://dx.doi.org/10.1093/jnci/djj231] [PMID: 16788158]

[68] Arrey Tarkang P, Franzoi KD, Lee S, et al. In vitro antiplasmodial activities and synergistic combinations of differential solvent extracts of the polyherbal product, Nefang. BioMed Res Int 2014; 2014: 835013.
[http://dx.doi.org/10.1155/2014/835013] [PMID: 24877138]

[69] Lawal IO, Grierson DS, Afolayan AJ. Phytotherapeutic information on plants used for the treatment of tuberculosis in eastern cape province, South Africa. Evid Based Complement Alternat Med 2014; 2014: 735423.
[http://dx.doi.org/10.1155/2014/735423] [PMID: 24864158]

[70] Sharma H, Yunus GY, Agrawal R, Kalra M, Verma S, Bhattar S. Antifungal efficacy of three medicinal plants *Glycyrrhiza glabra*, *Ficus religiosa*, and *Plantago major* against oral *Candida albicans*: A comparative analysis. Indian J Dent Res 2016; 27(4): 433-6.
[http://dx.doi.org/10.4103/0970-9290.191895] [PMID: 27723643]

[71] Luo KW, Yue GG, Ko CH, et al. In vivo and in vitro anti-tumor and anti-metastasis effects of *Coriolus versicolor* aqueous extract on mouse mammary 4T1 carcinoma. Phytomedicine 2014; 21(8-9): 1078-87.
[http://dx.doi.org/10.1016/j.phymed.2014.04.020] [PMID: 24856767]

[72] Cameron M, Chrubasik S. Oral herbal therapies for treating osteoarthritis. Cochrane Database Syst Rev 2014; (5): CD002947.
[PMID: 24848732]

[73] Li XZ, Zhang SN, Liu SM, Lu F. Recent advances in herbal medicines treating Parkinson's disease. Fitoterapia 2013; 84: 273-85.
[http://dx.doi.org/10.1016/j.fitote.2012.12.009] [PMID: 23266574]

[74] Liu LM, Cheng SF, Shieh PC, et al. The methanol extract of *Euonymus laxiflorus*, *Rubia lanceolata* and *Gardenia jasminoides* inhibits xanthine oxidase and reduce serum uric acid level in rats. Food Chem Toxicol 2014; 70: 179-84.
[http://dx.doi.org/10.1016/j.fct.2014.05.004] [PMID: 24845958]

[75] Dangarembizi R, Erlwanger KH, Moyo D, Chivandi E. Phytochemistry, pharmacology and ethnomedicinal uses of *Ficus thonningii* (Blume Moraceae): a review. Afr J Tradit Complement Altern Med 2012; 10(2): 203-12.
[PMID: 24146443]

[76] Duarte MC, Tavares GS, Valadares DG, et al. Antileishmanial activity and mechanism of action from a purified fraction of *Zingiber officinalis* Roscoe against *Leishmania amazonensis*. Exp Parasitol 2016; 166: 21-8.
[http://dx.doi.org/10.1016/j.exppara.2016.03.026] [PMID: 27013260]

[77] Bhalang K, Thunyakitpisal P, Rungsirisatean N. Acemannan, a polysaccharide extracted from *Aloe vera*, is effective in the treatment of oral aphthous ulceration. J Altern Complement Med 2013; 19(5): 429-34.

[http://dx.doi.org/10.1089/acm.2012.0164] [PMID: 23240939]

[78]   Lamorde M, Tabuti JR, Obua C, *et al.* Medicinal plants used by traditional medicine practitioners for the treatment of HIV/AIDS and related conditions in Uganda. J Ethnopharmacol 2010; 130(1): 43-53.
[http://dx.doi.org/10.1016/j.jep.2010.04.004] [PMID: 20451595]

[79]   Avijgan M, Kamran A, Abedini A. Effectiveness of Aloe Vera Gel in Chronic Ulcers in Comparison with Conventional Treatments. Iran J Med Sci 2016; 41(3): S30.
[PMID: 27516663]

[80]   Vemu B, Selvasubramanian S, Pandiyan V. Emu oil offers protection in Crohn's disease model in rats. BMC Complement Altern Med 2016; 16: 55.
[http://dx.doi.org/10.1186/s12906-016-1035-y] [PMID: 26852336]

[81]   Finberg MJ, Muntingh GL, van Rensburg CE. A comparison of the leaf gel extracts of *Aloe ferox* and *Aloe vera* in the topical treatment of atopic dermatitis in Balb/c mice. Inflammopharmacology 2015; 23(6): 337-41.
[http://dx.doi.org/10.1007/s10787-015-0251-2] [PMID: 26510768]

CHAPTER 10

Caspases and Phytochemicals: An Important Link in Cancer Chemoprevention

Manish Kumar[1], Sandeep Kaur[2], Varinder Kaur[2] and Satwinderjeet Kaur[2,*]

[1] *Department of Biology, S.D. College, Barnala, Punjab-148101, India*

[2] *Department of Botanical and Environmental Sciences, Guru Nanak Dev University, Amritsar-143005, Punjab, India*

Abstract: The use of natural phytochemicals over other treatment regimens offers a safe and broad-spectrum strategy to overcome the incidence of cancer. Phytochemicals play a pivotal role for the induction of apoptosis in cancerous cells by activation of pro-apoptotic proteins and initiator caspases that in turn cleave and execute the activation of down-stream effector caspases. The caspases belong to the family of proteases that exist as inactive pro-forms or zymogens that on cleavage forms an active enzyme mediating apoptosis. Since cancer cells become insensitive to anti-growth and apoptotic signals, thus these caspases remain inactive in them. Array of phytochemicals belonging to different categories *viz.* phenols, flavonoids, terpenoids, alkaloids, stilbenes *etc.* are reported to trigger the activation of caspases in cancer cells by modulating various cell-death stimulating pathways. In the present review, we highlighted the role of different phytochemicals that execute the cleavage of caspases in cancerous cells by activating pro-apoptotic enzymes and proteins involved in the induction of apoptosis. This will assist to select and identify the molecular mechanisms of some effective phytochemicals targeting caspases in their chemopreventive ability.

Keywords: Alkaloids, Apoptosis, Bcl-2 Family, Cancer, Caspases, Chemoprevention, Cytoplasmic Proteins, Cysteinyl Specific Proteases, Effector Caspases, Epigenetic, Extrinsic Pathway, Flavonoids, Initiator Caspases, Intrinsic Pathway, Nuclear Proteins, Phenols, Phytoconstituents, Stilbenes, Terpenoids, Zymogens.

INTRODUCTION

Cancer is a multi-stage patho-physiological condition occurring in response to genetic alteration, immune suppression and malignant transformation, and comprises of 3 main phases *viz.* initiation, promotion and progression [1, 2]. It is typified by altered or dys-regulated cellular processes such as proliferation,

---

* **Corresponding author Satwinderjeet Kaur:** Department of Botanical and Environmental Sciences, Guru Nanak Dev University, Amritsar-143005, Punjab, India; Tel: +91-9988195508; Email: sjkaur2011@gmail.com

Ashita Sharma, Manish Kumar, Satwinderjeet Kaur & Avinash Kaur Nagpal (Eds.)

inflammation, apoptosis, angiogenesis *etc.* [2, 3]. It was observed that targeting single pathway is not sufficient to manage cancer as there are multiple pathways involved in cancer growth and progression [4, 5]. Drug resistance is also one of the major obstacles in cancer treatment [6, 7]. Chemotherapy is amongst the vital strategies used for the treatment of cancer but it is associated with a number of side effects which limit its efficacy [8, 9]. Advances in the field of molecular biology and biochemistry have helped scientific community to gain deeper insight of the processes involved in carcinogenesis *viz.* initiation, promotion and progression [10]. During initiation phase, carcinogenic agents can either undergo metabolic activation or can get detoxified after entering the body. Metabolic activation leads to conversion of these substances into reactive species which can interact with DNA and may cause damage. The promotion phase of carcinogenesis is a reversible process which involves the accumulation of actively proliferating pre-neoplastic cells. In the progression stage of carcinogenesis, cancer cells show invasiveness and spread to various parts of the body [11].

Normally, cells in human body undergo death by two different mechanisms - 1) necrosis and 2) apoptosis. The necrotic cell death occurs in response to accidental injury leading to a traumatic death characterized by swelling, bursting and leakage of cellular elements into surroundings. It usually leads to inflammatory responses in the body. On the other hand, apoptotic cell death is well regulated and programmed. It is characterized by different morphological alterations such as cells' rounding off, blebbing of plasma membrane, nuclear condensation and fragmentation [12, 13]. Apoptosis is generally exploited by organisms to eradicate unwanted cells without affecting neighbouring cell population or without any inflammatory response [14]. Uncontrolled cell division and proliferation is the hallmark characteristic of cancer cells, achieved by subverting apoptotic cell death as one of the strategy. Induction of apoptosis in cancer cells is considered as a major strategy to manage cancer [15, 16]. Caspases (cysteine proteases) play an important role in the process of apoptosis *via* initiating the cascades of proteolytic cleavage events [17, 18]. It has been observed that caspases are found to be inactivated in cancer cells and their activation could bring back cells to the track of apoptosis.

Caspases

Caspases (called as enzymes of death) are cysteinyl specific proteases that cleave proteins at a site next to an aspartate (Asp) residue. Caspases exist as an inactive enzyme, procaspases (tripartite zymogens) in the cells and their activation requires the cleavage of their zymogenic forms (procaspases) at $Asp(P_1)$-$X(P_1')$ bonds to form active forms (caspases) [19]. The caspases were discovered in 1993 when the significance of *ced-3* gene was recognized by H. Robert Horvitz in

Caenorhabditis elegans as this gene was found to encode for a cysteine protease [20]. In humans, the family of caspases consists of 14 known enzymes that further come under two different sub-families. Out of all these enzymes, the initiator caspases are caspases-2,-8,-9,-10 which play an essential role in apoptosis and couple the cell death stimuli to effector caspases such as caspase-3,-6,-7. On the other hand, the caspases-1,-4,-5,-11,-12,-13,-14 are involved in the inflammatory responses mediated by pyroptosis [21 - 23]. The inflammatory caspases are a part of ICE (IL-1β converting enzyme) sub-family; while both apoptotic (initiator and effector) caspases constitute CED-3 sub-family. During apoptosis, the mechanism which is involved in the auto-activation of pro-caspases by other caspases in a chain reaction is the presence of aspartate (Asp) residue at their maturation cleavage sites [24]. Further, this apoptotic cell death is regulated by multiple and inter-connected pathways which ensures that caspases are activated only in the targeted cells [16]. There are two caspase-mediated apoptotic pathways *viz.* intrinsic and extrinsic [25]. Extrinsic pathway is activated upon binding death ligands to their respective death receptors such as Fas and tumour necrosis factor α (TNF-α) that in turn leads to formation of signalling complexes which activate caspase-8 from procaspase-8 leading to cleavage and activation of caspase-3 from procaspase-3. In some cases, caspase-8 also leads to cleavage of BH3 interacting domain (Bid) which can initiate mitochondrial or intrinsic pathway. Besides this, the alteration of inner mitochondrial membrane potential (MMP) in response to different stimuli like chemotherapeutic drugs, radiations, toxins, reactive oxygen species (ROS) *etc.* leads to cytochrome c release from mitochondria. Afterwards, the binding of cytochrome c to apoptotic protease activating factor (Apaf-1) and procaspase-9 results in the formation of apoptosome which in turn activates caspase-3 (Intrinsic Pathway) (Fig. **1**). Caspase-3 activation finally turns on the execution pathway which involves endonucleases' activation leading to DNA fragmentation and condensation along with degradation of cytoskeletal proteins causing apoptosis [25 - 33]. Besides caspases, B-cell lymphoma 2 (Bcl-2) family of proteins are also involved in apoptosis and decide the cell fate (live or die) by regulating the release of mitochondrial apoptogenic factors which further initiate the cascade of apoptotic signalling [25, 34 - 38]. Caspases are also regulated by various other proteins such as inhibitors of apoptosis (IAP's) [39 - 44], cellular FLICE inhibitory protein (cFLIP) [45, 46], second mitochondria-derived activator of caspase (Smac)/direct inhibitor of apoptosis-binding protein with low pI (DIABLO), Omi/high temperature requirement protein A2 (HtrA2) [47, 48] and heat shock proteins (HSPs) [49 - 54]. Thorough understanding apoptotic signalling mechanisms may result in the recognition of prospective targets for cancer management [55]. The present chapter highlights the different categories of phytochemicals in targeting caspases for induction of apoptosis.

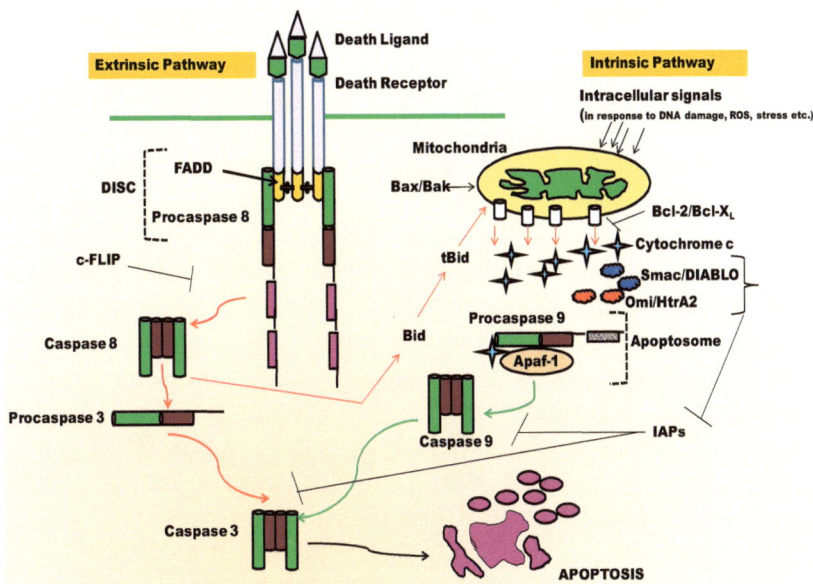

Fig. (1). Diagrammatic representation of Intrinsic and extrinsic pathways of apoptosis.

Caspases and Cancer

Caspases play a crucial role in stereotypical events and coordination occurring during apoptosis and thereby result in the biochemical and morphological changes like cell shrinkage, DNA fragmentation, loss of cell adhesion, membrane blebbing, chromatin condensation and formation of apoptotic bodies that stimulate their own phagocytic engulfment [19].

In case of cancer cells, caspases remain inactive either by somatic mutations, gene silencing or some other mechanisms. Moreover, the activation of caspases is disrupted by cellular endogenous caspase inhibitors such as X-linked inhibitor of apoptosis protein (XIAP) or c-FLIP and also, the up-stream pathways involved in stimulating the caspases' activation are de-regulated [19]. Alternatively, silencing of caspases in cancer cells is also governed by the generation of their alternatively spliced variants that are catalytically non-active or sometimes anti-apoptotic also [16, 56]. Various studies have also confirmed the existence of mutant forms of caspase-2,-3,-6,-7,-8,-9,-10 in certain types of cancer [19, 57, 58]. These mutations may possibly include mis-sense, non-sense, frame-shift, deletion mutations *etc*. Several cancer cells have also been identified to hold genetic polymorphism in caspase genes as demonstrated by a growing number of investigations [59 - 63]. Therefore, as a result of all above factors, the dys-regulation of caspases may lead to initiation of various degenerative ailments like cancer, inflammatory disorders and other related diseases [23].

Endogenous Factors Controlling the Activity of Caspases

The regulatory factors may either enhance or inhibit the enzymatic activity of caspases and also affect the activation of particular cellular substrates. Caspases are regulated by various proteins as well as other stimuli within cellular microenvironment such as transcription factors, ionic flux and regulatory nucleotides [64 - 66].

The Bcl-2 family of proteins regulates the permeability of mitochondrial membrane with the help of its pro-apoptotic and anti-apoptotic members. There are 25 genes which are identified in this family. Pro-apoptotic protein members are Bcl-10, Bcl-2 associated X protein (Bax), Bcl-2 antagonist killer (Bak), Bid, Bcl-2 associated death promoter (Bad), Bcl-2 interacting mediator of cell death (Bim), Bcl-2 interacting killer (Bik), B lymphocyte kinase (Blk) and Noxa while anti-apoptotic proteins include Bcl-2, myeloid leukaemia cell differentiation protein 1 (Mcl-1), Bcl extra large (Bcl-XL), Bcl extra small (Bcl-XS), Bcl-w and Bcl-2 associated athanogene (BAG). These proteins act as decider of the cell fate *i.e.* either cell undergoes apoptosis or subvert apoptosis. These proteins control the release of cytochrome c and formation of apoptosome which is comprised of cytochrome c, Apaf 1 and caspase-9 in mitochondrial dependent pathway [25,37.67]. These two types of proteins are found in the different sub-cellular locations. In the absence of any apoptotic signal, members of pro-apoptotic family are present in cytosol or cytoskeleton while that of anti-apoptotic ones are integral membrane proteins in mitochondria, endoplasmic reticulum or nuclear membrane [68 - 74]. Dys-regulation of Bcl-2 is frequently associated with the de-activation of caspases and thus progression of human malignant diseases by causing resistance to therapies [75]. Therefore, these proteins act as prime targets for novel specific anticancer therapeutics [76]. Recent study has reported mir-30a-5p as a novel direct target for Bcl-2 which increases the sensitivity of paclitaxel in non-small cell lung cancer for induction of apoptosis by inhibiting Bcl-2 expression [77].

Other proteins that inhibit caspase activation involve IAPs and cFLIP. In cancer cells, the frequent over-expression of these proteins occurs which may contribute to the survival of tumor cells, disease progression, chemo-resistance and poor prognosis [78]. Originally, IAP's were discovered in baculovirus where they were shown to prevent apoptosis of host cell [79]. These proteins play an important role in the regulation of apoptosis [39]. There are eight human IAP's known, out of which XIAP and survivin are extensively studied [80, 81]. Survivin also acts as cancer marker as its expression is high in many human cancers rather than normal tissues [82] and it is a known caspase inhibitor [44]. Other members such as XIAP, c-IAP1, and c-IAP2 have been found to bind and inhibit caspases-3,-7,-9

[40 - 43]. Kleinsimon *et al.* [83] have reported apoptosis inducing potential of European mistletoe (*Viscum album* L.) in osteosarcoma cell line through down-regulation of IAPs expression level. One of the other proteins *i.e.* cFLIP acts as competitive inhibitor of caspase-8 and -10. cFLIP prevents the attachment of caspase-8 and -10 to the adaptor protein Fas associated death domain (FADD) within death inducing signalling complex (DISC), thus inhibiting extrinsic pathway of apoptosis [45, 46].

Besides IAPs, Bcl-2 family and cFLIP, the prevention of caspase-dependent cell death can also take place in response to numerous other oncogenes *e.g.* Akt, retrovirus associated DNA sequences (Ras), epidermal growth factor receptor (EGFR) [84, 85].

Epigenetic mechanisms also come into play for regulating the activity of caspases at transcriptional as well as post-translational levels [86]. miRNAs which are considered both as oncogenes and tumour suppressor genes (TSGs) keep a watch on the activity of caspases under different conditions [87, 88]. *e.g.* knockout of some oncomir's including 20a, 21, 25, 96, 188, 590 in cultured cancer cell types has been shown to elicit apoptotic cell death [89 - 91]. Alternatively, the up-regulation of apoptotic pathway by some tumor suppressor miRNAs such as mir-15, 16, 34a, 204-5p, 451, 708 has also been apparent chiefly through extrinsic (death receptor-mediated) pathway [92 - 97]. Hyper-methylation of CpG islands of caspases (mainly caspase-3,-8,-9) and other apoptosis-related genes including TRAIL, Apaf-1, Fas also inhibits apoptotic process in cancerous tissues as apparent from various studies. Moreover, when de-methylation of these genes was carried out, the apoptosis was brought back on track which led to prevention of tumour progression [98 - 103].

Experimental support on the function of TSGs in the induction of caspase activity is also found in literature [86]. RBM5 (LUCA-15) was capable to up-regulate the expression of pro-apoptotic form of caspase-2 in HEK-293 and HeLa cancer cells by binding to pre-mRNA of caspase-2. It led to splicing of pre-mRNA in such a way that it eventually amplified the expression of pro-apoptotic form of caspase-2 [104]. Likewise, the potentiation of previously commenced apoptotic pathway has been shown to occur in some cellular models by interaction of *p*53 with caspase-6 and -10 directly which can modulate their transcriptional activity.

Orchestration of apoptosis by increased release of intracellular Ca^{2+} and cytochrome c is well known. Both act as inter-organellar messengers in apoptosis and play vital role in the execution of this process [105, 106]. Ca^{2+} enhances cytoplasmic cytochrome c flux and which in turn leads to the formation of apoptososme [107 - 110]. Inside the apoptososme, the molecular mechanism of caspase-9 activation by cytochrome c involves its binding to C-terminal WD40 repeats of Apaf-1 and in this response, the interaction of activated Apaf-1 with dATP/ATP becomes possible. In this manner, the fully activated apoptososme is formed thereby activating caspase-9 [111 - 114]. Certain inhibitors of Ca^{2+} are involved in the prevention of apoptosis *via* caspase repression. De-activation of caspase-8 by apoptosis repressor with caspase recruitment domain (ARC) has been reported in different cell types under *in vitro* conditions [115]. It was found to be mediated by binding of ARC with intracellular Ca^{2+} thereby decreasing its concentration.

Cytoplasmic and Nuclear Proteins as Substrates for Caspases

When activated, the caspases mediate their effector functions primarily *via* suppressing the protective pathways in the cell and execution of apoptosis by means of selective degradation of cytoplasmic as well as nuclear proteins involved in the cell survival. There are many examples of proteins which act as substrates for caspases [116, 117]:

- Cytoplasmic proteins (actin, growth arrest specific gene/Gas2, gelsolin, β-cateninkeratin-18, plasminogen activator inhibitor-2 *etc.*)
- Nuclear proteins (lamins, nuclear mitotic apparatus protein/NuMA, heterogeneous nuclear ribonucleoproteins/hnRNPs, scaffold attachment factor A/SAF-A *etc.*)
- Proteins involved in DNA metabolism and repair (poly ADP ribose polymerase/PARP, DNA dependent protein kinase catalytic subunit/DNA-PKcs, mini chromosome maintenance protein 3/MCM3, RNA polymerase I upstream binding factor, DNA polymerase II *etc.*)
- Protein kinases (PKC, protein kinase N/PKN, *p*21 activated kinase 2/PAK2, serine/threonine kinase 4/STK4, rapidly accelerated fibrosarcoma 1/Raf1, Akt *etc.*)
- Proteins involved in the signal transduction pathways (interleukins, Ras GTPase activating protein/RasGAP, GDP dissociation inhibitor/GDI, cytosolic phospholipase A2/cPLA2, nuclear factor κB/NF-κB *etc.*)
- Proteins involved in the regulation of cell cycle and proliferation (*p*21Cip1/Waf1, *p*RB, neural precursor cell expressed developmentally down regulated protein 4/nedd4 *etc.*)
- Permeability transition pore (a multi-protein complex formed at contact site of inner and outer mitochondrial membranes)

Caspases as Therapeutic Targets in Cancer

A number of side effects are associated with conventional cancer treatment regimen as many therapeutic agents possess very less selectivity towards cancer cells. This often results in the deterioration of life quality in cancer patients [118 - 120]. Although, there is great advancement in the field of modern medicine, the medicinal plants are still serving human race as source of prospective traditional medicine [121 - 124]. In the last few decades, phytoconstituents have achieved enormous interest due to their antiproliferative activities [125, 126]. Literature survey established that phytoconstituents embrace massive potential to induce apoptosis in cancer cells *via* targeting caspases [127 - 130].

Phenolic Acids Targeting Caspases

Phenolic acids are polyphenolic organic compounds consisting of hydroxy-cinnamic and hydroxybenzoic acids. Hydroxycinnamic acids include caffeic acid, *p*-coumaric acid and ferulic acid while hydroxybenzoic acids include ellagic acid, vanillic acid, 3,4-dihydroxybenzoic acid, *p*-hydroxybenzoic acid, gallic acid and syringic acid [131 - 133]. Phenolic acids are one of the most popular phytochemicals having a role as dietary agents. Phenolic acids induce apoptosis in various kinds of cancer cells, thus impart their anticancer activity *via* targeting caspases, cell cycle, Bcl-2 family proteins, phosphatidylinositol-3-kinase (PI3K)/Akt pathway *etc*. Herein, we have concluded various reports of phenolic acids inducing apoptosis *via* targeting caspases (Fig. **2**).

Ellagic acid was reported to induce apoptosis in human bladder cancer (T24) cells [134]. It was observed that ellagic acid treatment resulted in arresting of cells in G0/G1 phase activating caspase-3, up-regulating *p53* and *p21* and down-regulating CDK2 gene expression. Supporting studies in human osteogenic sarcoma (HOS) cells by Han *et al.* [135] also documented ellagic acid as a potential growth inhibitor and apoptosis inducer. Ellagic acid induced apoptosis in IIOS cell line *via* up-regulating Bax expression, activating caspase-3 and cleaving PARP. Apoptotic potential of another phenolic acid, protocatechuic acid was evaluated using various cancer cell lines *viz*. MCF-7, A549, HepG2, HeLa and LNCaP [136]. It was observed that protocatechuic acid reduced the growth of cancer cells and induced apoptosis *via* MMP loss, decreased Na^+-K^+-ATPase activity and up-regulated caspase-3 activity. Further, this compound also decreased the expression of interleukins and vascular endothelial growth factor (VEGF) in all cancer cells. Gallic acid caused apoptosis in human non-small-cell lung cancer NCI-H460 cells *via* G2/M phase arrest, intracellular Ca^{2+} generation, MMP loss, apoptosis inducing factor (AIF) release and activation of caspase-8,-9,-3 along with modulation of Bcl-2 family proteins [137]. In another study, gallic

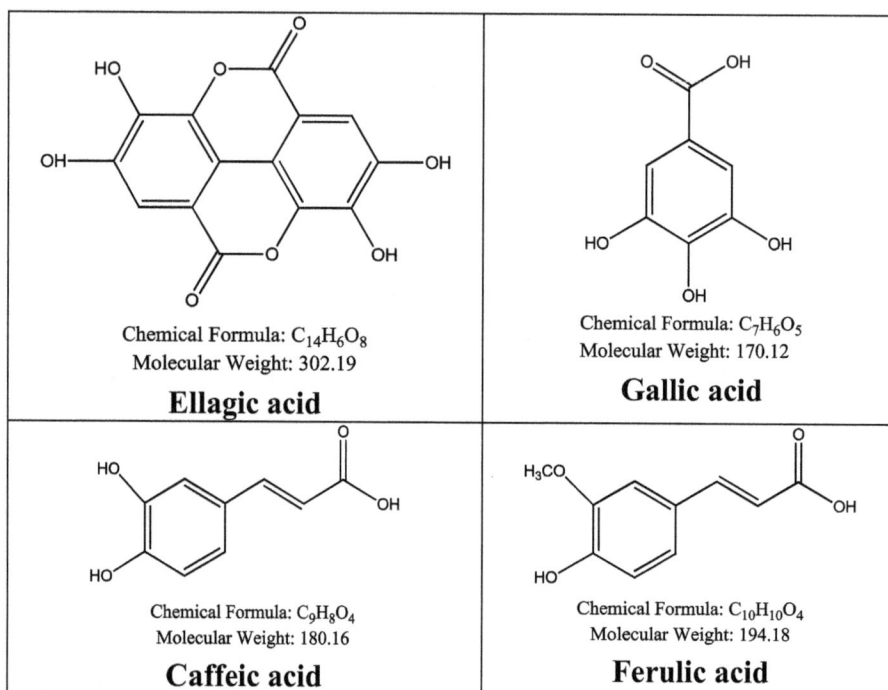

Chemical Formula: $C_{14}H_6O_8$
Molecular Weight: 302.19
Ellagic acid

Chemical Formula: $C_7H_6O_5$
Molecular Weight: 170.12
Gallic acid

Chemical Formula: $C_9H_8O_4$
Molecular Weight: 180.16
Caffeic acid

Chemical Formula: $C_{10}H_{10}O_4$
Molecular Weight: 194.18
Ferulic acid

Fig. (2). Chemical structures of some phenolics involved in caspases mediated apoptosis.

acid was reported to induce apoptosis in human melanoma (A375.S2) cells [138]. Gallic acid modulated the expression of Bcl-2 family proteins such as Bax and Bcl-2, induced ROS generation, MMP loss, cytochrome c release, cytosolic release of AIF and Endo G along with up-regulation of caspase-9 and -3 leading to apoptosis. Growth inhibitory potential of caffeic acid isolated from *Ocimum gratissimum* Linn. was evaluated in HeLa cells [139]. Caffeic acid dose-dependently inhibited the growth of HeLa cells and induced apoptosis *via* reducing Bcl-2 level, cytochrome c release into cytosol and increasing cleaved caspase-3 and *p*53 levels. Overall, it was concluded that caffeic acid induced apoptosis in HeLa cells *via* mitochondrial apoptotic pathway. Gallic acid (GA) and its derivative methyl gallate (MG) from seed coat extracts of *Givotia rottleriformis* inhibited the growth of human epidermoid carcinoma (A431) skin cancer cells by inducing apoptosis [140]. It was found that the treatment of A431 cells with both compounds resulted in reduced Bcl-2 and increased caspase-3 expression. Phenolic compound, ferulic acid was examined for its antiproliferative potency in osteosarcoma cell lines (143B and MG63) [141]. Treatment with ferulic acid caused apoptosis and arrest of cells in G_0/G_1 phase *via* reducing the expression of cell cycle-related proteins and Bcl-2 and increasing the expression of Bax and caspase-3 activity. Further, it also inhibited the activation of PI3K/Akt pathway.

Flavonoids Targeting Caspases

Flavonoids are secondary plant metabolites found in fruits, grains, vegetables and other plant parts including root, stem and leaves with variety of structural differences. Basic chemical structure of flavonoids is made up of fifteen-carbon skeleton having two benzene rings (A and B) linked through a heterocyclic pyran ring (C) [142]. There are more than 9,000 types of flavonoids known [143, 144]. Like phenolic acids, flavonoids are recognized to be potential inducers of apoptosis in cancer cells by targeting various cell signalling molecules implicated in apoptotic process such as caspases [145 - 148]. Fig. (**3**) shows structures of some flavonoids having modulatory effects on caspases activity.

Wang and coworkers [149] evaluated various flavonoids *viz*. apigenin, quercetin, myricetin and kaempferol for apoptosis induction in human leukaemia (HL-60) cell line. Treatment of HL-60 cells with flavonoids resulted in MMP loss, cytochrome c release, ROS generation, up-regulation of caspase-3 and cleavage of procaspase-9 and PARP. Apoptosis induction activity of different flavonoids was found to be in the order of apigenin > quercetin > myricetin > kaempferol. Lee *et al.* [150] reported that baicalein inhibited the growth of CH27 cells *via* inducing apoptosis. Apoptosis was caused by cell arrest in S-phase *via* reduction in the activity of Bcl-2 and procaspase-3. In another study, Apigenin was evaluated for apoptosis induction in leukemia (THP-1) cells by Vargo *et al.* [151]. Authors reported that apigenein treatment resulted in the activation of caspase-9,-3, enhancement in ROS generation, phosphorylation of mitogen activated protein kinases (MAPKs) including p38 and extracellular signal-regulated kinase (ERK) and activation of protein kinase c δ (PKC-δ) in leukemia cells. Over all from experimental results, the authors concluded that the mechanism responsible for apigenin induced apoptosis in THP-1 cells involves the activation of a PKC-δ dependent pathway. Granado-Serrano *et al.* [152] reported that quercetin induced apoptosis in HepG2 cells. It was found from results that apoptosis induction followed mitochondrial pathway *via* activation of caspase-3, -9, modulation of Bcl-2 family proteins and suppression of PI3K/Akt and ERK pathways. A dietary flavonoid, fisetin was reported to inhibit growth and induce apoptosis in prostate cancer (LNCaP) cells [153]. Fisetin arrested LNCaP cells in G1 phase along with modulation of various cell cycle proteins and modulating cyclin dependent kinase inhibitor (CKI)–cyclin–cyclin dependent kinase (CDK) network. Apoptosis induction by fisetin was associated with cytochrome c release, modulation of Bcl-2 family proteins and Smac/DIABLO, reduction in XIAP expression and caspase-3,-8,-9 activation. Apigenin, (-)-epigallocatechin, (-)-epigallocatechin-3-gallate (EGCG) and genistein were investigated by Das and coworkers [154] for apoptosis inducing potential in human glioblastoma (T98G and U87MG) cells. It was reported that all the four compounds induced apoptosis in glioblastoma cells

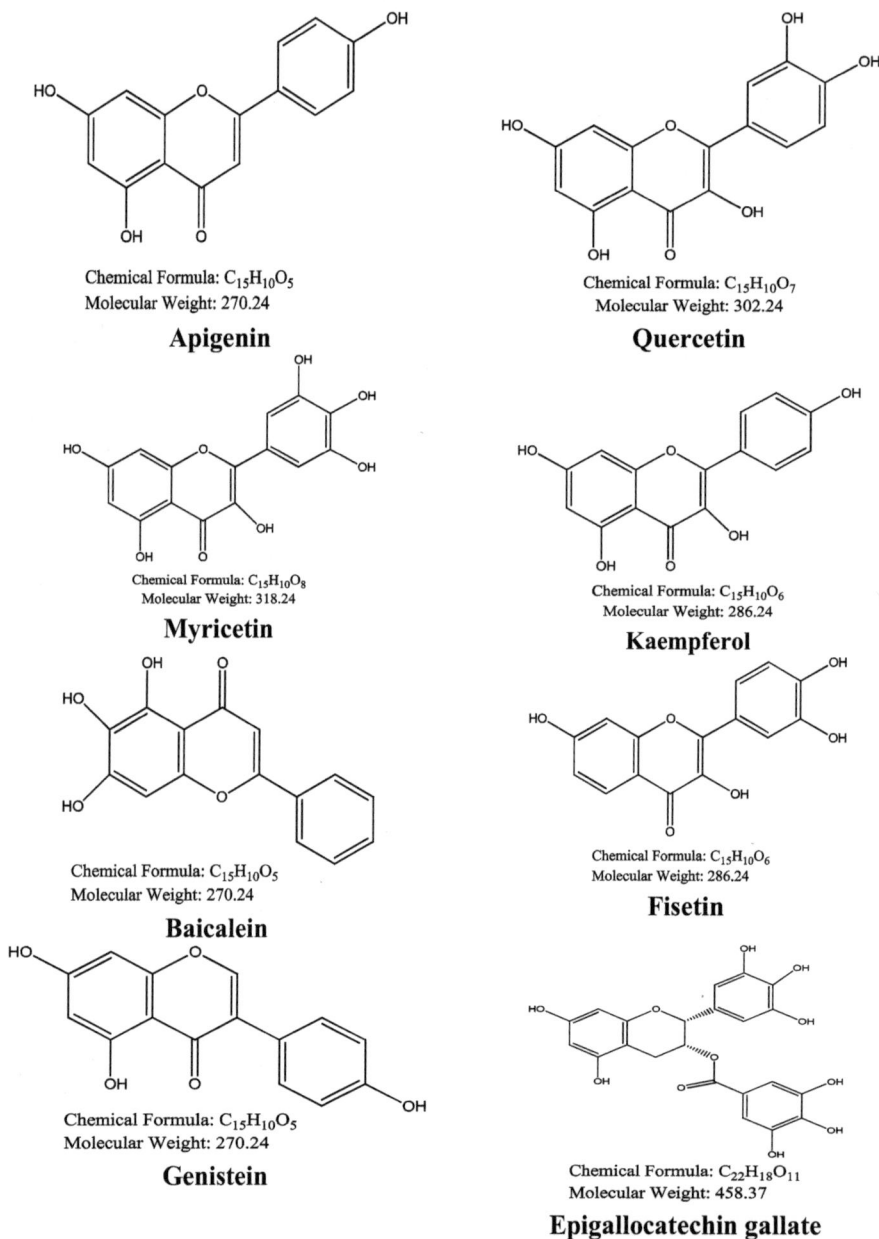

Chemical Formula: $C_{15}H_{10}O_5$
Molecular Weight: 270.24

Apigenin

Chemical Formula: $C_{15}H_{10}O_7$
Molecular Weight: 302.24

Quercetin

Chemical Formula: $C_{15}H_{10}O_8$
Molecular Weight: 318.24

Myricetin

Chemical Formula: $C_{15}H_{10}O_6$
Molecular Weight: 286.24

Kaempferol

Chemical Formula: $C_{15}H_{10}O_5$
Molecular Weight: 270.24

Baicalein

Chemical Formula: $C_{15}H_{10}O_6$
Molecular Weight: 286.24

Fisetin

Chemical Formula: $C_{15}H_{10}O_5$
Molecular Weight: 270.24

Genistein

Chemical Formula: $C_{22}H_{18}O_{11}$
Molecular Weight: 458.37

Epigallocatechin gallate

Fig. (3). Chemical structures of some flavonoids involved in caspases mediated apoptosis.

by enhancing ROS levels along with phosphorylation of *p*38 (MAPK) and activation of c-Jun N-terminal kinase-1 (JNK) pathway. Treatment of glioblastoma cells with flavonoids resulted in the increased intracellular free Ca^{2+}, increased Bax expression, mitochondrial membrane potential loss, cytochrome c and Smac/DIABLO release into cytosol, caspase-4,-9,-3 activation and modulation of some other proteins. (-)-epigallocatechin and EGCG also enhanced the activity of caspase-8. Further, it was observed that all the four flavonoids did not induce apoptosis in normal astrocytes (human). Overall apoptosis inducing activity was in the order (50 μM): apigenin < (-)-epigallocatechin < EGCG < genistein in both glioblastoma cells tested. Total flavonoids from *Polygonum amplexicaule* (TFPA) induced apoptosis in various cell lines (HepG2, Huh-7) of hepatocellular carcinoma *via* suppression of signal transducer and activator of transcription (STAT3) signaling. TFPA also enhanced the expression of SHP1 and caspase-3 activity and reduced the expression of Bcl-xL and Mcl-1 [155]. Water soluble total flavonoids (WSTF) from *Isodon lophanthoides* var. *gerardianus* were investigated for apoptosis inducing activity in HepG2 cells [156]. It was found that WSTF arrested HepG2 cells in G_0/G_1 phase and modulated the expression of various proteins including cytochrome c release, caspase-3 activation, decreased Bcl-2, survivin and Mcl-1 expression and increased Bax expression along with enhanced ROS generation. Authors also identified eight compounds in WSTF including caffeic acid, vicenin II, vicenin I, isoschaftoside, schaftoside, vitexin, 6,8-di-C-a-$_L$-arabinosylapigenin and rutin. Flavonoids *viz.* medicarpin and millepurpan were reported to induce apoptosis in leukaemia drug-sensitive (P388) multidrug resistant (P388/DOX) cells [157]. Both induced apoptosis *via* mitochondrial pathways through cleavage of PARP and reducing procaspase-3 activity. Expression of pro-apoptotic and anti-apoptotic Bcl-2 family proteins was also modulated on treatment with both compounds. The compounds also modulated P-gp-mediated efflux of drugs, thus potentiated the anticancer potential of chemotherapeutics in multidrug resistant cells.

Terpenoids Targeting Caspases

Terpenoids are largest group of plant secondary metabolites consisting of more than 40,000 compounds. These are made up of isoprenoid units and on the basis of these units; these are divided into different categories *viz.* monoterpenes, diterpenes, triterpenes, tetraterpenes and polyterpenes [158 - 162]. Many terpenoids are known to confer chemopreventive effects in breast, prostate and liver cancer [162 - 164]. Numerous mono- and triterpenoids have been identified to inhibit cancer cell growth *via* induction of apoptosis and differentiation in tumor cells, suppression of angiogenesis, cancer invasion and metastasis, targeting cell signaling molecules involved in apoptosis such as caspases, Bcl-2 family proteins, NF-κB *etc* [162 - 167]. In the next section, we summarized some of the reports of terpenoids in targeting caspases for apoptosis induction (Fig. **4**).

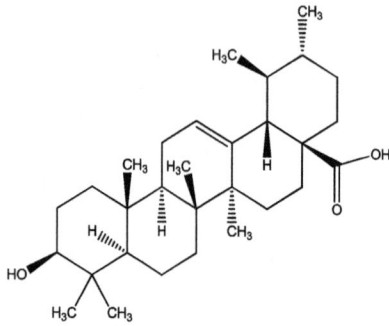

Chemical Formula: $C_{30}H_{48}O_3$
Molecular Weight: 456.70

Ursolic acid

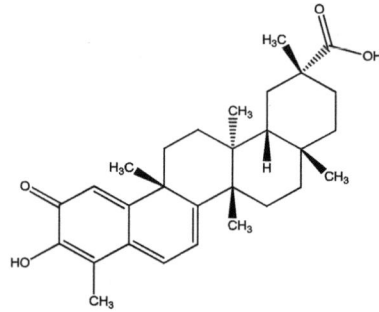

Chemical Formula: $C_{29}H_{38}O_4$
Molecular Weight: 450.61

Celastrol

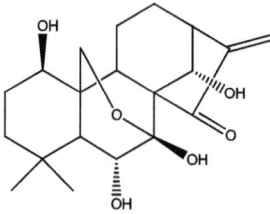

Chemical Formula: $C_{20}H_{28}O_6$
Molecular Weight: 364.43

Oridonin

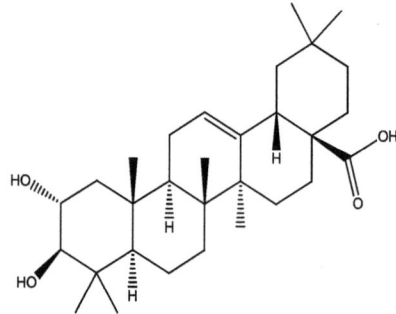

Chemical Formula: $C_{30}H_{48}O_4$
Molecular Weight: 472.70

Maslinic acid

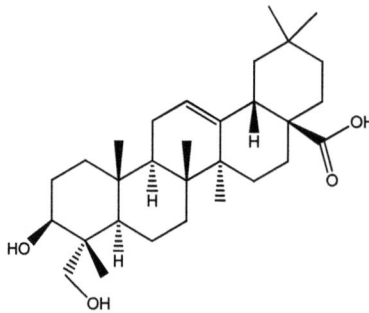

Chemical Formula: $C_{30}H_{48}O_4$
Molecular Weight: 472.70

Hederagenin

Fig. (4). Chemical structures of some terpenoids involved in caspases mediated apoptosis.

Ursolic acid caused apoptosis in human prostate epithelial cells *via* activating caspase-1,-3,-8,-9 and reducing the expression of cIAP family proteins [168]. In another study, ursolic acid was reported to inhibit the growth of HaCaT cells and induce apoptosis [169]. Apoptosis induction was mediated through cell cycle arrest in G1 phase and caspase-3 activation. Celastrol isolated from *Tripterygium regelii* was assessed for antiproliferative and apoptosis inducing activity in MCF-7 cells [170]. Compound treatment caused apoptosis through cytochrome c and AIF release, activation of caspase-7,-8,-9, cleavage of PARP and Bid, up-regulation of Bax and down-regulation of Bcl-2 in MCF-7 cells. Sun *et al.* [171] tested oridonin on gastric cancer (HGC-27) cells for anticancer activity and reported that oridonin induced mitochondrial dependent apoptosis *via* regulation of Apaf-1, Bcl-2 family proteins, caspase-3 and cytochrome c. Reyes-Zurita *et al.* [172] tested maslinic acid for its apoptosis inducing and antiproliferative activity in colon cancer (Caco-2) cells. Maslinic acid induced apoptosis in Caco-2 cells was associated with caspase-8,-3 activation, Bid cleavage, reduced Bcl-2 expression and activation of c-Jun N-terminal kinase (JNK). Hederagenin from *Hedera helix* L. leaves was evaluated for anticancer activity [173]. It was found that hederagenin treatment resulted ROS increase, MMP loss, down-regulation of Bcl-2, procaspase-9,-3 expressions and up-regulation of Bax, caspase-3,-9 expressions in LoVo cells (colon cancer). Terpenoid fraction from *Plectranthus hadiensis* shoots was investigated for apoptosis inducing activity on colon (HCT-15) cancer cells [174]. Apoptosis induction in HCT cells by terpenoid fraction was accompanied by increased activity of Bax and caspase-3 and decreased Bcl-2 and COX-2 activity. Results suggested the involvement of mitochondrial-dependent apoptotic pathway [175].

Alkaloids Targeting Caspases

Alkaloids are large group of plant secondary metabolites characterized by their basic nature and having one or more nitrogen atom in heterocyclic ring (Bhutani, 2007). Most popular anticancer agents of alkaloid nature include vinca alkaloids *viz.* vinblastine and vincristine isolated from *Catharanthus roseus* [176]. There are many reports in the literature about alkaloids in context to their anticancer properties. Mitochondria which are known to be implicated in tumorigenesis are target of many of the anticancer alkaloids [177 - 179]. Several reports in the literature have highlighted the activation of caspases in alkaloids-treated cancer cells (Fig. **5**).

Tetrandrine, a bis-benzylisoquinoline alkaloid induced apoptogenic cell death in leukaemia (U937) cells *via* induction of oxidative stress, activation of PKC-δ and caspase-3,-7,-8,-9 [180]. Yang *et al.* [181] reported α-chaconine, a potato glycoalkaloid as a potential inducer of apoptosis in colon (HT-29) cancer cells. Further, apoptosis induction by α-chaconine in HT-29 cells was mediated by inhibiting ERK and activating caspase-3. Berberine, an isoquinoline plant alkaloid was tested for apoptosis inducing activity in human promonocytic U937 cells

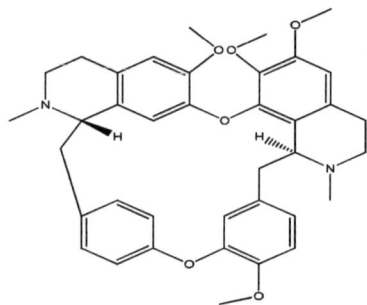

Chemical Formula: $C_{38}H_{42}N_2O_6$
Molecular Weight: 622.75
Tetrandrine

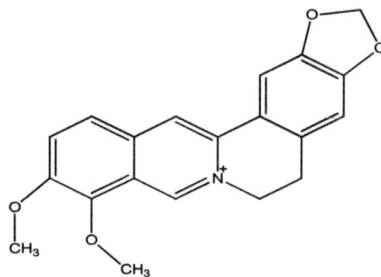

Chemical Formula: $C_{20}H_{18}NO_4^+$
Molecular Weight: 336.36
Berberine

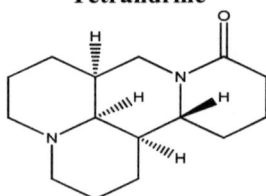

Chemical Formula: $C_{15}H_{24}N_2O$
Molecular Weight: 248.36
Matrine

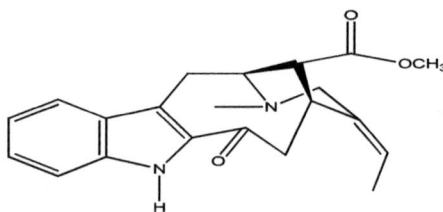

Chemical Formula: $C_{21}H_{24}N_2O_3$
Molecular Weight: 352.43
Vobasine

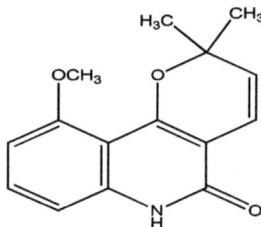

Chemical Formula: $C_{15}H_{15}NO_3$
Molecular Weight: 257.28
Haplophytin-A

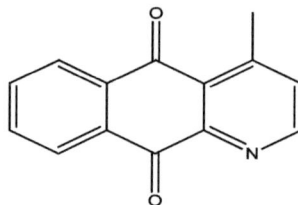

Chemical Formula: $C_{14}H_9NO_2$
Molecular Weight: 223.23
Cleistopholine

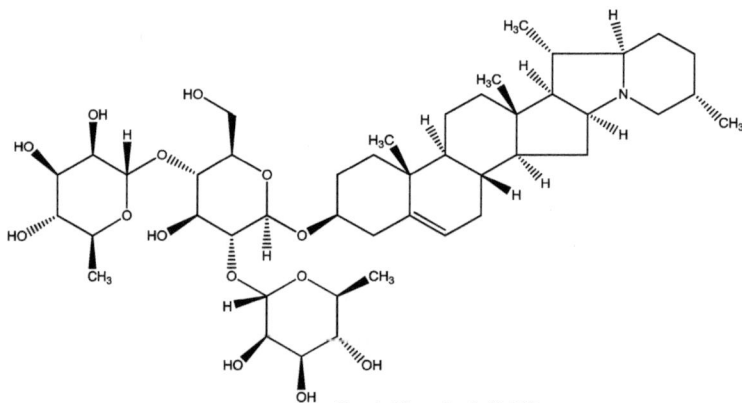

Chemical Formula: $C_{45}H_{73}NO_{14}$
Molecular Weight: 852.06
α- Chaconine

Fig. 5 cont.....

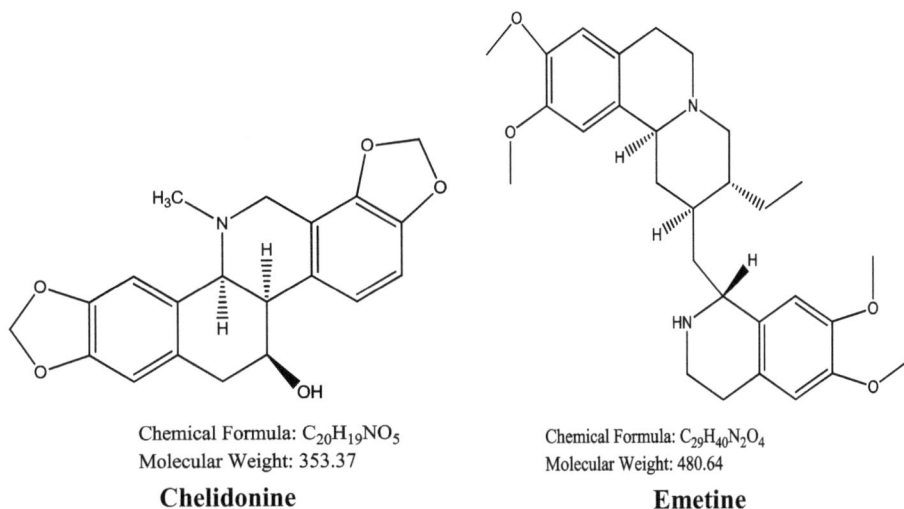

Chemical Formula: $C_{20}H_{19}NO_5$
Molecular Weight: 353.37

Chelidonine

Chemical Formula: $C_{29}H_{40}N_2O_4$
Molecular Weight: 480.64

Emetine

Fig. (5). Chemical structures of some alkaloids involved in caspases mediated apoptosis.

[182]. Berberine treatment induced apoptosis in U937 cells *via* decreasing MMP, increasing ROS, activating caspase-3, -9 pointing towards the involvement of mitochondrial/caspase-dependent pathway. Matrine was reported to incite apoptosis in gastric carcinoma (SGC-7901) cells *via* increasing the expression of Fas/Fas ligand (FasL) and caspase-3 activation [183]. Monoterpene indole alkaloids *viz.* tabernaemontanine and vobasine isolated from methanol extract of *Tabernaemontana elegans* leaves were found to induce apoptosis in HuH-7 cells *via* enhancing caspase-3 activity [184]. Won *et al.* [185] evaluated haplophytin-A (quinoline alkaloid) for apoptosis inducing potential in HL-60 cells and reported that it incited apoptosis in HL-60 by caspase-8-mediated extrinsic and mitochondrial-dependent intrinsic apoptotic pathways. Chelidonine and an alkaloid extract obtained from *Chelidonium majus* (harbouring protoberberine and benzo[c]phenanthridine alkaloids) were reported to regulate multidrug resistance in cancer cells *via* down-regulating ATP binding cassette (ABC)-transporter proteins such as P-glycoprotein (P-gp), multidrug resistance associated protein 1 (MRP1), breast cancer resistance protein (BCRP); metabolic genes including glutathione-S-transeferase (GST), CYP3A4, pregnane X receptor (PXR) and up-regulating the genes of caspase family [186]. Emetine (an alkaloid) was reported to increase the sensitization of ovarian carcinoma (SKOV3) cells to cisplatin [187]. Emetine/cisplatin (in combination) reduced colony formation and induced apoptosis *via* mediating caspase-3,-7,-8 activation and Bcl-xl down-regulation. Cathachunine isolated from *Catharanthus roseus* induced apoptosis in human Leukemia (HL-60) cells *via* ROS-mediated intrinsic pathway [188]. Treatment of HL-60 cells with cathachunine resulted in the modulation of Bcl-2/Bax ratio,

mitochondrial membrane potential loss, cytochrome c release and cleavage of caspase-3, PARP and increased ROS generation. An alkaloid, cleistopholine from *Enicosanthellum pulchrum* demonstrated apoptosis inducing activity in human ovarian (CAOV-3) cancer cells [189]. Cleistopholine treatment arrested the cells in G_0/G_1 phase and induced apoptosis *via* activation of caspase-9 & -3, MMP loss, cytochrome c release into cytosol, up-regulation of Bax, down-regulation of Bcl-2, Hsp70 and survivin proteins. *Aconitum szechenyianum* alkaloids (ASA) were investigated for anticancer activity against A549 cells by Fan *et al.* [190]. Authors concluded from results that ASA induced apoptosis in A549 cells *via* increasing Bax expression, decreasing Bcl-2 expression, cytochrome c release, activating caspase-3,-8,-9 and modulating *p*38 MAPK signaling pathway.

Polysaccharides Targeting Caspases

Polysaccharides are bio-molecules consisting of monosaccharide units joined together with the help of glycosidic bonds (in linear fashion or having branched side chains) [191]. These are found naturally in variety of living organisms including plants, microorganisms and animals [192]. Experimental studies have demonstrated that polysaccharides can diminish the growth of tumors by various mechanisms such as inducing apoptosis, suppressing cancer metastasis and enhancing the efficacy of conventional chemotherapeutics [191]. Survey of literature showed that polysaccharides possess potential to target caspases in inducing apoptosis in different types of cancer.

Water-soluble polysaccharide fraction (WACP; molecular weight 5.8×10^4 Da) from medicinal herb *Artemisia capillaries* induced apoptosis in nasopharyngeal carcinoma (CNE-2) cells [193]. Mechanism of WACP induced apoptosis involved MMP loss, cytochrome c release and caspase-3,-9 activation. A water-soluble polysaccharide (WPS-2-1) obtained from brown alga (*Laminaria japonica*) has been reported for antiproliferative activity [194]. Polysaccharide (molecular weight 80 kDa) induced apoptosis in A375 cells *via* MMP loss, modulation of Bax and Bcl-2 expression and activating caspase-3,-9. *Atractylodes macrocephala* polysaccharides (AMPs) were investigated for inducing apoptosis in C6 glioma cells [195]. It was reported that AMPs treatment to C6 cells resulted in MMP loss, cytochrome c release, activation of caspases-3,-9 along with PARP cleavage, thus suggesting the involvement of mitochondrial pathway in AMPs induced apoptosis. Ma *et al.* [196] studied a polysaccharide fraction (SFPSA; molecular weight 168 kDa) from *Stachys floridana* rhizomes for antiproliferative potency against HT-29 colon cancer cells. It was reported that SFPSA induced apoptosis and arrested the cells in G2/M phase of cell cycle along with reduction in Bcl-2 mRNA level and enhancement in Bax and *p*53 mRNA level and increased caspase-3 activity. A polysaccharide fraction namely CP-1 (molecular weight 12 kDa) was extracted from adlay seed *i.e. Coix lachryma-jobi* L. and investigated for apoptosis inducing activity [197]. CP-1 treatment resulted in the induction of apoptosis in A549 cells through MMP loss, activation of caspase-3,-9 and DNA damage (increased comet tail length). Overall, it was concluded that CP-1 induced apoptosis in A549 cells *via* intrinsic mitochondrial pathway. Polysaccharide fractions *viz.* F1 (molecular weight ~65 kDa) and F2 (molecular weight ~45 kDa) from *Cymbopogon citratus* was assessed for antiproliferative activity against Siha and LNCap cells [198]. Fractions demonstrated apoptotic potential on cancer cells *via* intrinsic pathway through MMP loss, cytochrome c release, modulating the expression of Bcl-2 family genes leading to activation of caspase-3. A polysaccharide isolated from a parasitic plant, *Boschniakia rossica* (BRP; molecular weight 22 kDa) was evaluated for anticancer activity in human larynx squamous carcinoma (Hep2) cells [199]. BRP induced apoptosis and arrested the cells in G_0/G_1 phase. Analysis of results demonstrated that BRP treatment resulted in the cleavage of procaspase-3,-8,-9 along with enhanced expression of death receptor (DR5) and pro-apoptotic protein (Bax) and down-regulated the expression of anti-apoptotic protein (Bcl-2). *Angelica sinensis* polysaccharide (ASP) from *Angelica sinensis* roots was evaluated for anticancer activity in breast cancer cells [200]. ASP induced cell death in breast cancer cells involved CREB signaling, caspase-3,-9 activation, MMP loss, release of cytochrome c, translocation of Bax to mitochondria, PARP cleavage and modulation of Apaf-1 and Bcl-2 family proteins. Apoptosis inducing activity of water-soluble polysaccharides extracted from *Sipunculus nudus* (SNP) in Hepg2.2.15 cells was evaluated by Su *et al.* [201]. SNP induced apoptosis was mediated by up-regulation of TNF-α, caspase-3 and Bax and down-regulation of survivin, Bcl-2 and VEGF.

Stilbenes Targeting Caspases

Stilbenes are secondary plant metabolites with 1,2-diphenylethylene nucleus derived from phenylpropanoid pathway [202]. More than 400 types of natural stilbene compounds are known and many of them provide health benefits such as resveratrol. As enzyme 'stilbene synthase' required for biosynthesis of stilbenes is not universally present, therefore stilbenes are found only in some plant families [202 - 204]. Stilbenes have potential to hit many cellular targets such as caspases, Bcl-2 family proteins, HSPs for inducing cell death (autophagy and apoptosis) [205]. Fig. (**6**) shows chemical structures of some stilbenes having modulatory effects on caspases activity.

Resveratrol (trans-3,4',5-trihydroxystilbene) inhibited the growth and induced apoptosis in U251 glioma cells *via* G_0/G_1 growth arrest, cytochrome c release, caspase-9,-3 activation, increased Bax expression and PARP cleavage [206].

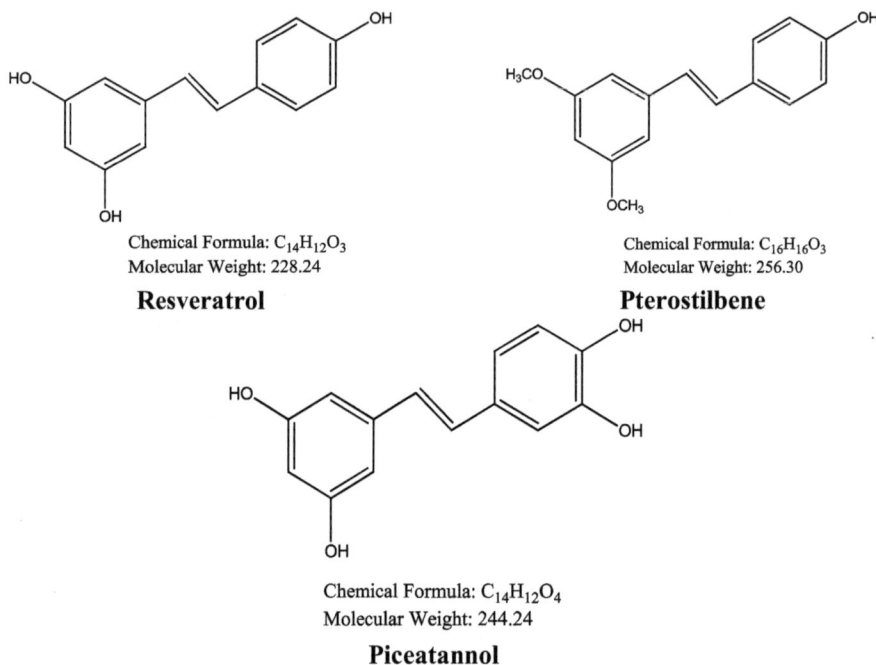

Chemical Formula: $C_{14}H_{12}O_3$
Molecular Weight: 228.24
Resveratrol

Chemical Formula: $C_{16}H_{16}O_3$
Molecular Weight: 256.30
Pterostilbene

Chemical Formula: $C_{14}H_{12}O_4$
Molecular Weight: 244.24
Piceatannol

Fig. (6). Chemical structures of some stilbenes involved in caspases mediated apoptosis.

Similarly, another stilbene, piceatannol treatment caused apoptosis in human leukaemia U937 cells *via* inducing accumulation of cells in subG1 phase along with modulation of Bcl-2 and cIAP-2 expression, caspase-3 activation and PARP cleavage [207]. In another study, piceatannol induced apoptotic cell death in human prostate (DU145) cancer cells [208]. Compound treatment caused up-regulation of caspase-8,-9,-7,-3, truncated Bid, Bax, Bik, Bcl-2 related ovarian killer (Bok), Fas and cleaved PARP along with cytochrome c release in cytoplasm and down-regulation of Mcl-1 and Bcl-xl. Resveratrol induced apoptosis in human oral (KB) cancer cells [209]. Experimental results depicted that resveratrol induced apoptosis was mediated through activation of caspases-3,-7. Pterostilbene treatment in MOLT4 (leukaemia) cells resulted in cell cycle arrest in S-phase and apoptosis mediated by loss of MMP and caspase-3 activation [210]. In a study by Hsiao *et al.* [211], it was reported that pterostilbene induced apoptotic cell death in acute myeloid leukaemia (AML) cells mediated through arrest of cells in G_0/G_1 phase and involvement of MAPK. Compound treatment also resulted in the activation of caspase -8,-9,-3.

SUMMARY

As caspases play a significant role in apoptotic mechanism of cell death, multitude of phytochemicals or crude plant extracts/fractions have been identified by different workers all over the world highlighting their potential to activate caspases and induce apoptosis in cancer cells. Further studies are needed to explore more signaling molecules along with already known ones which may have involvement in the activation of caspases or in the formation of signaling complexes in route of cell death pathways. This would help in better understanding and targeting molecular mechanisms involved in caspase-mediated apoptosis for management of cancer.

CONSENT FOR PUBLICATION

Not applicable.

CONFLICT OF INTEREST

The authors declare no conflict of interest, financial or otherwise.

ACKNOWLEDGEMENTS

This work was supported by S.D. College, Barnala, Punjab, India and Guru Nanak Dev University, Amritsar-143005, Punjab, India.

REFERENCES

[1]   Hanahan D, Weinberg RA. The hallmarks of cancer. Cell 2000; 100(1): 57-70.
       [http://dx.doi.org/10.1016/S0092-8674(00)81683-9] [PMID: 10647931]

[2]   Franco R, Schoneveld O, Georgakilas AG, Panayiotidis MI. Oxidative stress, DNA methylation and carcinogenesis. Cancer Lett 2008; 266(1): 6-11.
       [http://dx.doi.org/10.1016/j.canlet.2008.02.026] [PMID: 18372104]

[3]   Panayiotidis M. Reactive oxygen species (ROS) in multistage carcinogenesis. Cancer Lett 2008; 266(1): 3-5.
       [http://dx.doi.org/10.1016/j.canlet.2008.02.027] [PMID: 18359554]

[4]   Vogelstein B, Kinzler KW. Cancer genes and the pathways they control. Nat Med 2004; 10(8): 789-99.
       [http://dx.doi.org/10.1038/nm1087] [PMID: 15286780]

[5]   Dandawate P, Ahmad A, Deshpande J, *et al.* Anticancer phytochemical analogs 37: synthesis, characterization, molecular docking and cytotoxicity of novel plumbagin hydrazones against breast cancer cells. Bioorg Med Chem Lett 2014; 24(13): 2900-4.
       [http://dx.doi.org/10.1016/j.bmcl.2014.04.100] [PMID: 24835626]

[6]   Meguid RA, Hooker CM, Taylor JT, *et al.* Recurrence after neoadjuvant chemoradiation and surgery for esophageal cancer: does the pattern of recurrence differ for patients with complete response and those with partial or no response? J Thorac Cardiovasc Surg 2009; 138(6): 1309-17.
       [http://dx.doi.org/10.1016/j.jtcvs.2009.07.069] [PMID: 19931663]

[7]   Gomez-Veiga F, Mariño A, Alvarez L, *et al.* Brachytherapy for the treatment of recurrent prostate cancer after radiotherapy or radical prostatectomy. BJU Int 2012; 109 (Suppl. 1): 17-21.
       [http://dx.doi.org/10.1111/j.1464-410X.2011.10826.x] [PMID: 22239225]

[8]   Payne AS, James WD, Weiss RB. Dermatologic toxicity of chemotherapeutic agents. Semin Oncol

2006; 33(1): 86-97.
[http://dx.doi.org/10.1053/j.seminoncol.2005.11.004] [PMID: 16473647]

[9]   Rybak LP, Mukherjea D, Jajoo S, Ramkumar V. Cisplatin ototoxicity and protection: clinical and experimental studies. Tohoku J Exp Med 2009; 219(3): 177-86.
[http://dx.doi.org/10.1620/tjem.219.177] [PMID: 19851045]

[10]   Sporn MB. Approaches to prevention of epithelial cancer during the preneoplastic period. Cancer Res 1976; 36(7 PT 2): 2699-702.
[PMID: 1277177]

[11]   Surh YJ. Cancer chemoprevention with dietary phytochemicals. Nat Rev Cancer 2003; 3(10): 768-80.
[http://dx.doi.org/10.1038/nrc1189] [PMID: 14570043]

[12]   Goodsell DS. The molecular perspective: caspases. Oncologist 2000; 5(5): 435-6.
[http://dx.doi.org/10.1634/theoncologist.5-5-435] [PMID: 11040280]

[13]   Kroemer G, Galluzzi L, Vandenabeele P, et al. Classification of cell death: recommendations of the Nomenclature Committee on Cell Death 2009. Cell Death Differ 2009; 16(1): 3-11.
[http://dx.doi.org/10.1038/cdd.2008.150] [PMID: 18846107]

[14]   Reed JC. Mechanisms of apoptosis. Am J Pathol 2000; 157(5): 1415-30.
[http://dx.doi.org/10.1016/S0002-9440(10)64779-7] [PMID: 11073801]

[15]   Kerr JF, Wyllie AH, Currie AR. Apoptosis: a basic biological phenomenon with wide-ranging implications in tissue kinetics. Br J Cancer 1972; 26(4): 239-57.
[http://dx.doi.org/10.1038/bjc.1972.33] [PMID: 4561027]

[16]   Olsson M, Zhivotovsky B. Caspases and cancer. Cell Death Differ 2011; 18(9): 1441-9.
[http://dx.doi.org/10.1038/cdd.2011.30] [PMID: 21455218]

[17]   Shah S, Gapor A, Sylvester PW. Role of caspase-8 activation in mediating vitamin E-induced apoptosis in murine mammary cancer cells. Nutr Cancer 2003; 45(2): 236-46.
[http://dx.doi.org/10.1207/S15327914NC4502_14] [PMID: 12881019]

[18]   Li J, Yuan J. Caspases in apoptosis and beyond. Oncogene 2008; 27(48): 6194-206.
[http://dx.doi.org/10.1038/onc.2008.297] [PMID: 18931687]

[19]   Jäger R, Zwacka RM. The enigmatic roles of caspases in tumor development. Cancers (Basel) 2010; 2(4): 1952-79.
[http://dx.doi.org/10.3390/cancers2041952] [PMID: 24281211]

[20]   Yuan J, Shaham S, Ledoux S, Ellis HM, Horvitz HR. The C. elegans cell death gene ced-3 encodes a protein similar to mammalian interleukin-1 β-converting enzyme. Cell 1993; 75(4): 641-52.
[http://dx.doi.org/10.1016/0092-8674(93)90485-9] [PMID: 8242740]

[21]   Slee EA, Harte MT, Kluck RM, et al. Ordering the cytochrome c-initiated caspase cascade: hierarchical activation of caspases-2, -3, -6, -7, -8, and -10 in a caspase-9-dependent manner. J Cell Biol 1999; 144(2): 281-92.
[http://dx.doi.org/10.1083/jcb.144.2.281] [PMID: 9922454]

[22]   Boatright KM, Salvesen GS. Caspase activation. Biochem Soc Symp 2003; 70(70): 233-42.
[http://dx.doi.org/10.1042/bss0700233] [PMID: 14587296]

[23]   McIlwain DR, Berger T, Mak TW. Caspase functions in cell death and disease. Cold Spring Harb Perspect Biol 2013; 5(4): a008656.
[http://dx.doi.org/10.1101/cshperspect.a008656] [PMID: 23545416]

[24]   Shi Y. Caspase activation, inhibition, and reactivation: a mechanistic view. Protein Sci 2004; 13(8): 1979-87.
[http://dx.doi.org/10.1110/ps.04789804] [PMID: 15273300]

[25]   Elmore S. Apoptosis: a review of programmed cell death. Toxicol Pathol 2007; 35(4): 495-516.
[http://dx.doi.org/10.1080/01926230701320337] [PMID: 17562483]

[26]   Li H, Zhu H, Xu CJ, Yuan J. Cleavage of BID by caspase 8 mediates the mitochondrial damage in the Fas pathway of apoptosis. Cell 1998; 94(4): 491-501.
[http://dx.doi.org/10.1016/S0092-8674(00)81590-1] [PMID: 9727492]

[27]   Luo X, Budihardjo I, Zou H, Slaughter C, Wang X. Bid, a Bcl2 interacting protein, mediates cytochrome c release from mitochondria in response to activation of cell surface death receptors. Cell 1998; 94(4): 481-90.
[http://dx.doi.org/10.1016/S0092-8674(00)81589-5] [PMID: 9727491]

[28]   Fulda S, Debatin KM. Targeting apoptosis pathways in cancer therapy. Curr Cancer Drug Targets 2004; 4(7): 569-76.
[http://dx.doi.org/10.2174/1568009043332763] [PMID: 15578914]

[29]   Jin Z, El-Deiry WS. Overview of cell death signaling pathways. Cancer Biol Ther 2005; 4(2): 139-63.
[http://dx.doi.org/10.4161/cbt.4.2.1508] [PMID: 15725726]

[30]   Xu G, Shi Y. Apoptosis signaling pathways and lymphocyte homeostasis. Cell Res 2007; 17(9): 759-71.
[http://dx.doi.org/10.1038/cr.2007.52] [PMID: 17576411]

[31]   Gonzalvez F, Ashkenazi A. New insights into apoptosis signaling by Apo2L/TRAIL. Oncogene 2010; 29(34): 4752-65.
[http://dx.doi.org/10.1038/onc.2010.221] [PMID: 20531300]

[32]   Galluzzi L, Kepp O, Trojel-Hansen C, Kroemer G. Mitochondrial control of cellular life, stress, and death. Circ Res 2012; 111(9): 1198-207.
[http://dx.doi.org/10.1161/CIRCRESAHA.112.268946] [PMID: 23065343]

[33]   Shamas-Din A, Kale J, Leber B, Andrews DW. Mechanisms of action of Bcl-2 family proteins. Cold Spring Harb Perspect Biol 2013; 5(4): a008714.
[http://dx.doi.org/10.1101/cshperspect.a008714] [PMID: 23545417]

[34]   Oltvai ZN, Milliman CL, Korsmeyer SJ. Bcl-2 heterodimerizes *in vivo* with a conserved homolog, Bax, that accelerates programmed cell death. Cell 1993; 74(4): 609-19.
[http://dx.doi.org/10.1016/0092-8674(93)90509-O] [PMID: 8358790]

[35]   Gross A, McDonnell JM, Korsmeyer SJ. BCL-2 family members and the mitochondria in apoptosis. Genes Dev 1999; 13(15): 1899-911. a
[http://dx.doi.org/10.1101/gad.13.15.1899] [PMID: 10444588]

[36]   Gross A, Yin XM, Wang K, *et al.* Caspase cleaved BID targets mitochondria and is required for cytochrome c release, while BCL-XL prevents this release but not tumor necrosis factor-R1/Fas death. J Biol Chem 1999; 274(2): 1156-63. b
[http://dx.doi.org/10.1074/jbc.274.2.1156] [PMID: 9873064]

[37]   Zhang JH, Zhang Y, Herman B. Caspases, apoptosis and aging. Ageing Res Rev 2003; 2(4): 357-66.
[http://dx.doi.org/10.1016/S1568-1637(03)00026-6] [PMID: 14522240]

[38]   Green DR, Kroemer G. The pathophysiology of mitochondrial cell death. Science 2004; 305(5684): 626-9.
[http://dx.doi.org/10.1126/science.1099320] [PMID: 15286356]

[39]   Deveraux QL, Reed JC. IAP family proteins--suppressors of apoptosis. Genes Dev 1999; 13(3): 239-52.
[http://dx.doi.org/10.1101/gad.13.3.239] [PMID: 9990849]

[40]   Deveraux QL, Takahashi R, Salvesen GS, Reed JC. X-linked IAP is a direct inhibitor of cell-death proteases. Nature 1997; 388(6639): 300-4.
[http://dx.doi.org/10.1038/40901] [PMID: 9230442]

[41]   Deveraux QL, Roy N, Stennicke HR, *et al.* IAPs block apoptotic events induced by caspase-8 and cytochrome c by direct inhibition of distinct caspases. EMBO J 1998; 17(8): 2215-23.

[http://dx.doi.org/10.1093/emboj/17.8.2215] [PMID: 9545235]

[42] Roy N, Deveraux QL, Takahashi R, Salvesen GS, Reed JC. The c-IAP-1 and c-IAP-2 proteins are direct inhibitors of specific caspases. EMBO J 1997; 16(23): 6914-25.
[http://dx.doi.org/10.1093/emboj/16.23.6914] [PMID: 9384571]

[43] Salvesen GS, Duckett CS. IAP proteins: blocking the road to death's door. Nat Rev Mol Cell Biol 2002; 3(6): 401-10.
[http://dx.doi.org/10.1038/nrm830] [PMID: 12042762]

[44] Tamm I, Wang Y, Sausville E, *et al.* IAP-family protein survivin inhibits caspase activity and apoptosis induced by Fas (CD95), Bax, caspases, and anticancer drugs. Cancer Res 1998; 58(23): 5315-20.
[PMID: 9850056]

[45] Stennicke HR, Salvesen GS. Properties of the caspases. Biochim Biophys Acta 1998; 1387(1-2): 17-31.
[http://dx.doi.org/10.1016/S0167-4838(98)00133-2] [PMID: 9748481]

[46] Ghavami S, Hashemi M, Ande SR, *et al.* Apoptosis and cancer: mutations within caspase genes. J Med Genet 2009; 46(8): 497-510.
[http://dx.doi.org/10.1136/jmg.2009.066944] [PMID: 19505876]

[47] Du C, Fang M, Li Y, Li L, Wang X. Smac, a mitochondrial protein that promotes cytochrome c-dependent caspase activation by eliminating IAP inhibition. Cell 2000; 102(1): 33-42.
[http://dx.doi.org/10.1016/S0092-8674(00)00008-8] [PMID: 10929711]

[48] Verhagen AM, Vaux DL. Cell death regulation by the mammalian IAP antagonist Diablo/Smac. Apoptosis 2002; 7(2): 163-6.
[http://dx.doi.org/10.1023/A:1014318615955] [PMID: 11865200]

[49] Concannon CG, Orrenius S, Samali A. Hsp27 inhibits cytochrome c-mediated caspase activation by sequestering both pro-caspase-3 and cytochrome c. Gene Expr 2001; 9(4-5): 195-201.
[http://dx.doi.org/10.3727/000000001783992605] [PMID: 11444529]

[50] Pandey P, Farber R, Nakazawa A, *et al.* Hsp27 functions as a negative regulator of cytochrome c-dependent activation of procaspase-3. Oncogene 2000; 19(16): 1975-81.
[http://dx.doi.org/10.1038/sj.onc.1203531] [PMID: 10803458]

[51] Bruey JM, Ducasse C, Bonniaud P, *et al.* Hsp27 negatively regulates cell death by interacting with cytochrome c. Nat Cell Biol 2000; 2(9): 645-52.
[http://dx.doi.org/10.1038/35023595] [PMID: 10980706]

[52] Creagh EM, Carmody RJ, Cotter TG. Heat shock protein 70 inhibits caspase-dependent and -independent apoptosis in Jurkat T cells. Exp Cell Res 2000; 257(1): 58-66.
[http://dx.doi.org/10.1006/excr.2000.4856] [PMID: 10854054]

[53] Samali A, Cai J, Zhivotovsky B, Jones DP, Orrenius S. Presence of a pre-apoptotic complex of pro-caspase-3, Hsp60 and Hsp10 in the mitochondrial fraction of jurkat cells. EMBO J 1999; 18(8): 2040-8.
[http://dx.doi.org/10.1093/emboj/18.8.2040] [PMID: 10205158]

[54] Xanthoudakis S, Roy S, Rasper D, *et al.* Hsp60 accelerates the maturation of pro-caspase-3 by upstream activator proteases during apoptosis. EMBO J 1999; 18(8): 2049-56.
[http://dx.doi.org/10.1093/emboj/18.8.2049] [PMID: 10205159]

[55] Bold RJ, Termuhlen PM, McConkey DJ. Apoptosis, cancer and cancer therapy. Surg Oncol 1997; 6(3): 133-42.
[http://dx.doi.org/10.1016/S0960-7404(97)00015-7] [PMID: 9576629]

[56] Lan Q, Zheng T, Chanock S, *et al.* Genetic variants in caspase genes and susceptibility to non-Hodgkin lymphoma. Carcinogenesis 2007; 28(4): 823-7.
[http://dx.doi.org/10.1093/carcin/bgl196] [PMID: 17071630]

[57]   Soung YH, Lee JW, Kim HS, *et al.* Inactivating mutations of CASPASE-7 gene in human cancers. Oncogene 2003; 22(39): 8048-52.
[http://dx.doi.org/10.1038/sj.onc.1206727] [PMID: 12970753]

[58]   Shalini S, Dorstyn L, Dawar S, Kumar S. Old, new and emerging functions of caspases. Cell Death Differ 2015; 22(4): 526-39.
[http://dx.doi.org/10.1038/cdd.2014.216] [PMID: 25526085]

[59]   Hosgood HD III, Baris D, Zhang Y, *et al.* Caspase polymorphisms and genetic susceptibility to multiple myeloma. Hematol Oncol 2008; 26(3): 148-51.
[http://dx.doi.org/10.1002/hon.852] [PMID: 18381704]

[60]   Lee SY, Choi YY, Choi JE, *et al.* Polymorphisms in the caspase genes and the risk of lung cancer. J Thorac Oncol 2010; 5(8): 1152-8.
[http://dx.doi.org/10.1097/JTO.0b013e3181e04543] [PMID: 20661084]

[61]   Liamarkopoulos E, Gazouli M, Aravantinos G, *et al.* Caspase 8 and caspase 9 gene polymorphisms and susceptibility to gastric cancer. Gastric Cancer 2011; 14(4): 317-21.
[http://dx.doi.org/10.1007/s10120-011-0045-1] [PMID: 21461653]

[62]   Theodoropoulos GE, Michalopoulos NV, Panoussopoulos SG, Taka S, Gazouli M. Effects of caspase-9 and survivin gene polymorphisms in pancreatic cancer risk and tumor characteristics. Pancreas 2010; 39(7): 976-80.
[http://dx.doi.org/10.1097/MPA.0b013e3181d705d4] [PMID: 20357690]

[63]   Yilmaz SG, Yencilek F, Yildirim A, Yencilek E, Isbir T. Effects of caspase 9 gene polymorphism in patients with prostate cancer. In Vivo 2017; 31(2): 205-8.
[http://dx.doi.org/10.21873/invivo.11046] [PMID: 28358701]

[64]   Skurk C, Maatz H, Kim HS, *et al.* The Akt-regulated forkhead transcription factor FOXO3a controls endothelial cell viability through modulation of the caspase-8 inhibitor FLIP. J Biol Chem 2004; 279(2): 1513-25.
[http://dx.doi.org/10.1074/jbc.M304736200] [PMID: 14551207]

[65]   Shabala S. Salinity and programmed cell death: unravelling mechanisms for ion specific signalling. J Exp Bot 2009; 60(3): 709-12.
[http://dx.doi.org/10.1093/jxb/erp013] [PMID: 19269993]

[66]   Hou YM, Yang X. Regulation of cell death by transfer RNA. Antioxid Redox Signal 2013; 19(6): 583-94.
[http://dx.doi.org/10.1089/ars.2012.5171] [PMID: 23350625]

[67]   Kominami K, Nakabayashi J, Nagai T, *et al.* The molecular mechanism of apoptosis upon caspase-8 activation: quantitative experimental validation of a mathematical model. Biochim Biophys Acta 2012; 1823(10): 1825-40.
[http://dx.doi.org/10.1016/j.bbamcr.2012.07.003] [PMID: 22801217]

[68]   Hockenbery D, Nuñez G, Milliman C, Schreiber RD, Korsmeyer SJ. Bcl-2 is an inner mitochondrial membrane protein that blocks programmed cell death. Nature 1990; 348(6299): 334-6.
[http://dx.doi.org/10.1038/348334a0] [PMID: 2250705]

[69]   Krajewski S, Tanaka S, Takayama S, Schibler MJ, Fenton W, Reed JC. Investigation of the subcellular distribution of the bcl-2 oncoprotein: residence in the nuclear envelope, endoplasmic reticulum, and outer mitochondrial membranes. Cancer Res 1993; 53(19): 4701-14.
[PMID: 8402648]

[70]   de Jong D, Prins FA, Mason DY, Reed JC, van Ommen GB, Kluin PM. Subcellular localization of the bcl-2 protein in malignant and normal lymphoid cells. Cancer Res 1994; 54(1): 256-60.
[PMID: 8261449]

[71]   Zhu W, Cowie A, Wasfy GW, Penn LZ, Leber B, Andrews DW. Bcl-2 mutants with restricted subcellular location reveal spatially distinct pathways for apoptosis in different cell types. EMBO J

1996; 15(16): 4130-41.
[http://dx.doi.org/10.1002/j.1460-2075.1996.tb00788.x] [PMID: 8861942]

[72] Hsu YT, Wolter KG, Youle RJ. Cytosol-to-membrane redistribution of Bax and Bcl-X(L) during apoptosis. Proc Natl Acad Sci USA 1997; 94(8): 3668-72.
[http://dx.doi.org/10.1073/pnas.94.8.3668] [PMID: 9108035]

[73] Gross A, Jockel J, Wei MC, Korsmeyer SJ. Enforced dimerization of BAX results in its translocation, mitochondrial dysfunction and apoptosis. EMBO J 1998; 17(14): 3878-85.
[http://dx.doi.org/10.1093/emboj/17.14.3878] [PMID: 9670005]

[74] Puthalakath H, Huang DC, O'Reilly LA, King SM, Strasser A. The proapoptotic activity of the Bcl-2 family member Bim is regulated by interaction with the dynein motor complex. Mol Cell 1999; 3(3): 287-96.
[http://dx.doi.org/10.1016/S1097-2765(00)80456-6] [PMID: 10198631]

[75] Frenzel A, Grespi F, Chmelewskij W, Villunger A. Bcl2 family proteins in carcinogenesis and the treatment of cancer. Apoptosis 2009; 14(4): 584-96.
[http://dx.doi.org/10.1007/s10495-008-0300-z] [PMID: 19156528]

[76] Akl H, Vervloessem T, Kiviluoto S, et al. A dual role for the anti-apoptotic Bcl-2 protein in cancer: mitochondria versus endoplasmic reticulum. Biochim Biophys Acta 2014; 1843(10): 2240-52.
[http://dx.doi.org/10.1016/j.bbamcr.2014.04.017] [PMID: 24768714]

[77] Xu X, Jin S, Ma Y, et al. miR-30a-5p enhances paclitaxel sensitivity in non-small cell lung cancer through targeting BCL-2 expression. J Mol Med (Berl) 2017; 95(8): 861-71.
[http://dx.doi.org/10.1007/s00109-017-1539-z] [PMID: 28487996]

[78] Silke J, Meier P. Inhibitor of apoptosis (IAP) proteins-modulators of cell death and inflammation. Cold Spring Harb Perspect Biol 2013; 5(2): a008730.
[http://dx.doi.org/10.1101/cshperspect.a008730] [PMID: 23378585]

[79] Roulston A, Marcellus RC, Branton PE. Viruses and apoptosis. Annu Rev Microbiol 1999; 53: 577-628.
[http://dx.doi.org/10.1146/annurev.micro.53.1.577] [PMID: 10547702]

[80] Silke J, Hawkins CJ, Ekert PG, et al. The anti-apoptotic activity of XIAP is retained upon mutation of both the caspase 3- and caspase 9-interacting sites. J Cell Biol 2002; 157(1): 115-24.
[http://dx.doi.org/10.1083/jcb.200108085] [PMID: 11927604]

[81] Colnaghi R, Connell CM, Barrett RM, Wheatley SP. Separating the anti-apoptotic and mitotic roles of survivin. J Biol Chem 2006; 281(44): 33450-6.
[http://dx.doi.org/10.1074/jbc.C600164200] [PMID: 16950794]

[82] Ambrosini G, Adida C, Altieri DC. A novel anti-apoptosis gene, survivin, expressed in cancer and lymphoma. Nat Med 1997; 3(8): 917-21.
[http://dx.doi.org/10.1038/nm0897-917] [PMID: 9256286]

[83] Kleinsimon S, Kauczor G, Jaeger S, Eggert A, Seifert G, Delebinski C. ViscumTT induces apoptosis and alters IAP expression in osteosarcoma in vitro and has synergistic action when combined with different chemotherapeutic drugs. BMC Complement Altern Med 2017; 17(1): 26.
[http://dx.doi.org/10.1186/s12906-016-1545-7] [PMID: 28061770]

[84] Beere HM, Wolf BB, Cain K, et al. Heat-shock protein 70 inhibits apoptosis by preventing recruitment of procaspase-9 to the Apaf-1 apoptosome. Nat Cell Biol 2000; 2(8): 469-75.
[http://dx.doi.org/10.1038/35019501] [PMID: 10934466]

[85] Shortt J, Johnstone RW. Oncogenes in cell survival and cell death. Cold Spring Harb Perspect Biol 2012; 4(12): a009829.
[http://dx.doi.org/10.1101/cshperspect.a009829] [PMID: 23209150]

[86] Hauser AT, Jung M. Targeting epigenetic mechanisms: potential of natural products in cancer chemoprevention. Planta Med 2008; 74(13): 1593-601.

[http://dx.doi.org/10.1055/s-2008-1081347] [PMID: 18704881]

[87]   Park SM, Peter ME. microRNAs and death receptors. Cytokine Growth Factor Rev 2008; 19(3-4): 303-11.
[http://dx.doi.org/10.1016/j.cytogfr.2008.04.011] [PMID: 18490189]

[88]   Garofalo M, Condorelli GL, Croce CM, Condorelli G. MicroRNAs as regulators of death receptors signaling. Cell Death Differ 2010; 17(2): 200-8.
[http://dx.doi.org/10.1038/cdd.2009.105] [PMID: 19644509]

[89]   Yang L, Belaguli N, Berger DH. MicroRNA and colorectal cancer. World J Surg 2009; 33(4): 638-46.
[http://dx.doi.org/10.1007/s00268-008-9865-5] [PMID: 19123024]

[90]   Visani M, Acquaviva G, Marucci G, Ragazzi M, Fraceschi E. microRNA in brain neoplasia: A review Int J Brain Disord Treat 2015; 1(002)

[91]   Gozuacik D, Akkoc Y, Ozturk DG, Kocak M. Autophagy-Regulating microRNAs and Cancer. Front Oncol 2017; 7: 65.
[http://dx.doi.org/10.3389/fonc.2017.00065] [PMID: 28459042]

[92]   Cimmino A, Calin GA, Fabbri M, *et al.* miR-15 and miR-16 induce apoptosis by targeting BCL_2. Proc Natl Acad Sci USA 2005; 102(39): 13944-9.
[http://dx.doi.org/10.1073/pnas.0506654102] [PMID: 16166262]

[93]   Welch C, Chen Y, Stallings RL. MicroRNA-34a functions as a potential tumor suppressor by inducing apoptosis in neuroblastoma cells. Oncogene 2007; 26(34): 5017-22.
[http://dx.doi.org/10.1038/sj.onc.1210293] [PMID: 17297439]

[94]   Zhang B, Pan X, Cobb GP, Anderson TA. microRNAs as oncogenes and tumor suppressors. Dev Biol 2007; 302(1): 1-12.
[http://dx.doi.org/10.1016/j.ydbio.2006.08.028] [PMID: 16989803]

[95]   Saini S, Yamamura S, Majid S, *et al.* MicroRNA-708 induces apoptosis and suppresses tumorigenicity in renal cancer cells. Cancer Res 2011; 71(19): 6208-19.
[http://dx.doi.org/10.1158/0008-5472.CAN-11-0073] [PMID: 21852381]

[96]   Wang R, Wang ZX, Yang JS, Pan X, De W, Chen LB. MicroRNA-451 functions as a tumor suppressor in human non-small cell lung cancer by targeting ras-related protein 14 (RAB14). Oncogene 2011; 30(23): 2644-58.
[http://dx.doi.org/10.1038/onc.2010.642] [PMID: 21358675]

[97]   Lin YC, Lin JF, Tsai TF, Chou KY, Chen HE, Hwang TI. Tumor suppressor miRNA-204-5p promotes apoptosis by targeting BCL2 in prostate cancer cells. Asian J Surg 2017; 40(5): 396-406.
[http://dx.doi.org/10.1016/j.asjsur.2016.07.001] [PMID: 27519795]

[98]   Hopkins-Donaldson S, Ziegler A, Kurtz S, *et al.* Silencing of death receptor and caspase-8 expression in small cell lung carcinoma cell lines and tumors by DNA methylation. Cell Death Differ 2003; 10(3): 356-64.
[http://dx.doi.org/10.1038/sj.cdd.4401157] [PMID: 12700635]

[99]   Gonzalez-Gomez P, Bello MJ, Inda MM, *et al.* Deletion and aberrant CpG island methylation of Caspase 8 gene in medulloblastoma. Oncol Rep 2004; 12(3): 663-6.
[PMID: 15289853]

[100]   Bello MJ, De Campos JM, Isla A, Casartelli C, Rey JA. Promoter CpG methylation of multiple genes in pituitary adenomas: frequent involvement of caspase-8. Oncol Rep 2006; 15(2): 443-8.
[PMID: 16391867]

[101]   Hervouet E, Cheray M, Vallette FM, Cartron PF. DNA methylation and apoptosis resistance in cancer cells. Cells 2013; 2(3): 545-73.
[http://dx.doi.org/10.3390/cells2030545] [PMID: 24709797]

[102]   Borhani N, Manoochehri M, Gargari SS, Novin MG, Mansouri A, Omrani MD. Decreased expression

of proapoptotic genes caspase-8-and BCL2-associated agonist of cell death (BAD) in ovarian cancer Clin Ovarian other Gynecol Cancer 2014; 7: 18-23.

[103] Shenoy N, Vallumsetla N, Zou Y, *et al.* Role of DNA methylation in renal cell carcinoma. J Hematol Oncol 2015; 8: 88.
[http://dx.doi.org/10.1186/s13045-015-0180-y] [PMID: 26198328]

[104] Fushimi K, Ray P, Kar A, Wang L, Sutherland LC, Wu JY. Up-regulation of the proapoptotic caspase 2 splicing isoform by a candidate tumor suppressor, RBM5. Proc Natl Acad Sci USA 2008; 105(41): 15708-13.
[http://dx.doi.org/10.1073/pnas.0805569105] [PMID: 18840686]

[105] Norberg E, Gogvadze V, Ott M, *et al.* An increase in intracellular Ca_2^+ is required for the activation of mitochondrial calpain to release AIF during cell death. Cell Death Differ 2008; 15(12): 1857-64.
[http://dx.doi.org/10.1038/cdd.2008.123] [PMID: 18806756]

[106] Kim KY, Cho HJ, Yu SN, *et al.* Interplay of reactive oxygen species, intracellular Ca^{2+} and mitochondrial homeostasis in the apoptosis of prostate cancer cells by deoxypodophyllotoxin. J Cell Biochem 2013; 114(5): 1124-34.
[http://dx.doi.org/10.1002/jcb.24455] [PMID: 23192945]

[107] Wang S, El-Deiry WS. Cytochrome c: a crosslink between the mitochondria and the endoplasmic reticulum in calcium-dependent apoptosis. Cancer Biol Ther 2004; 3(1): 44-6.
[http://dx.doi.org/10.4161/cbt.3.1.740] [PMID: 14726686]

[108] Contreras L, Drago I, Zampese E, Pozzan T. Mitochondria: the calcium connection. Biochim Biophys Acta 2010; 1797(6-7): 607-18.
[http://dx.doi.org/10.1016/j.bbabio.2010.05.005] [PMID: 20470749]

[109] Giorgi C, Baldassari F, Bononi A, *et al.* Mitochondrial $Ca^{(2+)}$ and apoptosis. Cell Calcium 2012; 52(1): 36-43.
[http://dx.doi.org/10.1016/j.ceca.2012.02.008] [PMID: 22480931]

[110] Vygodina T, Kirichenko A, Konstantinov AA. Direct regulation of cytochrome c oxidase by calcium ions. PLoS One 2013; 8(9): e74436.
[http://dx.doi.org/10.1371/journal.pone.0074436] [PMID: 24058566]

[111] Li P, Nijhawan D, Budihardjo I, *et al.* Cytochrome c and dATP-dependent formation of Apaf-1/caspase-9 complex initiates an apoptotic protease cascade. Cell 1997; 91(4): 479-89.
[http://dx.doi.org/10.1016/S0092-8674(00)80434-1] [PMID: 9390557]

[112] Jiang X, Wang X. Cytochrome c promotes caspase-9 activation by inducing nucleotide binding to Apaf-1. J Biol Chem 2000; 275(40): 31199-203.
[http://dx.doi.org/10.1074/jbc.C000405200] [PMID: 10940292]

[113] Hu Q, Wu D, Chen W, *et al.* Molecular determinants of caspase-9 activation by the Apaf-1 apoptosome. Proc Natl Acad Sci USA 2014; 111(46): 16254-61.
[http://dx.doi.org/10.1073/pnas.1418000111] [PMID: 25313070]

[114] Li Y, Zhou M, Hu Q, *et al.* Mechanistic insights into caspase-9 activation by the structure of the apoptosome holoenzyme. Proc Natl Acad Sci USA 2017; 114(7): 1542-7.
[http://dx.doi.org/10.1073/pnas.1620626114] [PMID: 28143931]

[115] Jo DG, Jun JI, Chang JW, *et al.* Calcium binding of ARC mediates regulation of caspase 8 and cell death. Mol Cell Biol 2004; 24(22): 9763-70.
[http://dx.doi.org/10.1128/MCB.24.22.9763-9770.2004] [PMID: 15509781]

[116] Earnshaw WC, Martins LM, Kaufmann SH. Mammalian caspases: structure, activation, substrates, and functions during apoptosis. Annu Rev Biochem 1999; 68: 383-424.
[http://dx.doi.org/10.1146/annurev.biochem.68.1.383] [PMID: 10872455]

[117] Poreba M, Strózyk A, Salvesen GS, Drąg M. Caspase substrates and inhibitors. Cold Spring Harb Perspect Biol 2013; 5(8): a008680.

[http://dx.doi.org/10.1101/cshperspect.a008680] [PMID: 23788633]

[118] Cella D, Fallowfield LJ. Recognition and management of treatment-related side effects for breast cancer patients receiving adjuvant endocrine therapy. Breast Cancer Res Treat 2008; 107(2): 167-80.
[http://dx.doi.org/10.1007/s10549-007-9548-1] [PMID: 17876703]

[119] Mans DR, da Rocha AB, Schwartsmann G. Anti-cancer drug discovery and development in Brazil: targeted plant collection as a rational strategy to acquire candidate anti-cancer compounds. Oncologist 2000; 5(3): 185-98.
[http://dx.doi.org/10.1634/theoncologist.5-3-185] [PMID: 10884497]

[120] Eschenhagen T, Force T, Ewer MS, *et al.* Cardiovascular side effects of cancer therapies: a position statement from the Heart Failure Association of the European Society of Cardiology. Eur J Heart Fail 2011; 13(1): 1-10.
[http://dx.doi.org/10.1093/eurjhf/hfq213] [PMID: 21169385]

[121] Valerio LG Jr, Gonzales GF. Toxicological aspects of the South American herbs cat's claw (Uncaria tomentosa) and Maca (Lepidium meyenii) : a critical synopsis. Toxicol Rev 2005; 24(1): 11-35.
[http://dx.doi.org/10.2165/00139709-200524010-00002] [PMID: 16042502]

[122] Choudhary K, Singh M, Pillai U. Ethnobotanical survey of Rajasthan—an update. Am Eurasian J Bot 2008; 1: 38-45.

[123] McClatchey WC, Mahady GB, Bennett BC, Shiels L, Savo V. Ethnobotany as a pharmacological research tool and recent developments in CNS-active natural products from ethnobotanical sources. Pharmacol Ther 2009; 123(2): 239-54.
[http://dx.doi.org/10.1016/j.pharmthera.2009.04.002] [PMID: 19422851]

[124] Ji HF, Li XJ, Zhang HY. Natural products and drug discovery. Can thousands of years of ancient medical knowledge lead us to new and powerful drug combinations in the fight against cancer and dementia? EMBO Rep 2009; 10(3): 194-200.
[http://dx.doi.org/10.1038/embor.2009.12] [PMID: 19229284]

[125] Aravindaram K, Yang NS. Anti-inflammatory plant natural products for cancer therapy. Planta Med 2010; 76(11): 1103-17.
[http://dx.doi.org/10.1055/s-0030-1249859] [PMID: 20432202]

[126] Tosetti F, Noonan DM, Albini A. Metabolic regulation and redox activity as mechanisms for angioprevention by dietary phytochemicals. Int J Cancer 2009; 125(9): 1997-2003.
[http://dx.doi.org/10.1002/ijc.24677] [PMID: 19551861]

[127] Firoozinia M, Moghadamtousi SZ, Sadeghilar A, Karimian H, Noordin MIB. Golden natural plant compounds activate apoptosis via both mitochondrial and death receptor pathways: a review eJBio 2015; 11: 126-37.

[128] Nie J, Zhao C, Deng LI, *et al.* Efficacy of traditional Chinese medicine in treating cancer. Biomed Rep 2016; 4(1): 3-14.
[http://dx.doi.org/10.3892/br.2015.537] [PMID: 26870326]

[129] Kumar M, Kaur V, Kumar S, Kaur S. Phytoconstituents as apoptosis inducing agents: strategy to combat cancer. Cytotechnology 2016; 68(4): 531-63.
[http://dx.doi.org/10.1007/s10616-015-9897-2] [PMID: 26239338]

[130] Kaur V, Kumar M, Kumar A, Kaur K, Dhillon VS, Kaur S. Pharmacotherapeutic potential of phytochemicals: Implications in cancer chemoprevention and future perspectives. Biomed Pharmacother 2018; 97: 564-86.
[http://dx.doi.org/10.1016/j.biopha.2017.10.124] [PMID: 29101800]

[131] Clifford MN. Chlorogenic acids and other cinnamates–nature, occurrence and dietary burden. J Sci Food Agric 1999; 79: 362-72.
[http://dx.doi.org/10.1002/(SICI)1097-0010(19990301)79:3<362::AID-JSFA256>3.0.CO;2-D]

[132] Tomás☐Barberán FA, Clifford MN. Dietary hydroxybenzoic acid derivatives–nature, occurrence and

dietary burden. J Sci Food Agric 2000; 80: 1024-32.
[http://dx.doi.org/10.1002/(SICI)1097-0010(20000515)80:7<1024::AID-JSFA567>3.0.CO;2-S]

[133]   Manach C, Scalbert A, Morand C, Rémésy C, Jiménez L. Polyphenols: food sources and bioavailability. Am J Clin Nutr 2004; 79(5): 727-47.
[http://dx.doi.org/10.1093/ajcn/79.5.727] [PMID: 15113710]

[134]   Li TM, Chen GW, Su CC, *et al.* Ellagic acid induced p53/p21 expression, G1 arrest and apoptosis in human bladder cancer T24 cells. Anticancer Res 2005; 25(2A): 971-9.
[PMID: 15868936]

[135]   Han DH, Lee MJ, Kim JH. Antioxidant and apoptosis-inducing activities of ellagic acid. Anticancer Res 2006; 26(5A): 3601-6.
[PMID: 17094489]

[136]   Yin MC, Lin CC, Wu HC, Tsao SM, Hsu CK. Apoptotic effects of protocatechuic acid in human breast, lung, liver, cervix, and prostate cancer cells: potential mechanisms of action. J Agric Food Chem 2009; 57(14): 6468-73.
[http://dx.doi.org/10.1021/jf9004466] [PMID: 19601677]

[137]   Ji BC, Hsu WH, Yang JS, *et al.* Gallic acid induces apoptosis *via* caspase-3 and mitochondrion-dependent pathways *in vitro* and suppresses lung xenograft tumor growth *in vivo*. J Agric Food Chem 2009; 57(16): 7596-604.
[http://dx.doi.org/10.1021/jf901308p] [PMID: 20349925]

[138]   Lo C, Lai TY, Yang JH, *et al.* Gallic acid induces apoptosis in A375.S2 human melanoma cells through caspase-dependent and -independent pathways. Int J Oncol 2010; 37(2): 377-85.
[PMID: 20596665]

[139]   Chang WC, Hsieh CH, Hsiao MW, Lin WC, Hung YC, Ye JC. Caffeic acid induces apoptosis in human cervical cancer cells through the mitochondrial pathway. Taiwan J Obstet Gynecol 2010; 49(4): 419-24.
[http://dx.doi.org/10.1016/S1028-4559(10)60092-7] [PMID: 21199742]

[140]   Kamatham S, Kumar N, Gudipalli P. Isolation and characterization of gallic acid and methyl gallate from the seed coats of *Givotia rottleriformis* Griff. and their anti-proliferative effect on human epidermoid carcinoma A431 cells. Toxicol Rep 2015; 2: 520-9.
[http://dx.doi.org/10.1016/j.toxrep.2015.03.001] [PMID: 28962387]

[141]   Wang T, Gong X, Jiang R, Li H, Du W, Kuang G. Ferulic acid inhibits proliferation and promotes apoptosis *via* blockage of PI3K/Akt pathway in osteosarcoma cell. Am J Transl Res 2016; 8(2): 968-80.
[PMID: 27158383]

[142]   Middleton E Jr. Effect of plant flavonoids on immune and inflammatory cell function. Adv Exp Med Biol 1998; 439: 175-82.
[http://dx.doi.org/10.1007/978-1-4615-5335-9_13] [PMID: 9781303]

[143]   Williams CA, Grayer RJ. Anthocyanins and other flavonoids. Nat Prod Rep 2004; 21(4): 539-73.
[http://dx.doi.org/10.1039/b311404j] [PMID: 15282635]

[144]   Xiao ZP, Peng ZY, Peng MJ, Yan WB, Ouyang YZ, Zhu HL. Flavonoids health benefits and their molecular mechanism. Mini Rev Med Chem 2011; 11(2): 169-77.
[http://dx.doi.org/10.2174/138955711794519546] [PMID: 21222576]

[145]   Watson WH, Cai J, Jones DP. Diet and apoptosis. Annu Rev Nutr 2000; 20: 485-505.
[http://dx.doi.org/10.1146/annurev.nutr.20.1.485] [PMID: 10940343]

[146]   Manson MM. Cancer prevention -- the potential for diet to modulate molecular signalling. Trends Mol Med 2003; 9(1): 11-8.
[http://dx.doi.org/10.1016/S1471-4914(02)00002-3] [PMID: 12524205]

[147]   Yang CS, Landau JM, Huang MT, Newmark HL. Inhibition of carcinogenesis by dietary polyphenolic

compounds. Annu Rev Nutr 2001; 21: 381-406.
[http://dx.doi.org/10.1146/annurev.nutr.21.1.381] [PMID: 11375442]

[148]   Ramos S. Effects of dietary flavonoids on apoptotic pathways related to cancer chemoprevention. J Nutr Biochem 2007; 18(7): 427-42.
[http://dx.doi.org/10.1016/j.jnutbio.2006.11.004] [PMID: 17321735]

[149]   Wang I-K, Lin-Shiau S-Y, Lin J-K. Induction of apoptosis by apigenin and related flavonoids through cytochrome c release and activation of caspase-9 and caspase-3 in leukaemia HL-60 cells. Eur J Cancer 1999; 35(10): 1517-25.
[http://dx.doi.org/10.1016/S0959-8049(99)00168-9] [PMID: 10673981]

[150]   Lee HZ, Leung HW, Lai MY, Wu CH. Baicalein induced cell cycle arrest and apoptosis in human lung squamous carcinoma CH27 cells. Anticancer Res 2005; 25(2A): 959-64.
[PMID: 15868934]

[151]   Vargo MA, Voss OH, Poustka F, Cardounel AJ, Grotewold E, Doseff AI. Apigenin-induced-apoptosis is mediated by the activation of PKCdelta and caspases in leukemia cells. Biochem Pharmacol 2006; 72(6): 681-92.
[http://dx.doi.org/10.1016/j.bcp.2006.06.010] [PMID: 16844095]

[152]   Granado-Serrano AB, Martín MA, Bravo L, Goya L, Ramos S. Quercetin induces apoptosis *via* caspase activation, regulation of Bcl-2, and inhibition of PI-3-kinase/Akt and ERK pathways in a human hepatoma cell line (HepG2). J Nutr 2006; 136(11): 2715-21.
[http://dx.doi.org/10.1093/jn/136.11.2715] [PMID: 17056790]

[153]   Khan N, Afaq F, Syed DN, Mukhtar H. Fisetin, a novel dietary flavonoid, causes apoptosis and cell cycle arrest in human prostate cancer LNCaP cells. Carcinogenesis 2008; 29(5): 1049-56.
[http://dx.doi.org/10.1093/carcin/bgn078] [PMID: 18359761]

[154]   Das A, Banik NL, Ray SK. Flavonoids activated caspases for apoptosis in human glioblastoma T98G and U87MG cells but not in human normal astrocytes. Cancer 2010; 116(1): 164-76.
[PMID: 19894226]

[155]   Xiang M, Su H, Hong Z, Yang T, Shu G. Chemical composition of total flavonoids from *Polygonum amplexicaule* and their pro-apoptotic effect on hepatocellular carcinoma cells: Potential roles of suppressing STAT3 signaling. Food Chem Toxicol 2015; 80: 62-71.
[http://dx.doi.org/10.1016/j.fct.2015.02.020] [PMID: 25754378]

[156]   Feng CP, Tang HM, Huang S, *et al.* Evaluation of the effects of the water-soluble total flavonoids from *Isodon lophanthoides* var.gerardianus (Benth.) H. Hara on apoptosis in HepG2 cell: Investigation of the most relevant mechanisms. J Ethnopharmacol 2016; 188: 70-9.
[http://dx.doi.org/10.1016/j.jep.2016.04.042] [PMID: 27132715]

[157]   Gatouillat G, Magid AA, Bertin E, *et al.* Medicarpin and millepurpan, two flavonoids isolated from Medicago sativa, induce apoptosis and overcome multidrug resistance in leukemia P388 cells. Phytomedicine 2015; 22(13): 1186-94.
[http://dx.doi.org/10.1016/j.phymed.2015.09.005] [PMID: 26598918]

[158]   Sacchettini JC, Poulter CD. Creating isoprenoid diversity. Science 1997; 277(5333): 1788-9.
[http://dx.doi.org/10.1126/science.277.5333.1788] [PMID: 9324768]

[159]   Peñuelas J, Munné-Bosch S. Isoprenoids: an evolutionary pool for photoprotection. Trends Plant Sci 2005; 10(4): 166-9.
[http://dx.doi.org/10.1016/j.tplants.2005.02.005] [PMID: 15817417]

[160]   Rohdich F, Bacher A, Eisenreich W. Isoprenoid biosynthetic pathways as anti-infective drug targets. Biochem Soc Trans 2005; 33(Pt 4): 785-91.
[http://dx.doi.org/10.1042/BST0330785] [PMID: 16042599]

[161]   Withers ST, Keasling JD. Biosynthesis and engineering of isoprenoid small molecules. Appl Microbiol Biotechnol 2007; 73(5): 980-90.

[http://dx.doi.org/10.1007/s00253-006-0593-1] [PMID: 17115212]

[162] Rabi T, Bishayee A. Terpenoids and breast cancer chemoprevention. Breast Cancer Res Treat 2009; 115(2): 223-39.
[http://dx.doi.org/10.1007/s10549-008-0118-y] [PMID: 18636327]

[163] Yang H, Dou QP. Targeting apoptosis pathway with natural terpenoids: implications for treatment of breast and prostate cancer. Curr Drug Targets 2010; 11(6): 733-44.
[http://dx.doi.org/10.2174/138945010791170842] [PMID: 20298150]

[164] Thoppil RJ, Bishayee A. Terpenoids as potential chemopreventive and therapeutic agents in liver cancer. World J Hepatol 2011; 3(9): 228-49.
[http://dx.doi.org/10.4254/wjh.v3.i9.228] [PMID: 21969877]

[165] Sagar SM, Yance D, Wong RK. Natural health products that inhibit angiogenesis: a potential source for investigational new agents to treat cancer-Part 1. Curr Oncol 2006; 13(1): 14-26.
[PMID: 17576437]

[166] Kuttan G, Pratheeshkumar P, Manu KA, Kuttan R. Inhibition of tumor progression by naturally occurring terpenoids. Pharm Biol 2011; 49(10): 995-1007.
[http://dx.doi.org/10.3109/13880209.2011.559476] [PMID: 21936626]

[167] Huang M, Lu JJ, Huang MQ, Bao JL, Chen XP, Wang YT. Terpenoids: natural products for cancer therapy. Expert Opin Investig Drugs 2012; 21(12): 1801-18.
[http://dx.doi.org/10.1517/13543784.2012.727395] [PMID: 23092199]

[168] Choi YH, Baek JH, Yoo MA, Chung HY, Kim ND, Kim KW. Induction of apoptosis by ursolic acid through activation of caspases and down-regulation of c-IAPs in human prostate epithelial cells. Int J Oncol 2000; 17(3): 565-71.
[PMID: 10938399]

[169] Harmand PO, Duval R, Liagre B, et al. Ursolic acid induces apoptosis through caspase-3 activation and cell cycle arrest in HaCat cells. Int J Oncol 2003; 23(1): 105-12.
[PMID: 12792782]

[170] Yang HS, Kim JY, Lee JH, et al. Celastrol isolated from *Tripterygium regelii* induces apoptosis through both caspase-dependent and -independent pathways in human breast cancer cells. Food Chem Toxicol 2011; 49(2): 527-32.
[http://dx.doi.org/10.1016/j.fct.2010.11.044] [PMID: 21134410]

[171] Sun KW, Ma YY, Guan TP, et al. Oridonin induces apoptosis in gastric cancer through Apaf-1, cytochrome c and caspase-3 signaling pathway. World J Gastroenterol 2012; 18(48): 7166-74.
[http://dx.doi.org/10.3748/wjg.v18.i48.7166] [PMID: 23326121]

[172] Reyes-Zurita FJ, Rufino-Palomares EE, Medina PP, et al. Antitumour activity on extrinsic apoptotic targets of the triterpenoid maslinic acid in p53-deficient Caco-2 adenocarcinoma cells. Biochimie 2013; 95(11): 2157-67.
[http://dx.doi.org/10.1016/j.biochi.2013.08.017] [PMID: 23973282]

[173] Liu BX, Zhou JY, Li Y, et al. Hederagenin from the leaves of ivy (*Hedera helix* L.) induces apoptosis in human LoVo colon cells through the mitochondrial pathway. BMC Complement Altern Med 2014; 14: 412.
[http://dx.doi.org/10.1186/1472-6882-14-412] [PMID: 25342273]

[174] Menon DB, Gopalakrishnan VK. Terpenoids isolated from the shoot of *Plectranthus hadiensis* induces apoptosis in human colon cancer cells *via* the mitochondria-dependent pathway. Nutr Cancer 2015; 67(4): 697-705.
[http://dx.doi.org/10.1080/01635581.2015.1019631] [PMID: 25837437]

[175] Bhutani SP. Chapter 6: Alkaloids. Organic Chemistry: Selected Topics 2007; p. 287.

[176] Newman DJ, Cragg GM. Natural products as sources of new drugs over the last 25 years. J Nat Prod 2007; 70(3): 461-77.

[http://dx.doi.org/10.1021/np068054v] [PMID: 17309302]

[177]   Kaminskyy V, Kulachkovskyy O, Stoika R. A decisive role of mitochondria in defining rate and intensity of apoptosis induction by different alkaloids. Toxicol Lett 2008; 177(3): 168-81.
[http://dx.doi.org/10.1016/j.toxlet.2008.01.009] [PMID: 18325696]

[178]   Rovini A, Savry A, Braguer D, Carré M. Microtubule-targeted agents: when mitochondria become essential to chemotherapy. Biochim Biophys Acta 2011; 1807(6): 679-88.
[http://dx.doi.org/10.1016/j.bbabio.2011.01.001] [PMID: 21216222]

[179]   Urra FA, Cordova-Delgado M, Pessoa-Mahana H, *et al.* Mitochondria: a promising target for anticancer alkaloids. Curr Top Med Chem 2013; 13(17): 2171-83.
[http://dx.doi.org/10.2174/15680266113139990150] [PMID: 23978135]

[180]   Jang BC, Lim KJ, Paik JH, *et al.* Tetrandrine-induced apoptosis is mediated by activation of caspases and PKC-delta in U937 cells. Biochem Pharmacol 2004; 67(10): 1819-29.
[http://dx.doi.org/10.1016/j.bcp.2004.01.018] [PMID: 15130759]

[181]   Yang SA, Paek SH, Kozukue N, Lee KR, Kim JA. Alpha-chaconine, a potato glycoalkaloid, induces apoptosis of HT-29 human colon cancer cells through caspase-3 activation and inhibition of ERK 1/2 phosphorylation. Food Chem Toxicol 2006; 44(6): 839-46.
[http://dx.doi.org/10.1016/j.fct.2005.11.007] [PMID: 16387404]

[182]   Jantova S, Cipak L, Letasiova S. Berberine induces apoptosis through a mitochondrial/caspase pathway in human promonocytic U937 cells. Toxicol. *In Vitro* 2007; 21(1): 25-31.
[http://dx.doi.org/10.1016/j.tiv.2006.07.015] [PMID: 17011159]

[183]   Dai ZJ, Gao J, Ji ZZ, *et al.* Matrine induces apoptosis in gastric carcinoma cells *via* alteration of Fas/FasL and activation of caspase-3. J Ethnopharmacol 2009; 123(1): 91-6.
[http://dx.doi.org/10.1016/j.jep.2009.02.022] [PMID: 19429345]

[184]   Mansoor TA, Ramalho RM, Mulhovo S, Rodrigues CM, Ferreira MJ. Induction of apoptosis in HuH-7 cancer cells by monoterpene and beta-carboline indole alkaloids isolated from the leaves of Tabernaemontana elegans. Bioorg Med Chem Lett 2009; 19(15): 4255-8.
[http://dx.doi.org/10.1016/j.bmcl.2009.05.104] [PMID: 19525111]

[185]   Won KJ, Chung KS, Lee YS, *et al.* Haplophytin-A induces caspase-8-mediated apoptosis *via* the formation of death-inducing signaling complex in human promyelocytic leukemia HL-60 cells. Chem Biol Interact 2010; 188(3): 505-11.
[http://dx.doi.org/10.1016/j.cbi.2010.09.001] [PMID: 20833157]

[186]   El-Readi MZ, Eid S, Ashour ML, Tahrani A, Wink M. Modulation of multidrug resistance in cancer cells by chelidonine and *Chelidonium majus* alkaloids. Phytomedicine 2013; 20(3-4): 282-94.
[http://dx.doi.org/10.1016/j.phymed.2012.11.005] [PMID: 23238299]

[187]   Sun Q, Yogosawa S, Iizumi Y, Sakai T, Sowa Y. The alkaloid emetine sensitizes ovarian carcinoma cells to cisplatin through downregulation of bcl-xL. Int J Oncol 2015; 46(1): 389-94.
[http://dx.doi.org/10.3892/ijo.2014.2703] [PMID: 25310746]

[188]   Wang XD, Li CY, Jiang MM, *et al.* Induction of apoptosis in human leukemia cells through an intrinsic pathway by cathachunine, a unique alkaloid isolated from Catharanthus roseus. Phytomedicine 2016; 23(6): 641-53.
[http://dx.doi.org/10.1016/j.phymed.2016.03.003] [PMID: 27161405]

[189]   Nordin N, Majid NA, Mohan S, *et al.* Cleistopholine isolated from *Enicosanthellum pulchrum* exhibits apoptogenic properties in human ovarian cancer cells. Phytomedicine 2016; 23(4): 406-16.
[http://dx.doi.org/10.1016/j.phymed.2016.02.016] [PMID: 27002411]

[190]   Fan Y, Jiang Y, Liu J, Kang Y, Li R, Wang J. The anti-tumor activity and mechanism of alkaloids from *Aconitum szechenyianum* Gay. Bioorg Med Chem Lett 2016; 26(2): 380-7.
[http://dx.doi.org/10.1016/j.bmcl.2015.12.006] [PMID: 26711147]

[191]   Zong A, Cao H, Wang F. Anticancer polysaccharides from natural resources: a review of recent

research. Carbohydr Polym 2012; 90(4): 1395-410.
[http://dx.doi.org/10.1016/j.carbpol.2012.07.026] [PMID: 22944395]

[192] Caliceti P, Salmaso S, Bersani S. Polysaccharide-based anticancer prodrugs.Macromolecular anticancer therapeutics. New York, NY: Humana Press Inc 2010; pp. 163-219.
[http://dx.doi.org/10.1007/978-1-4419-0507-9_5]

[193] Feng G, Wang X, You C, *et al.* Antiproliferative potential of Artemisia capillaris polysaccharide against human nasopharyngeal carcinoma cells. Carbohydr Polym 2013; 92(2): 1040-5.
[http://dx.doi.org/10.1016/j.carbpol.2012.10.024] [PMID: 23399126]

[194] Peng Z, Liu M, Fang Z, Chen L, Wu J, Zhang Q. *In vitro* antiproliferative effect of a water-soluble *Laminaria japonica* polysaccharide on human melanoma cell line A375. Food Chem Toxicol 2013; 58: 56-60.
[http://dx.doi.org/10.1016/j.fct.2013.04.026] [PMID: 23612001]

[195] Li X, Liu F, Li Z, Ye N, Huang C, Yuan X. *Atractylodes macrocephala* polysaccharides induces mitochondrial-mediated apoptosis in glioma C6 cells. Int J Biol Macromol 2014; 66: 108-12.
[http://dx.doi.org/10.1016/j.ijbiomac.2014.02.019] [PMID: 24556116]

[196] Ma L, Qin C, Wang M, *et al.* Preparation, preliminary characterization and inhibitory effect on human colon cancer HT-29 cells of an acidic polysaccharide fraction from *Stachys floridana* Schuttl. ex Benth. Food Chem Toxicol 2013; 60: 269-76.
[http://dx.doi.org/10.1016/j.fct.2013.07.060] [PMID: 23911803]

[197] Lu X, Liu W, Wu J, *et al.* A polysaccharide fraction of adlay seed (*Coixlachryma-jobi* L.) induces apoptosis in human non-small cell lung cancer A549 cells. Biochem Biophys Res Commun 2013; 430(2): 846-51.
[http://dx.doi.org/10.1016/j.bbrc.2012.11.058] [PMID: 23200838]

[198] Thangam R, Sathuvan M, Poongodi A, *et al.* Activation of intrinsic apoptotic signaling pathway in cancer cells by Cymbopogon citratus polysaccharide fractions. Carbohydr Polym 2014; 107: 138-50.
[http://dx.doi.org/10.1016/j.carbpol.2014.02.039] [PMID: 24702929]

[199] Wang Z, Lu C, Wu C, *et al.* Polysaccharide of *Boschniakia rossica* induces apoptosis on laryngeal carcinoma Hep2 cells. Gene 2014; 536(1): 203-6.
[http://dx.doi.org/10.1016/j.gene.2013.11.090] [PMID: 24334128]

[200] Zhou WJ, Wang S, Hu Z, Zhou ZY, Song CJ. *Angelica sinensis* polysaccharides promotes apoptosis in human breast cancer cells *via* CREB-regulated caspase-3 activation. Biochem Biophys Res Commun 2015; 467(3): 562-9.
[http://dx.doi.org/10.1016/j.bbrc.2015.09.145] [PMID: 26431878]

[201] Su J, Jiang L, Wu J, Liu Z, Wu Y. Anti-tumor and anti-virus activity of polysaccharides extracted from *Sipunculus nudus*(SNP) on Hepg2.2.15. Int J Biol Macromol 2016; 87: 597-602.
[http://dx.doi.org/10.1016/j.ijbiomac.2016.03.022] [PMID: 26987430]

[202] Chong J, Poutaraud A, Hugueney P. Metabolism and roles of stilbenes in plants. Plant Sci 2009; 177: 143-55.
[http://dx.doi.org/10.1016/j.plantsci.2009.05.012]

[203] Shen T, Wang X-N, Lou H-X. Natural stilbenes: an overview. Nat Prod Rep 2009; 26(7): 916-35.
[http://dx.doi.org/10.1039/b905960a] [PMID: 19554241]

[204] Rivière C, Pawlus AD, Mérillon J-M. Natural stilbenoids: distribution in the plant kingdom and chemotaxonomic interest in Vitaceae. Nat Prod Rep 2012; 29(11): 1317-33.
[http://dx.doi.org/10.1039/c2np20049j] [PMID: 23014926]

[205] Sirerol JA, Rodríguez ML, Mena S, Asensi MA, Estrela JM, Ortega AL. Role of natural stilbenes in the prevention of cancer. Oxid Med Cell Longev 2016; 2016: 3128951.
[http://dx.doi.org/10.1155/2016/3128951] [PMID: 26798416]

[206] Jiang H, Zhang L, Kuo J, *et al.* Resveratrol-induced apoptotic death in human U251 glioma cells. Mol

Cancer Ther 2005; 4(4): 554-61.
[http://dx.doi.org/10.1158/1535-7163.MCT-04-0056] [PMID: 15827328]

[207]　Kim YH, Park C, Lee JO, *et al.* Induction of apoptosis by piceatannol in human leukemic U937 cells through down-regulation of Bcl-2 and activation of caspases. Oncol Rep 2008; 19(4): 961-7.
[PMID: 18357382]

[208]　Kim EJ, Park H, Park SY, Jun JG, Park JH. The grape component piceatannol induces apoptosis in DU145 human prostate cancer cells *via* the activation of extrinsic and intrinsic pathways. J Med Food 2009; 12(5): 943-51.
[http://dx.doi.org/10.1089/jmf.2008.1341] [PMID: 19857055]

[209]　Kim S-H, Kim H-J, Lee M-H, *et al.* Resveratrol induces apoptosis of KB human oral cancer cells. J Korean Soc Appl Biol Chem 2011; 54: 966-71.
[http://dx.doi.org/10.1007/BF03253187]

[210]　Siedlecka-Kroplewska K, Jozwik A, Kaszubowska L, Kowalczyk A, Boguslawski W. Pterostilbene induces cell cycle arrest and apoptosis in MOLT4 human leukemia cells. Folia Histochem Cytobiol 2012; 50(4): 574-80.
[http://dx.doi.org/10.5603/FHC.2012.0080] [PMID: 23264221]

[211]　Hsiao P-C, Chou Y-E, Tan P, *et al.* Pterostilbene simultaneously induced G0/G1-phase arrest and MAPK-mediated mitochondrial-derived apoptosis in human acute myeloid leukemia cell lines. PLoS One 2014; 9(8): e105342.
[http://dx.doi.org/10.1371/journal.pone.0105342] [PMID: 25144448]

CHAPTER 11

Understanding Molecular Diagnosis in Cancer

Trupti N. Patel[*] and **Priyanjali Bhattacharya**

Department of Integrative Biology, Vellore Institute of Technology, Vellore, India

Abstract: The term diagnosis refers to a set of tests or investigations involving various biological samples such as tissue, body fluids and/or other irregularities in the body. While diagnosis is a broad term, examining the abnormalities in macromolecules - DNA, RNA or proteins at the level of nucleic acids and amino acids makes it a molecular diagnosis. As the world of diagnosis is rapidly expanding and advancing, newer minimal invasive techniques are being researched to spare the patients from various discomforts. Healthcare professionals are also aiming towards personalized medicine, making the treatment more selective and accurate, at the same time ensuring patients' safety. This is being achieved by high throughput analysis, though currently still economically very challenging. Technologies such as Histopathology, Polymerase Chain Reaction (PCR), Flow Cytometry (FC), Immunohistochemistry (IHC), Capillary Electrophoresis (CE) and Fluorescent *In Situ* Hybridization (FISH) are already in use for routine diagnosis in cancer. Though, high throughput technologies like, Next Generation Sequencing (NGS), Microarray studies, miRNA detection, ChIP (Chromatin Immune-Precipitation) and MeDIP-seq (Methyl-DNA Immuno-precipitation), SELDI-TOF MS (Surface-Enhanced Laser Desorption/Ionization Time-of Flight Mass Spectrometry), quantitative proteomic analysis and many more are triumphing into the diagnostic avenue, they are restrictively catering to high socio-economic groups in developing countries. An average citizen is still missing out on these forms of diagnosis and personalized treatments, thus devaluing the very aim of diagnosis and treatment in cancer. It becomes the pursuit of diagnostic community to come up with cost effective investigations and affordable treatments that could be efficiently validated. In this chapter, we briefly discuss the differences between diagnostic, prognostic and predictive tests along with a detailed understanding of genes and gene products that form the basis of molecular diagnosis in cancer. We also focus on currently available high-throughput methods and the challenges associated with them.

Keywords : Capillary Electrophoresis, ChIP, Flow Cytometry, Fluorescence *in situ* Hybridization, Gene Aberration, Histopathology, Immunohistochemistry, MeDIP-seq, Microarray, Molecular Diagnosis, Next Generation Sequencing, Polymerase Chain Reaction, SELDI-TOF MS.

[*] **Corresponding author Trupti N Patel:** Department of Integrative Biology, Vellore Institute of Technology University, Vellore, India; Tel: +919597873669; E-mails: tnpatel@vit.ac.in, dr.tnpatel@gmail.com

Ashita Sharma, Manish Kumar, Satwinderjeet Kaur & Avinash Kaur Nagpal (Eds.)

MOLECULAR DIAGNOSIS IN RELATION WITH PUBLIC HEALTH

"Health is a state of complete physical, mental and social well-being and not merely the absence of disease or infirmity." – **World Health Organization (WHO)**

With the fast-changing sciences and availability of rural internet, awareness about diseases and well-being has led to a demand for quality healthcare. This has also surged the number of insured lives and development of *In-Vitro* Diagnostic (IVD) market worldwide allowing corporate presence, which provides medical devices and means to examine biological specimen, identify the disease and monitor its progression. IVD market targets hospitals and clinics, pathology laboratories, research institutes, consulting firms for a wide range of genetic, cardiovascular, infectious and autoimmune diseases as well as with highest returns being netted from diabetes care and cancer. In India, around 60%-70% of IVD market covers clinical biochemistry, immunoassays and hematology and by 2020, the IVD share in India is expected to reach double [1]. Most of the diagnoses presently include hematological and biochemical profiles, microbial assays, urine analysis, histopathological and immuno- assays, with molecular assays progressing from conventional PCR (Polymerase chain Reaction) and cytogenetics to using microarrays, running gene panel analysis, and massive parallel sequencing, as a part of new world diagnosis. All the global leading companies in diagnostics have pitched in India with routine clinical tests and beyond, with Indian companies competing equally hard to strive and survive. Haematologic diagnosis is the third largest segment in Indian IVD market valued at 800-900 Cr with 62% value share and anticipated to surpass $2.6 billion by 2020 growing at 5% CAGR (Compound Annual Growth Rate). It is strongly driven by intense desire to advance in less invasive molecular diagnostic tests and improve technology related clinical outcomes due to radical increase in genetic disorders, cancers (solid tumor and haematologic malignancies) as well as viral related infections like swine flu, dengue and others [1, 2].'Molecular Oncodiagnositcs' is rapidly evolving field working towards identifying and targeting molecular alterations in cancer cells thereby generating 'Omics' such as- genome, exome, transcriptome, proteome, methylome, metabolome, topome and microbiome in the diagnostic arena. This has directed patient diagnosis to a high throughput level apart from routine histopathological examination and translating our knowledge from a clinical context to a molecular perspective. According to 2016 ICMR report- *"India is likely to have over 17.3 lakh new cases of cancer and over 8.8 lakh deaths due to the disease by 2020 with cancers of breast, lung and cervix topping the list"* [3]. International Cancer Genome Consortium (ICGC) has identified a total of 63,480,214 mutations and 57,753 genes contributing to mutations for a wide range of cancers [4]. Nevertheless, this signifies the necessity for an improved cancer diagnosis and management across the globe. India has higher mortality rate due to lack of awareness and expensive treatments and diagnosis often not accessible to patients due to financial challenges and skewed geographic distribution of clinical laboratories. The unavailability of mass cancer screening program in hospitals also limits the accurate diagnosis of patients.

Predictive, Prognostic and Diagnostic Tests- A Brief Understanding

Predictive Test

Predictive test looks for any kind of gene aberration (mutation) which could be linked with a disease condition prior to appearance of symptoms. These tests are dependent on the pedigree analysis and are used to evaluate the susceptibility of patients with no illness but family history of treatable or non-treatable hereditary conditions. Predictive examinations help understand the risk of diseases such as cancer and genetic conditions not restricted to but including myotonic dystrophy, Huntington disease, Ataxia, long QT syndrome, spastic paraplegia, hypertrophic cardiomyopathy, fascio-scapulo-humeral muscular dystrophy [5]. Presymptomatic and Predispositional test are two different types of predictive tests that reveal the means through which the patients develop symptoms and are more likely to develop disease respectively. Positive predictive test leads to genetic counseling of patient and family members carrying genetic abnormalities that can predispose them to develop the disease in future. Genetic counseling has a major role to curb psychological disturbances caused due to complications with a degree of uncertainty about whether and/or when the condition can develop, how fast symptoms can progress and how severely they can alter their lifestyle and bring about wellbeing of individual and family members [6]. Hence a predictive biomarker provides information related to the disease and therapeutic interventions that could be fabricated for the patients.

Prognostic Test

Prognostic test utilizes a set of pathological, molecular, biochemical or genetic markers to ascertain patient with various disease risks of specific future outcomes irrespective of therapy, further aiding intervention choice and proper counselling. Thus, presence or absence of such prognostic marker can be a measure in patient selection for ongoing trails and treatments but indeed not helpful in predicting treatment response [7]. Prognostic tests involve biomarkers that give information on recurrence in patients who receive curative treatment and biomarkers that correlate with duration of survival in patients with metastases post cancer onset. In cancers, two different types of prognostic biomarkers are known - one provides information on disease relapse for patients who are receiving treatments or progression free survival and the second studies of the biomarkers linked to metastatic advancement in patients [8]. According to a NIH Consensus Conference, '*a clinical useful prognostic marker must be a proven independent, significant factor that is easy to determine and interpret and has therapeutic consequences.*'

Diagnostic Test

Diagnostic tests assist physicians to discriminate patients with or without disease and to determine disease probability in clinically detected patients. Diagnostic tests are performed to establish diagnosis in symptomatic and asymptomatic patients, provide prognostic information in patients with disease, monitor therapy for good or adverse effects, and to confirm absence of disease in patients post therapy. There can be two different types of test probability- pre-test and post-test probability. Pre-test probability enables physicians to make decision on which test a patient should undergo and post-test probability is the cumulative result of diagnostic tests conferring the line of treatment by physicians. There are predictive values that can differ for same diagnostic test in various populations although these values are highly dependent on disease prevalence among population to be examined. Therefore, without having any prior knowledge on disease prevalence such values are impossible to be assessed accurately [9, 10].

Genes and Chromosomes - The Basis of Molecular Diagnosis

Every human cell contains 22 pairs of chromosomes known as autosomes (chromosome number 1-22) and 1 pair of sex chromosome (XX or XY). We often hear that we look like one of our family members. This is because of the genes that decide how we will look outside and how our system will work – the phenotype and the genotype. Genes reside on chromosomes and code for proteins, having multiple cellular and metabolic functions in human body. Proteins are not only the building blocks in our system but they also synthesize enzymes to control nearly all the biochemical processes. A chromosome encompasses hundreds to thousands of genes that are made up of deoxyribonucleic acids (DNA) which are double stranded, helical in structure and packed multi-folds. The four nucleotides in DNA molecule are- adenine (A), guanine (G), thymine (T) and cytosine (C). Surprisingly these nucleotides are arranged in different length and sequences creating a variety of genes thus helping to pass the genetic information from one generation to other. A healthy individual possesses different 'traits' which are specific characteristics determined by specific genes, giving rise to physiological diversity in our race [11]. Variation in traits is considered to be normal or polymorphic until otherwise seen to be altered due to mutation in genes of a single cell/individual thereby hindering normal functions, making a person sick and often triggering diseases. The 'Central Dogma' of molecular biology deals with DNA to RNA to Protein expression, a process by which DNA is transcribed into RNA and then translated into protein. There are somatic, de-novo and hereditary mutations that can happen spontaneously during an individual's lifetime, making cellular functionality frequently irregular to nonfunctional. Hereditary mutation passes from generation to generation whereas de-novo mutation arises from sperm or egg cell shortly after fertilization. Somatic mutation, the most common ones, occurs in regular cells as a result of environmental exposures and altered DNA replication [12]. A number of repair mechanisms exist in human body which helps to rectify such genetic errors and provide shelter from disease prevalence. When these systems lose control to correct the errors at molecular levels, genetically complex conditions like diabetes and cancers arise. Therefore, it is essential to adapt the overall diagnosis at molecular level and reveal the diversifying effect of mutations, likelihood of developing diseases and the symptoms associated with it along with its severity. In conditions like cancer, usually the clinicians start by collecting family history and physical examination leading to screening of blood, urine or any other body fluids and biopsy, which in turn is followed by any of the or combination of imaging techniques like CT scan, MRI, Ultrasound, PET scan, X-ray to determine the exact location of tumor in body, if present. By performing simple diagnostics test like hematology and histopathological examination of biopsy reveals the stage and grade of disease and overall chances of survival, simultaneously planning for best treatment options that can be provided to patients with minimal pain and an affordable cost. Healthcare professionals are trying to make cancer treatment 100% safe with an approach of 'personalized medicine' where patients receive treatment specific for major (driver) mutations that are diagnosed in their genomes. Although considering current scenario, there exist no treatment without severe side effects, molecular diagnostics in combination with targeted therapy is evolving and making the picture considerably ideal. The diagnosis of the new world involves studying DNA, RNA and protein expression with technologies not restricted to Polymerase Chain reaction (PCR), Next Generation Sequencing (NGS), Methylation, ChIP-seq, MALDI, SELDI and many more, in order to promote wellness and advance public health with effective interventions [13, 14].

Techniques in Cancer Diagnosis- Current Practices

Histopathology

Histopathology examines cellular structure (cytology) and tissues (histology) often retrieved from the biopsy samples at the site of abnormal growth. The technique uses light microscopy to generate a pathology report containing size, shape, growth rate and colour of the tissue which plays an important role in cancer staging and diagnosis. The tumor, node and metastasis (TNM) staging has been a significant method of understanding the tumor progression and still remains the method of choice across the globe. In both histology and cytology, there is a visible examination carried out on the patients followed by microscopic observation of the biopsy using either chemical fixation or frozen section processing [15]. Before undergoing microscopic examination, the tissues are collected and fixed in 10% neutral buffered formaldehyde to resist decay. Hematoxylin and Eosin are widely used dyes which form the gold standard for cancer diagnosis. Other dyes such as Oil red O, silver salt, congo red, safranin are also available to colour tissue sections [16]. Histopathological protocols vary slightly in various laboratory conditions though there is not much deviation in the reports generated for cancer examination.

Flow Cytometric Analysis

Flowcytometry provides a qualitative and quantitative analysis of multiple characteristics of single normal *versus* the tumor cell such as cellular size and complexity, live to dead ratio, DNA-RNA content, intracellular proteins and membrane bound makers. This technique is less expensive and used for immuno-phenotyping of cerebrospinal and serous cavity fluid, solid tumor cells, and cells extracted from whole blood, urine or bone marrow. Flow cytometry can classify, separate and quantify cells by means of cell surface antigens, DNA quantity and DNA proliferative activity by S-phase fraction (SPF). SPF is defined by the percentage or proportion of cells that are preparing for mitosis. An increase in SPF activity links with poor cellular differentiation, large tumor size and tumor aggressiveness. In flow cytometry, when cell surface antigens are tagged with fluorescent monoclonal antibodies, a stream of cells pass across the path of laser beam and the computer distinguishes normal cells from the abnormal ones and also helps in subgrouping of same cell types. During the course of work, the laser light traverses cell suspension that passes through flow cytometer, emitting light in all direction that are detected by photomultiplier tubes and digitized by computer analysis. Lymphocytes which are of two types- B and T cell lymphocytes and T cells are further categorized as helper or inducer, cytotoxic or suppressor and NK cells can be studied at various phases and with multiple CD markers in case of malignancies particularly blood related cancers. Flow cytometry is best understood by study of abnormal surface antigens on malignant cells that can contribute in identification of low grade tumor to high grade tumor. The human CD (Cluster of Differentiation) markers, which are widely used in flow cytometry for identification and characterization of leukocytes include- CD45 (leukocyte), CD3 (T-cells), CD4 (T-helper cells), CD8 (Cytotoxic T-cells), CD56 (NK cells), CD19 and CD 20 (B-cells), CD11c (Dendritic cells), CD14 and CD33 (Monocytes/ Macrophages), CD66b (granulocytes), CD34 (haematopoietic stem cells), CD41 and CD61 (platelets), CD235a (erythrocytes), CD146 (endothelial cells) and many more [17]. In oncology, S phase tumors and proliferative markers like Ki-67, PCNA (Proliferating Cell Nuclear Antigen) can be identified along with phenotyping of leukemia-lymphoma by characterizing leukocyte surface by measuring CD25, CD69. Propidium iodide, fluorescein, phycoerythrin are some of the routinely used fluorochromes which when conjugated with antibodies, bind to surface proteins, allowing accurate identification of several cellular properes [18, 19]. Flow cytometry allows identification of restricted chromosomal anomalies though not all of abnormality in karyotype. Due to these reasons, it is used as a prognostic indicator of solid tumors, the results of which indicate presence of multiple peaks on DNA histogram for aneuploidy tumors in comparison with control (diploid tumors) having single peak. However, such concept of prognostic indicator is controversial since a number of researches with a varying degree of results suggest diploid carcinomas are better prognostic factor than the aneuploid tumors whereas other studies do not consider ploidy as a significant risk factor [20].

Conventional and Molecular Cytogenetics

In conventional cytogenetics, chromosomal anomalies are detected using tumor cells culturing and harvesting them when maximum numbers of cells are in metaphase. Conventionally cytogenetics has been a method of choice for hematologic cancers especially the history relates chronic myeloid leukaemia and presence of Philadelphia chromosome [t(9;22)]. In 1960, the Philadelphia chromosome was first discovered to be genetically abnormal and was seen to be associated with chronic myeloid leukaemia (CML) which opened new age diagnostics in cancer cytogenetics. Cells pretreated with colchicine will result in mitotic arrest and addition of hypotonic solution in these cells significantly improves the quality and yield of metaphase spread. The metaphase cells are identified under the microscope for the presence of specific banding patterns post treatment called karyotype. One of the fundamental techniques that brought diagnostic advancement in field of cytogenetics is Fluorescence *In Situ* Hybridization (FISH). FISH is highly sensitive and specific, simple yet powerful technique sensing chromosomal abnormalities by molecular cytogenetics in general genetic disorders and in cancers. It uses fluorescent labelled DNA probes to target specific DNA sequences precisely to visualize one or more specific regions of genome. Discovery of FISH has enabled researcher to detect and report anomalies in interphase as well as metaphase chromosomes. An advanced and more expensive version of FISH allows simultaneous painting of each chromosome with different fluorescent dyes and computational pseudo multicolouring known as spectral karyotyping (SKY) in order to screen whole genome and rapidly evaluate deletions, duplication, inversions and translocations of metaphase chromosomes [21]. A rapid advancement in molecular cytogenetics related technology has reached another landmark by introducing Comparative Genome Hybridization – Array (CGH), for the detection of chromosomal copy number changes on a genome wide and high-resolution scale. Molecular cytogenetics is not only restricted to screening of complex karyotypic abnormalities but has a major role in identification of newer fusion genes which has high diagnosis value and precious to treatment of haematological malignancies facilitating use of personalized medicine in clinical practice [22].

Immunohistochemistry (IHC)

Immunohistochemistry (IHC), a qualitative and to some extent quantitative method, targets specific antigens in cells of a tissue with the help of primary monoclonal antibodies and visualizing antigen-antibody interaction through staining. IHC involves two stages- tissue fixation and its processing followed by result interpretation for the obtained expression. It is extensively used in the field of clinical diagnosis and research to enable physician differentiate and diagnose healthy and diseased tissue, characterizing primary sites of tumor and subtyping different cancer types, simultaneously developing therapeutic intervention [23]. One of the examples include Human Protein Atlas Project (HPA), funded by Knut and Alice Wallenberg Foundation, aimed to analytically discover human proteome using antibody-based proteomics [24]. IHC methods have brought about a revolution in approach to diagnosis of tumors of uncertain origin and discriminating primary from metastatic tumors. A panel of antibodies made on the basis of clinical history, morphological features, and results of other relevant investigations, is chosen to resolve these diagnostic dilemmas.

Electron Microscopy

Before Sanger sequencing, DNA sequencing approach was proposed with the concept of transmission electron microscopy. The target DNA molecule is linearized followed by complementary strand synthesis. In this technique, three bases of DNA are labeled with heavy isotopes and the other base remains unlabeled. Since natural DNA is transparent under TEM, the bases that are labeled with heavy atoms make DNA heavier and visible under microscope. Thus, the resulting complementary strand can be made observable and the four bases can be differentiated by size and intensity of dots exemplified by the bases [25, 26]. However, the conceptual idea diminished with advent of more feasible concepts and ideas in molecular diagnostics.

Western Blot Analysis, Two-Dimensional Gel Electrophoresis and Proteomics

Western blot was developed in the year 1979 and since then it has been widely used in research to identify, separate and analyze proteins on the basis of molecular weight and their ability to bind with specific antibodies *via* SDS-PAGE and immunoblotting techniques. Western blot is a rapid analytical technique that utilizes cost effective equipment and chemicals [27]. Once the antibody interacts with its protein of interest, a single band will be produced and the band thickness will determine the amount of protein present. The big question here is the validity of Western Blotting as a means to cancer diagnostics. Though for prostate oncology, there may be validated Western Blot assays in Europe but it does depend on use of control-quality antigen and a commercially available cGMP-produced antibody globally.

In the year 1995, the terminology 'Proteome' was first suggested to understand the 'protein complement of a genome'. In proteomics, the extracted proteins are first separated by means of difference in their isoelectric pH on immobilized gradient strip (IPGs - immobilized pH gradients) and then by molecular weight using SDS- gel electrophoresis. This technique allows detection of novel peptide spots or differentially expressed proteins on the gel which can be further identified using mass spectrometry. Inside a mass spectrometer, proteins are first converted into molecular ions and these ions are then resolved on the basis of mass to charge ratio thus identifying the amino acid sequence. Proteomics has come a long way with technological advances like MALDI-TOF, MS/MS and LC-GC/MS aiding biomarker discovery and disease-susceptibility studies [28, 29].

PCR (Polymerase Chain Reaction), Sanger Sequencing and Genomics

The basic idea behind 'Polymerase Chain Reaction' (PCR) was to produce millions and billions of copies of DNA segment of interest with the help of thermostable Taq polymerase, primers, dNTPs (Deoxyribonucleotide triphosphate) and buffer, through multiple cycles with a series of change in reaction temperatures. The three major steps of PCR include- denaturation, annealing and extension. The annealing temperature is solely calculated on melting temperature of primers and also varies from one primer to another. The application of PCR covers genetic testing, tissue typing, forensic science (paternity testing and fingerprinting), cancer diagnostics, identification of slow growing microorganisms, gene expression, DNA cloning, development of hybridization probes, DNA sequencing [30 - 33]. The continuous improvements in the PCR techniques have led to its enormous contribution in cancer biomarker identification and diagnosis of bcr (Breakpoint Cluster Region protein)/abl (Abelson Murine Leukaemia Viral Oncogene) translocations, G-Protein studies and oncogene expression. PCR technique is not a standalone technique and it works in combination with other downstream techniques like Southern blotting or direct sequencing to detect genomic anomalies in cancerous tissues [34]. Table **1** des-

Table 1. Different PCR techniques used in Cancer Diagnosis currently and in past.

PCR Types	Definition	Advantages	Disadvantages
Reverse transcriptase PCR (RT-PCR)	In direct RNA expression or cDNA technique	RNA transcript expressions, exons and introns mapping, sensitive and specific	Cannot determine functional proteins, high cost
Real Time PCR (qPCR)	• Intercalation of non-specific fluorescent dyes • Dual-labelled fluorogenic probe technique	No post PCR analysis, Higher sensitivity, resolution and precision, less time consuming	Non-specific product with non-specific dyes, costly
Nested PCR	Nested PCR involves two set of primers. The first primer set amplifies single locus and the second primer pair binds with the first PCR product	Reduces non-specific binding and increases specificity	Increases chances of product contamination
Methylation specific PCR	This technique determines CpG island methylation using methylated-specific primers (MSP or BSP)	Study of methylation pattern in disease and development	Only few sites situated within the template sequence can be investigated
Touchdown PCR	This technique works by applying higher annealing temperature than the primer's melting temperature and gradually reduces with the number of cycles until it has reached an optimum melting temperature	High sensitivity and specificity, increases product yield without requirement of redesign of primer or length optimization	Primer- dimer, non-specific products
Multiplex PCR	Multiplex PCR works by adding multiple primers in a single PCR reaction producing a varying length of amplicons that are specific for different DNA sequences. It requires optimization to ensure primer pairs are compatible for the reaction	Reduces amplification cost for each sample	Primer-dimer formation, less specific and less yield of product
Hot-start PCR	It uses specific antibodies in order to block the activity of Taq-polymerase at low temperature. The initial temperature is set at 95-degree C to denature the antibodies linked with enzyme active site. Once denatured, the amplification continues with greater specificity	Reduces non-specific priming	Increases product yield

cribes different types of PCR available and their advantages-disadvantages in current science.

Real time PCR also called qPCR was an answer to the biggest challenge for pathologist, a DNA based technique that can aid diagnosis and classification of the huge oncologic spectrum. qPCR in totality helps solid-tumor profiling in the clinical laboratory, identification of molecular markers which can guide to predict response, resistance, or toxicity to therapy. The end of qPCR allows the technician to derive a melting curve that identifies single base changes, copy number variations and microsatellite instabilities. Multiplexing PCR techniques have given us a major leap in time oriented report delivery to the critically significant cancer cases [35].

Sanger sequencing uses DNA polymerase, primer, dNTPs and dideoxy or chain terminating ddNTPs (Dideoxyribonucleotide triphosphate), which are labelled with different fluorescent tags. Dideoxy nucleotides are analogous to deoxynucleotides but the major difference is dideoxynucleotides lack hydroxyl group on 3' carbon of sugar ring whereas in deoxynucleotides, hydroxyl groups facilitate addition of new nucleotides in the existing chain. Therefore, when dideoxynucleotides are being incorporated in the chain it ends with ddNTPs. The entire process is repeated for a number of cycles generating fragments of different lengths and fluorescent labelled chain terminator nucleotides allow determining the sequence by marking end of each fragment. Finally, these fragments are run through capillary gel electrophoresis and the result contains series of peaks, an electropherogram, from which the DNA sequence is read [36]. Capillary Electrophoresis (CE) contributes largely in DNA sequencing and fragment analysis, a process by which ionic fragments are separated by size. As the fluorescent labelled DNA fragments are separated by size, they move across the path of laser beam, before reaching positive electrode where the laser light causes the dye to fluoresce and the (charge coupled device) CCD camera detects this fluorescence signal [37, 38]. Sanger sequencing is convenient to identify DNA variants with clinical significance. It has been applied to patients with non-small cell lung cancer (NSCLC) in selective treatment for EGFR (Epidermal Growth Factor Receptor) tyrosine kinase inhibitors whereas in colorectal cancer for identifying both KRAS (Kirsten ras oncogene homolog from the mammalian ras *gene* family) and BRAF mutation and also examining the resistance to anti-EGFR antibody therapy such as Cetuximab and Pantimumab [39 - 41].

Gene Panel Analysis

Gene Panel comprises of a set of genes that analyzes for specific gene mutations *viz.* structural variations, point mutations, indels, copy number variations and/or gene fusion associated with particular disease phenotype. These gene panels focus on candidate genes of diagnostic value simultaneously providing a cost-effective and simpler way of data analysis without high diagnostic expertise. It also generates a prognostic perception about tumors and enables clinicians to decide therapeutic interventions. According to American Society of Human Genetics, depending upon clinical symptoms of patient group, gene panel needs to be designed and evaluated before applying it to practice in routine diagnosis. The set of genes used to construct gene panel is important as it reflects the mutational hotspots and subsequent therapy. At the same time, the design of gene panel differs among laboratories where either, genes with diagnostic and therapeutic importance are preferred to be chosen according to literature survey or genes that are under clinical trial are considered. In order to fabricate and validate a diagnostic test it is necessary to state the importance and interpretation of gene panel analysis so that the quality of test conducted can be further improved [42, 43].

DNA Microarray

DNA microarray uses collection of micro - spots or probes of DNA, cDNA or oligonucleotides, attached to an optical solid surface such as microscopic slides or silicon chips to name a few. This technique helps in determining complementary binding of unknown samples and simultaneously allowing gene expression analysis. Microarray based genotyping covers array CGH (Comparative Genomic Hybridization), phenotype specific SNP (Single Nucleotide Polymorphism) panel and genome wide SNP panel interrogating multiple genes and evaluating their expression to offer high quality molecular diagnosis. This technique also enables analysis of binding of transcription factor in association with chromatin immune-precipitation referred to as 'ChIP-on-chip' method [44]. Tumor gene expression profiling with DNA microarray has been developed to determine primary sites in metastatic cancers and to sub-classify squamous cell carcinomas, soft tissue sarcomas and osteomas. It shows inter-tumor variations in histologically similar tumor, alleviating the challenges of cancer classification and increasing the precision of diagnosis. This technique has helped in depth classification of various leukemia and hereditary breast cancer with BRCA1 and BRCA2 status variations. There are certain limitations to this technique in that, it does not always provide accurate diagnosis of tumors and the data of gene expression significantly vary in same tumor under different experiment conditions. Due to unavailability of the early gene expressions in cancers, the profiling is still not possible thus highlighting that this technique is only useful with prior knowledge of gene expression [45]. Nonetheless, DNA microarray has made its mark in the diagnostic market and is currently used as a standalone method or in combination with qPCR or NGS techniques.

Next Generation Sequencing

In 1977, Scientists Sanger and Coulson developed a technology for studying the DNA sequence known as the, 'First Generation Sequencing'. The first-generation

DNA sequencing machines produced approximately 1KB reads but to enhance this, the scientist came up with 'shotgun sequencing' where overlapping DNA fragments were cloned and sequenced separately, and assembled into a contig, in silico [46]. Maxam and Gilbert improved upon this method by reducing the use of toxic chemicals and radioisotopes and their method is being followed till date with little to moderate advancement. The second-generation DNA sequencing, developed alongside dideoxy sequencing markedly differed wherein, the scientist used luminescent method for measuring pyrophosphate synthesis [47]. This method formed the basis of the current Third Generation Sequencing also known as the Next Generation Sequencing (NGS). NGS is a DNA sequencing technology, modernizing the entire genome research decreasing the time of decoding the known or unknown genome though researchers are still striving towards the cost reduction. This technology is slowly approaching towards clinical practice due to its ability to produce large number of accurate information [48, 49]. There are various NGS platforms that sequences millions of DNA fragments paralelly and identify the abnormal genomic regions and its related mutations (Table **2**). Whole Genome Sequencing (WGS), exome sequencing, denovo sequencing, targeted sequencing, mRNA and noncoding RNA sequencing, methylation sequencing, ChIP sequencing, ribosome profiling are the NGS methods that allow scientists to elucidate the early changes associated with genome, proteome, transcriptome, exome and epigenome thus, diagnosing the cancers at the primary stages [50]. It is necessary to choose an appropriate platform for data analysis depending on the requirement of sequencing coverage and depth [51, 52]. The table below shows the available NGS platforms, currently dominating the market, developed between 2000- 2013 (Table **3**).

Table 2. Next generation sequencing approaches – present and future of onco-diagnosis.

Whole Genome Sequencing (WGS)	Whole Genome Sequencing (WGS) is a high resolution, powerful method to detect small and large variation across the genome thereby identifying disease and its related mutation. The large volume of data produced by WGS and the gradually lowering cost is making WGS widely acceptable for clinical practice
Exomics	Exome are the protein-coding region of genome that contains approximately 80- 85% of disease causing variants. This makes exome sequencing a cost-effective alternative to WGS
Transcriptomics	This analyze the mRNA transcripts across the genome in normal and altered conditions
Dehovo sequencing	This type of NGS approach deals with novel genome sequencing without having any reference sequence for alignment and accumulating individual sequence reads into longer contiguous sequences. This helps in identification of complex rearrangements and structural variants

(Table 2) cont.....

Targeted sequencing	Targeted sequencing technique comprises of specific set of genes that are known or suspected with disease association in order to analyze gene mutation for a given sample
Methylome sequencing	Changing methylation patterns at cytosine residue can alter gene expression, single cell methylome sequencing is a major tool to study embryonic development and pre-malignant and malignant conditions
Chromatin Immunoprecipitation-ChIP sequencing	ChIP sequencing, which is often compatible for a wide range of DNA samples, is a powerful method to identify DNA binding sites for transcription factors and other proteins by combining chromatin immunoprecipitation and sequencing. In this technique, the DNA bound protein is immune-precipitated with the use of specific antibody and further co-precipitated, purified and sequenced
Ribosome profiling	This NGS technique offers profound sequencing of ribosome protected mRNA fragments, giving an insight about how many ribosomes are active in cell at a particular point of time, further predicting the actively translated proteins in order to monitor cellular translation. This helps in estimating protein abundance in a cell
DNA nanoball sequencing	Implicates rolling circle replication to amplify small genomic DNA fragments into DNA nanoballs allowing large amount of nanoballs to be sequenced per run at low cost

Next generation sequencing led to a collaborative effort, The Cancer Genome Atlas (TCGA) that has created a genomic data analysis pipeline that can effectively collect, select, and analyze human tissues for genomic alterations on a very large-scale aiding precision medicine.

Fourth Generation Sequencing Techniques in Cancer Diagnosis

SMRT

Single Molecule Real Time (SMRT) has been developed for long read sequencing, a challenge that is addressed *via* novel arenas of Zero Mode Waveguides (ZMV) and phospholinked nucleotides. Real time DNA replication detection is carried out while a phosphate chain is cleaved and nucleotide is incorporated by polymerase enzyme. The fluorescence emitted uses zero mode waveguide (ZMW), a nano-photonic confinement structure with a circular hole of approximately 70nm in diameter and 100nm in depth and within which DNA polymerase activity can be detected [53, 54]. For the purpose of sequencing, SMRT does not require a PCR, uses minimal amount of reagent and data output and analysis is much faster. The technology certainly provides sequencing-aided clinical cancer detection, thereby monitoring cancer diagnostics and treatment, refining concepts of precision medicine. SMRT has been used to establish FLT3 (Fms Like Tyrosine Kinase 3) gene related kinase domain mutation and sequenced BCR-ABL fusion gene transcript in order to detect complex mutations and splice isoforms [55].

Table 3. Next generation sequencing platforms in the diagnostic market and approximate cost per run.

Sequencing platforms	Description	Approximate cost per run
Roche GS-FLX 454 Sequencer	Based on pyrosequencing method using an Emulsion PCR with a speed of 13Mbp/h and can read up to 400- 700bp. It has an accuracy of 99.9%	Rs. 384180.00per run
Ion-torrent	Instead of using an optical signal, ion torrent sequencer detects the released H^+ ion (proton) upon addition of dNTPs to DNA polymerase. It can read 100- 400 bp and uses emulsion PCR with a speed of 25Mb- 16Gbp/h and has an accuracy of 99%	Rs. 14406.75 for Ion Torrent PGM - 314 Chip Rs. 27212.75 for Ion Torrent PGM- 316 Chip Rs. 40018.75 for Ion Torrent PGM- 318 Chip
Illumina/Solexa Genome Analyzer	Illumina employs reverse terminator sequencing technology with a sequence speed of 25Mbp and reading length of 100-300bp with an accuracy of 99.9%. It utilizes dideoxynucleotides to irreversibly terminate primer extension which differs from Sanger sequencing, as modified nucleotide analogues are used to terminate the same reversibly	Rs. 1.28 /million bases
ABI SOLiD	SOLiD, uses emulsion PCR in order to perform 'Sequencing by Oligonucleotide Ligation and Detection' with a speed of 21- 28Mbp/h, generating 80- 360Gbp as an output per run.	Rs. 2561.20×10^{-9}/ base

Helicos Sequencing

Helicos single molecule fluorescent sequencing reads a strand of DNA about 100-200 base pairs by the addition of poly-A sequence to 3' end of each DNA fragment. The polyA-polyT hybridized billion fragments of 50-55bp are read in a single go. Simultaneously location of each hybridized template can be detected due to its fluorescence label [53, 54]. Higher number of bases in the sequence leads to decreased percentage of strand read. Due to certain limitations this method is yet not validated in the diagnostic market though, it has been successfully applied to many omics study.

FRET Sequencing

Fluorescence Resonance Energy Transfer (FRET) is sequencing by synthesis approach that uses a polymerase containing donor fluorophore in combination with four different acceptor fluorophores for the corresponding four nucleotide bases. A FRET signal is generated whenever a polymerase encounters a nucleotide and providing sequence information based on the labels used for each base. FRET biosensors are best in detecting kinase inhibitors and/ or drugs in live cells and visualizing subcellular molecular signaling events in real time [56]. Bain *et al.,* (2007) showed that drug-resistant cancer cells can be determined using FRET in combination with flow cytometry since recurrence of any type of cancer and therapeutic failures could be a reason for cells becoming drug resistant. Hence, this technique is promising in assessing biopsy samples and predicting the probability of specific drug resistance in patients [53, 54, 57].

MinION - NanoporeTechnology

MinION – DNA nanopore sequencing is a real time, long-read, low-cost and portable system with an advantage of DNA, direct RNA or cDNA sequencing. The disruption in the current while a nano molecule passes through electrically resistant polymer membrane with protein nanopore reads the molecule. These pores can be either biological, solid-state *viz.* α-Hemolysin, MspA, Phi29, Al_2O_3or silicon-based. The technology has a vast application in analysis of both biological and chemical molecules. Although nanopore DNA sequencing has evolved fast but a rapid translocation speed of DNA and high error rate limits its performance [53, 54]. Norrisa *et al.* used nanopore sequencing technology MinION to detect structural variants that inactivate CDKN2A(Cyclin-dependent Kinase Inhibitor 2A)/p16 and SMAD4/DPC4 (Deleted in Pancreatic Carcinoma, locus 4) tumor suppressor genes in pancreatic cancer [58].

Danaher/Dover/AzcoPolonator G.007

This is a new platform in market that utilizes sequencing by ligation approach with the help of random bead based arrayed emulsion PCR in order to amplify DNA for parallel sequencing and to generate 8- 10 Gbp of reads per run. This bead-based array is replaced with patented rolling circle colonies for an increased efficiency and improved read length [53, 54].

PROMISES AND CHALLENGES OF MOLECULAR DIAGNOSTICS IN CLINICAL FIELD- CORRELATING MEDICINE AND BIOLOGY

Since the discovery of two-hit model of cancer by Knudson in 1971, a number of multifactorial, sophisticated models have been designed in the following decades to understand the contribution of nearly 20,000 human genes in cancer that arises from low-penetrance mutations in combination with environmental factors. Next Generation Sequencing technologies has evolved medical diagnostics by analyzing cause of disease, to revealing highly diversified mutations across various cancers and drafting targeted therapies. NGS can sequence multiple makers for a large set of samples in a single run thus decreasing the overall cost per run and using minimum sample concentration. These techniques are not only useful for confirmed diagnosis in terms of pre-symptomatic screening and prognosis prediction but also recently giving a boost to Cancer Genome Projects and Human Microbiome Projects. The medical field is still largely dependent on invasive painful technique of biopsy sampling for detection of cancers [59]. However, NGS techniques requires minimum sampling which can be obtained non-invasively from blood, serum or circulating tumor DNA which can disclose major information of primary cancer and the markers related to progression.

The major challenges for implementing NGS in clinical diagnostics are assay validation, gene panel selection, choice of sequencing technologies, result interpretation, data analysis and management, clinical reporting and the cost incurred per run. Before executing next generation sequencing as routine diagnostic tests in clinical laboratories, it requires stringent justification as per regulatory guidelines defined by The New York State Department of Health, The United States Centers for Disease Control and Prevention, American College of Medical Genetics, Association for Molecular Pathology for validation, testing and reporting of NGS results. In diagnostic laboratories, Sanger sequencing carries out mutational assays involved in sequencing of tumors with known aberrations. This in turn can aid in establishing capacity of new protocols using NGS to identify such mutations. At present, validation of NGS assays apply set of tumor samples with diverse mutations unlike similar tumors, which are clinically reported for most of the genes. Since NGS can screen several markers at a time, the selection of genes to be used in a panel becomes vital. There are a number of predesigned gene panels available commercially with prognostic and diagnostic relevance. Such gene panels can be customized as per the requirement and genetic history of the patients. NGS generates high volume, complex data which are massive compared to other low to moderate throughput tests. Laboratories require storing and backup systems for these data so that the patient test results can be traced back in compliance with Health Insurance Portability and Accountability Act [60]. Other challenges deal with comprehensive coverage and accurate representation of data. With consistent improvement in third generation technologies and with development of fourth generation sequencing techniques, it has been possible to eradicate certain limitations of cancer diagnosis. However, cancer diagnosis has a long way to go before the assays and test start reaching the poorly developed countries and citizens of the world irrespective of their socio-economic status.

CONSENT FOR PUBLICATION

Not applicable.

CONFLICT OF INTEREST

The authors declare no conflict of interest, financial or otherwise.

ACKNOWLEDGEMENTS

Declared none.

REFERENCES

[1]　SchenellaMenda (Business Insights and Analytics), BCC Research, IVD Market in India. (2016, August 27). Retrieved from: https://www.slideshare.net/ (Accessed on 8.11.2017)

[2]　Aggarwal S, Phadke SR. Medical genetics and genomic medicine in India: current status and opportunities ahead. Mol Genet Genomic Med 2015; 3(3): 160-71.
[http://dx.doi.org/10.1002/mgg3.150] [PMID: 26029702]

[3]　Jain M. Past, present and future of molecular oncology in India. IJMIO 2017; 2: 1-3.

[4]　Futreal PA, Coin L, Marshall M, *et al.* A census of human cancer genes. Nat Rev Cancer 2004; 4(3): 177-83.
[http://dx.doi.org/10.1038/nrc1299] [PMID: 14993899]

[5]　Marzuillo C, De Vito C, D'Andrea E, Rosso A, Villari P. Predictive genetic testing for complex diseases: a public health perspective. QJM 2014; 107(2): 93-7.
[http://dx.doi.org/10.1093/qjmed/hct190] [PMID: 24049051]

[6]　Evans JP, Skrzynia C, Burke W. The complexities of predictive genetic testing. BMJ 2001; 322(7293): 1052-6.
[http://dx.doi.org/10.1136/bmj.322.7293.1052] [PMID: 11325775]

[7]　Italiano A. Prognostic or predictive? It's time to get back to definitions! J Clin Oncol 2011; 29(35): 4718.
[http://dx.doi.org/10.1200/JCO.2011.38.3729] [PMID: 22042948]

[8]　Rector TS, Taylor BC, Wilt TJ. Systematic Review of Prognostic Tests.Methods Guide for Medical Test Reviews. Rockville, MD: Agency for Healthcare Research and Quality (US) 2012; pp. 1-13. [Internet]

[9]　Manikandan R, Dorairajan LN. How to appraise a diagnostic test. Indian J Urol 2011; 27(4): 513-9.
[http://dx.doi.org/10.4103/0970-1591.91444] [PMID: 22279321]

[10]　Stacey D, Bennett CL, Barry MJ, *et al.* Decision aids for people facing health treatment or screening decisions. Cochrane Database Syst Rev 2011; 5(10): CD001431.
[PMID: 21975733]

[11]　Griffiths AJF, Miller JH, Suzuki DT, Lewontin RC, Gelbart WM. Genetics and the Organism.NewYorkAn Introduction to Genetic Analysis 2000; pp. 1-26.

[12]　Griffiths AJF, Miller JH, Suzuki DT, Lewontin RC, Gelbart WM. Chromosomal Basis of Heredity.NewYorkAn Introduction to Genetic Analysis 2000; pp. 73-113.

[13]　Tajima F. Evolutionary relationship of DNA sequences in finite populations. Genetics 1983; 105(2): 437-60.
[PMID: 6628982]

[14]　Goode EL, Ulrich CM, Potter JD. Polymorphisms in DNA repair genes and associations with cancer risk. Cancer Epidemiol Biomarkers Prev 2002; 11(12): 1513-30.
[PMID: 12496039]

[15]　Ramaswamy S, Tamayo P, Rifkin R, *et al.* Multiclass cancer diagnosis using tumor gene expression

signatures. Proc Natl Acad Sci USA 2001; 98(26): 15149-54.
[http://dx.doi.org/10.1073/pnas.211566398] [PMID: 11742071]

[16]   Leong ASY, Zhuang Z. The changing role of pathology in breast cancer diagnosis and treatment. Pathobiology 2011; 78(2): 99-114.
[http://dx.doi.org/10.1159/000292644] [PMID: 21677473]

[17]   CD Marker Panel- Resources. 2017. Retrieved from: http://www.antibodies-online.com/resources/18/1540/cd-marker-panel/ (Accessed on 5.11.2017)

[18]   Brown M, Wittwer C. Flow cytometry: principles and clinical applications in hematology. Clin Chem 2000; 46(8 Pt 2): 1221-9.
[PMID: 10926916]

[19]   Colozza M, Azambuja E, Cardoso F, Sotiriou C, Larsimont D, Piccart MJ. Proliferative markers as prognostic and predictive tools in early breast cancer: where are we now? Ann Oncol 2005; 16(11): 1723-39.
[http://dx.doi.org/10.1093/annonc/mdi352] [PMID: 15980158]

[20]   Darzynkiewicz Z, Halicka HD, Zhao H. Analysis of cellular DNA content by flow and laser scanning cytometry. Adv Exp Med Biol 2010; 676: 137-47.
[http://dx.doi.org/10.1007/978-1-4419-6199-0_9] [PMID: 20687474]

[21]   Bishop R. Applications of fluorescence *in situ* hybridization (FISH) in detecting genetic aberrations of medical significance. Biosci Horiz 2010; 3: 85-95.
[http://dx.doi.org/10.1093/biohorizons/hzq009]

[22]   Wan TSK. Cancer cytogenetics: methodology revisited. Ann Lab Med 2014; 34(6): 413-25.
[http://dx.doi.org/10.3343/alm.2014.34.6.413] [PMID: 25368816]

[23]   Brandtzaeg P. The increasing power of immunohistochemistry and immunocytochemistry. J Immunol Methods 1998; 216(1-2): 49-67.
[http://dx.doi.org/10.1016/S0022-1759(98)00070-2] [PMID: 9760215]

[24]   Matos LL, Trufelli DC, de Matos MG, da Silva Pinhal MA. Immunohistochemistry as an important tool in biomarkers detection and clinical practice. Biomark Insights 2010; 5: 9-20.
[http://dx.doi.org/10.4137/BMI.S2185] [PMID: 20212918]

[25]   Zhang J, Chiodini R, Badr A, Zhang G. The impact of next-generation sequencing on genomics. J Genet Genomics 2011; 38(3): 95-109.
[http://dx.doi.org/10.1016/j.jgg.2011.02.003] [PMID: 21477781]

[26]   Ozsolak F. Third-generation sequencing techniques and applications to drug discovery. Expert Opin Drug Discov 2012; 7(3): 231-43.
[http://dx.doi.org/10.1517/17460441.2012.660145] [PMID: 22468954]

[27]   Mahmood T, Yang PC. Western blot: technique, theory, and trouble shooting. N Am J Med Sci 2012; 4(9): 429-34.
[http://dx.doi.org/10.4103/1947-2714.100998] [PMID: 23050259]

[28]   Verrills NM. Clinical proteomics: present and future prospects. Clin Biochem Rev 2006; 27(2): 99-116.
[PMID: 17077880]

[29]   Doustjalali SR, Bhuiyan M, Al-Jashamy K, *et al.* Two Dimensional Gel Electrophoresis: An Overview of Proteomic Technique in Cancer Research J Proteomics Bioinform 2014; 7: 077-81.

[30]   Johnson WM. The polymerase chain reaction: An overview and development of diagnostic PCR protocols at the LCDC. Can J Infect Dis 1991; 2(2): 89-91.
[http://dx.doi.org/10.1155/1991/580478] [PMID: 22529715]

[31]   Abath FG, Melo FL, Werkhauser RP, Montenegro L, Montenegro R, Schindler HC. Single-tube nested PCR using immobilized internal primers. Biotechniques 2002; 33(6): 1210-1212, 1214.

[http://dx.doi.org/10.2144/02336bm05] [PMID: 12503300]

[32]   Derks S, Lentjes MH, Hellebrekers DM, de Bruïne AP, Herman JG, van Engeland M. Methylation-specific PCR unraveled. Cell Oncol 2004; 26(5-6): 291-9.
       [PMID: 15623939]

[33]   Garibyan L, Avashia N. Polymerase chain reaction. J Invest Dermatol 2013; 133(3): 1-4.
       [http://dx.doi.org/10.1038/jid.2013.1] [PMID: 23399825]

[34]   Lyons J. The polymerase chain reaction and cancer diagnostics. Cancer 1992; 69(6) (Suppl.): 1527-31.
       [http://dx.doi.org/10.1002/1097-0142(19920315)69:6+<1527::AID-CNCR2820691304>3.0.CO;2-N]
       [PMID: 1540890]

[35]   Bernard PS, Wittwer CT. Real-time PCR technology for cancer diagnostics. Clin Chem 2002; 48(8): 1178-85.
       [PMID: 12142370]

[36]   Ross JS, Cronin M. Whole cancer genome sequencing by next-generation methods. Am J Clin Pathol 2011; 136(4): 527-39.
       [http://dx.doi.org/10.1309/AJCPR1SVT1VHUGXW] [PMID: 21917674]

[37]   Whatley H. Basic Principles and Modes of Capillary Electrophoresis.Clinical and Forensic Applications of Capillary Electrophoresis Pathology and Laboratory Medicine. Totowa, NJ: Humana Press 2001; pp. 21-58.

[38]   Karger BL, Guttman A. DNA Sequencing by Capillary Electrophoresis. Electrophoresis 2009; 30: S196-202.
       [http://dx.doi.org/10.1002/elps.200900218] [PMID: 19517496]

[39]   Kobayashi S, Boggon TJ, Dayaram T, *et al.* EGFR mutation and resistance of non-small-cell lung cancer to gefitinib. N Engl J Med 2005; 352(8): 786-92.
       [http://dx.doi.org/10.1056/NEJMoa044238] [PMID: 15728811]

[40]   Sequist LV, Joshi VA, Jänne PA, *et al.* Response to treatment and survival of patients with non-small cell lung cancer undergoing somatic EGFR mutation testing. Oncologist 2007; 12(1): 90-8.
       [http://dx.doi.org/10.1634/theoncologist.12-1-90] [PMID: 17285735]

[41]   Monzon FA, Ogino S, Hammond ME, Halling KC, Bloom KJ, Nikiforova MN. The role of KRAS mutation testing in the management of patients with metastatic colorectal cancer. Arch Pathol Lab Med 2009; 133(10): 1600-6.
       [PMID: 19792050]

[42]   Lim ECP, Brett M, Lai AH, *et al.* Next-generation sequencing using a pre-designed gene panel for the molecular diagnosis of congenital disorders in pediatric patients. Hum Genomics 2015; 9: 33.
       [http://dx.doi.org/10.1186/s40246-015-0055-x] [PMID: 26666243]

[43]   Jennings LJ, Arcila ME, Corless C, *et al.* Guidelines for Validation of Next-Generation Sequencing-Based Oncology Panels: A Joint Consensus Recommendation of the Association for Molecular Pathology and College of American Pathologists. J Mol Diagn 2017; 19(3): 341-65.
       [http://dx.doi.org/10.1016/j.jmoldx.2017.01.011] [PMID: 28341590]

[44]   Bumgarner R. DNA microarrays: Types, Applications and their future Curr Protoc Mol Biol 2013; 0 22 Unit–22.1

[45]   Kim IJ, Kang HC, Park JG. Microarray applications in cancer research. Cancer Res Treat 2004; 36(4): 207-13.
       [http://dx.doi.org/10.4143/crt.2004.36.4.207] [PMID: 20368836]

[46]   Heather JM, Chain B. The sequence of sequencers: The history of sequencing DNA. Genomics 2016; 107(1): 1-8.
       [http://dx.doi.org/10.1016/j.ygeno.2015.11.003] [PMID: 26554401]

[47]   Nyrén P, Lundin A. Enzymatic method for continuous monitoring of inorganic pyrophosphate

synthesis. Anal Biochem 1985; 151(2): 504-9.
[http://dx.doi.org/10.1016/0003-2697(85)90211-8] [PMID: 3006540]

[48]   Behjati S, Tarpey PS. What is next generation sequencing? Arch Dis Child Educ Pract Ed 2013; 98(6): 236-8.
[http://dx.doi.org/10.1136/archdischild-2013-304340] [PMID: 23986538]

[49]   Buermans HPJ, den Dunnen JT. Next generation sequencing technology: Advances and applications. Biochim Biophys Acta 2014; 1842(10): 1932-41.
[http://dx.doi.org/10.1016/j.bbadis.2014.06.015] [PMID: 24995601]

[50]   Barba M, Czosnek H, Hadidi A. Historical perspective, development and applications of next-generation sequencing in plant virology. Viruses 2014; 6(1): 106-36.
[http://dx.doi.org/10.3390/v6010106] [PMID: 24399207]

[51]   van Dijk EL, Auger H, Jaszczyszyn Y, Thermes C. Ten years of next-generation sequencing technology. Trends Genet 2014; 30(9): 418-26.
[http://dx.doi.org/10.1016/j.tig.2014.07.001] [PMID: 25108476]

[52]   Ansorge WJ. Next Generation DNA Sequencing (II): Techniques, Applications Next Generat Sequenc & Applic 2016; S1: 005: 1-10.

[53]   Myllykangas S, Buenrostro J, Ji HP. Overview of Sequencing Technology Platforms.Bioinformatics for High Throughput Sequencing. Stanford, CA, USA 2012; pp. 11-25.
[http://dx.doi.org/10.1007/978-1-4614-0782-9_2]

[54]   Blackmore JK, Karmakar S, Gu G, *et al.* The SMRT coregulator enhances growth of estrogen receptor-α-positive breast cancer cells by promotion of cell cycle progression and inhibition of apoptosis. Endocrinology 2014; 155(9): 3251-61.
[http://dx.doi.org/10.1210/en.2014-1002] [PMID: 24971610]

[55]   LeBlanc VG, Marra MA. Next-Generation Sequencing Approaches in Cancer: Where Have They Brought Us and Where Will They Take Us? Cancers (Basel) 2015; 7(3): 1925-58.
[http://dx.doi.org/10.3390/cancers7030869] [PMID: 26404381]

[56]   Lu S, Wang Y. FRET Biosensors for Cancer Detection and Evaluation of Drug Efficacy. Clin Cancer Res 2010; 16: 3822-4.
[http://dx.doi.org/10.1158/1078-0432.CCR-10-1333] [PMID: 20670948]

[57]   Bain J, Plater L, Elliott M, *et al.* The selectivity of protein kinase inhibitors: a further update. Biochem J 2007; 408(3): 297-315.
[http://dx.doi.org/10.1042/BJ20070797] [PMID: 17850214]

[58]   Norris AL, Workman RE, Fan Y, Eshleman JR, Timp W. Nanopore sequencing detects structural variants in cancer. Cancer Biol Ther 2016; 17(3): 246-53.
[http://dx.doi.org/10.1080/15384047.2016.1139236] [PMID: 26787508]

[59]   Shen T, Pajaro-Van de Stadt SH, Yeat NC, Lin JC-H. Clinical applications of next generation sequencing in cancer: from panels, to exomes, to genomes. Front Genet 2015; 6: 215.
[http://dx.doi.org/10.3389/fgene.2015.00215] [PMID: 26136771]

[60]   Singh RR, Luthra R, Routbort MJ, Patel KP, Medeiros LJ. Implementation of next generation sequencing in clinical molecular diagnostic laboratories: advantages, challenges and potential. Expert Rev Precis Med Drug Dev 2016; 1: 109-20.
[http://dx.doi.org/10.1080/23808993.2015.1120401]

SUBJECT INDEX

Apoptosis-related genes 216
Apoptosome 213, 214, 215
Apoptotic cell death 212, 213
Aqueous extracts 200, 203
Arachidonic acid (AA) 29
Aroclor 120, 123
Aromatic amines 27, 174, 175, 177
Aromatic hydrocarbons 14, 116, 117, 118
Aromatic hydroxylamines 174, 175, 176
Aromatics 3, 9, 145, 150, 170, 172, 176
Arsenic contamination 24
Arthrobacter sp 141, 149
Arthropods 99
Asbestos fibers 17, 25, 39
Asthma 10, 17, 102, 197
ATP binding cassette (ABC) 226
Atractylodes macrocephala polysaccharides
        (AMPs) 227

B

Bacteria 18, 53, 54, 97, 99, 100, 105, 107, 110,
        123, 130, 141, 143, 144, 149, 153
    toxic chemical pesticides 105
Bacterial bioassays 47, 53
Bacterial reverse mutation assays 128, 130
Baicalein 220, 221
Bax, up-regulation of 224, 227
Berberine 224, 225
Beryllium 5, 9, 12
Bioassays 47, 48, 50, 51, 52, 54, 57, 58, 59, 130
    battery of 58, 59
    plant-based 52
Bioaugmentation 94, 106
Biocatalysts 142, 143
Biochanin 166, 182, 183
Biodegradable contaminants 3
Biofilm 144
Bioluminescence 55, 56
Biopesticides 94, 96, 99
Biorationals 99
Bioremediation 94, 110, 138, 141, 143, 148, 151,
        152
Bioremediation approaches 106, 138
Bioremediation of contaminated soil 94
Bioremediation of heavy metal contaminated
        sites 146

Bioremediation strategies for pesticide
        degradation 103
Biosphere 2, 3, 53, 77, 153
Biostimulation 94, 105
Biotoxin 56
Biotransformation 53, 138, 148, 162, 168, 171,
        172, 173, 174, 175
Biotransformation reactions 168, 171, 172
Bioventing, process of 105, 106
Biphenyls 116, 117, 118, 119, 120, 121, 127,
        140, 151
    applications of 119, 120
    chemical properties of 117, 118
    polybrominated 116, 119
    polychlorinated 116, 117, 119, 127, 140, 151
    sources of 117, 121
Bisphenol A 24, 26, 34, 35
Bladder cancer 25, 177
Breast cancer resistance protein (BCRP) 226
Breast carcinoma cells 37
Bridges 68, 69, 70, 71, 72, 73, 74, 75, 76, 77, 78,
        80, 81, 82, 83
    anaphase 72, 78
    telophasic 68
Bromine atoms 118, 119

C

Cadmium compounds 25, 39
CAMP response element modulator (CREM) 29
Canaphases 69, 75, 76
Cancer 124, 26 97, 102, 220, 224, 260
    colon 197, 224
    pancreatic 28, 260
    prostate 24, 26, 102, 220
Cancer cell growth 218, 222
Cancer cells 31, 211, 212, 214, 215, 218, 219,
        220, 224, 226, 227, 228, 229, 230, 246,
        260
    alkaloids-treated 224
    breast 228
    drug-resistant 260
    skin 219
The cancer genome atlas (TCGA) 258
Cancer-related genes 32
Cancer risk assessment 40
Cancer treatment regimen, conventional 217